Lecture Notes
in Control and Information Sciences   361

Editors: M. Thoma, M. Morari

Hugo Oscar Méndez-Acosta, Ricardo Femat,
Victor González-Álvarez (Eds.)

Selected Topics in Dynamics and Control of Chemical and Biological Processes

Springer

Editors

Hugo Oscar Méndez-Acosta

UdG-Department of Chemical Engineering and
IPICYT-Department of Applied Mathematics
Camino a la presa San José # 2055
Col. Lomas 4 sección, CP 78126
San Luis Potosí, San Luis Potosí
Mexico

Victor González-Álvarez

UdG-Department of Chemical Engineering
Bvld. Marcelino García Barragán y Calzada
Olímpica (s/n), CP 44860
Guadalajara, Jalisco
Mexico

Ricardo Femat

IPICyT-Department of Applied Mathematics
Camino a la presa San José # 2055
Col. Lomas 4 sección, CP 78126
San Luis Potosí, San Luis Potosí
Mexico

Library of Congress Control Number: 2007929542

ISSN print edition: 0170-8643
ISSN electronic edition: 1610-7411
ISBN-10 3-540-73187-3 Springer Berlin Heidelberg New York
ISBN-13 978-3-540-73187-0 Springer Berlin Heidelberg New York

Springer is a part of Springer Science+Business Media
springer.com

© Springer-Verlag Berlin Heidelberg 2007
Printed in Germany

Typesetting: by the authors and SPS using a Springer LaTeX macro package

Printed on acid-free paper   SPIN: 12066124   89/SPS   5 4 3 2 1 0

To Laura, Valeria, Camila, Beatriz and Cande
H.O. Méndez-Acosta

To cow-eyes lady and my parents
R. Femat

To my lovely family
V. González-Álvarez

Preface

In view of the rapid changes in requirements, it has became necessary to place at the reader's disposal a book dealing with basic and advanced concepts and techniques for the monitoring and control of chemical and biochemical processes, as well as with the aspects of the implementation of these different robust techniques. To make the ideas covered in this book accessible to a larger audience, we have attempted to present a balanced view of the theoretical and practical issues of control systems. Different cases are presented to illustrate the controller and observer design procedures and their dynamic effects in the closed-loop. The book consists of peer-reviewed contributions and aims the integration of recent research by a group of colleagues working on theory and practice of the process control. All authors are members on educative institutions and are the new talent entering the field of process control. The coverage in contributions is from the pure theoretically analysis to the experimental implementation passing through physical interpretation. Thus, process modeling topics, theoretical conditions for controlling and physical interpretations of the controllers are involved in the book. In order to embrace as chemical as control engineering themes, authors have proposed different kind of processes and distinct control scopes. The themes have been chosen by accounting that contributions go beyond the academic custom "give me a model and I can find a controller" and the industry concept "the control is working, do not known why but do not touch it". This book is an effort to incorporate and link both theory-practice and chemical-control engineering.

Concerning the variety of processes, authors have contributed with distillation columns and alkylation, biological and exothermic reactors. In regard distillation columns and alkylation reactors, the control approaches have a physical meaning because the model governing their dynamics is to large to formally study them. Besides, in complementary manner, biological and exothermic reactors are studied in a more formal way; which, in combination with the practical approaches proposed for distillation and alkylation problems, can render to reader a balance on theory-practice. This wide orientation obeys to the diversity of control problems on chemical and biological processes.

In regard dynamics and control scopes, the contributions address analysis of open and closed-loop systems, fault detection and the dynamical behavior of controlled processes. Concerning control design, the contributors have exploited fuzzy and neuro-fuzzy techniques for control design and fault detection. Moreover, robust approaches to dynamical output feedback from geometric control are also included. In addition, the contributors have also enclosed results concerning the dynamics of controlled processes, such as the study of homoclinic orbits in controlled CSTR and the experimental evidence of how feedback interconnection in a recycling bioreactor can induce unpredictable (possibly chaotic) oscillations.

First, we would like to acknowledge the individuals who wrote chapters to this book for taking time out of their busy schedule to contribute to this book. Their hardwork and patience are gratefully appreciated. We also which to express our gratitude to the international reviewers whose selfless service to the profession is truly appreciated.

The book comprises three parts as follows:

Part 1 *Control of Chemical Processes.* Some common problems in chemical processes are presented and either classical solutions or physical interpretation of controllers are discussed. Thus, the first chapter includes modeling and local control whereas the second chapter is focussed on nonlinear control design from heat balance on chemical reactors. The three first chapters deal with regulation problems while the last one is devoted to a tracking one.

Part 2 *Control and Diagnosis of Biological Processes.* This part is oriented on biological wastewater treatment processes. From fuzzy-based designs to geometric control approaches, the wastewater processes are controlled to regulate the concentration of polluting agents, to guarantee the process stability or for fault diagnosis purposes. Control of wastewater treatment plants is a quite active research area and this chapter provides a scope on alternative solutions. Experimental results are included along this part.

Part 3 *Dynamics of Controlled Reactors.* The first chapter shows how the classical control can induce oscillatory behavior even chaotic in a chemical process. This chapter includes a theoretical treatment whereas the last one shows experimental evidence of how feedback control can induce oscillatory (possibly chaotic) behavior. Thus, this part includes a rarely found topic in control books.

San Luis Potosí, S.L.P., Mexico,   *H.O. Méndez-Acosta*
November 2006   *R. Femat*
                                                               V. González-Álvarez

Contents

Part I

Control of Chemical Processes

Nonisothermal Stirred-Tank Reactor with Irreversible Exothermic Reaction $A \rightarrow B$: 1. Modeling and Local Control

P. Albertos[1] and M. Pérez-Polo[2]

[1] Department of Systems Engineering and Control, ETSII
pedro@aii.upv.es
[2] Department of Physics, System Engineering and Signal Theory, EPS
manolo@dfists.ua.es

Summary. In this chapter, the nonlinear and linear mathematical model of a CSTR where an irreversible exothermic reaction takes place is considered. From the analysis of the steady states, the curves of removed and generated heat are deduced in order to determine the equilibrium point and the transfer function of the reactor. Considering a set of reactors ranging from small to high volume, it is shown that the open-loop transfer function can be conditionally stable, and the effect of the scaling-up is investigated by means of the root locus. Reactor and jacket equilibrium temperatures vs. reactor volume are deduced in order to obtain an intuitive notion of the reactor controllability. The problem of the local control of a single loop concentration and cascade temperature control from the linearized model previously considered is analyzed by using PI controllers. Several examples considering step responses to changes in the concentration and temperature references are considered. Finally, the decoupling and the feedback control of the reactor by using the pole-placement technique have been also considered.

1 Introduction

Chemical reactors intended for use in different processes differ in size, geometry and design. Nevertheless, a number of common features allows to classify them in a systematic way [3], [4], [9]. Aspects such as, flow pattern of the reaction mixture, conditions of heat transfer in the reactor, mode of operation, variation in the process variables with time and constructional features, can be considered. This work deals with the classification according to the flow pattern of the reaction mixture, the conditions of heat transfer and the mode of operation. The main purpose is to show the utility of a Continuous Stirred Tank Reactor (CSTR) both from the point of view of control design and the study of nonlinear phenomena.

According to the flow pattern reactors can be stirred-tank and tubular type. A stirred-tank reactor is a tank whose contents or reaction mixture is agitated

H.O. Méndez-Acosta et al. (Eds.): Dyn. & Ctrl. of Chem. & Bio. Proc., LNCIS 361, pp. 3–32, 2007.
springerlink.com

by a mechanical stirred or a recirculating pump. In the theory of chemical reactors it is customary first to consider perfect or complete mixing, in which all the reaction conditions are completely the same at any point throughout the reactor's volume. A tubular reactor is a tube or a pipe in which the reaction progresses as the reactants move down the conduit. In an ideal tubular reactor (or plug flow), the reactants and products move down the reactor in the form of plugs, so that a distinct distribution of individual concentrations, temperature and other variables is established over the reactor's length. In a fluidized bed reactor, a suspension layer is formed when a stream of a fluid is passing upward through a layer of a solid granular material. The value of the fluid speed is the appropriate to obtain a bed where the particles of the granular material are suspended. These reactors are characterized by a high efficiency in heterogeneous processes involving mass and heat transfer.

Related to the heat transfer conditions, the release or absorption of heat of the reaction as well as that of the associated physical processes, such as dissolution, crystallization, evaporation etc. taking place within the chemical reactor, must be considered. As heat is released or absorbed, the temperature in the reactor changes and a temperature gradient is established between the reactor and the surroundings. In an adiabatic reactor all the released or absorbed heat is used in internal heat transfer and, consequently, the reaction mixture is either heated or cooled. In an isothermal reactor the heat transfer to or from the surroundings maintains a constant temperature inside the reactor. Moreover, in a heat regulated reactor, the heat evolved or taken up by the chemical reaction is partly balanced by the heat exchanged with the surroundings, and partly causes a change in the temperature of the reaction mixture.

The classification taking into account the operation mode considers how the reactants are feed to and products are discharged from a reactor. Thus, there are batch reactors, semibatch reactors and continuous reactors. In batch reactors all reactants are feed into a tank prior to starting the reaction and the product mixture is discharged once the process is finished, so the process variables of a batch reactor vary with time. In a continuous reactor, the admission of the reactants and discharge of the products proceed in parallel, and so there is no time wasted on nonproductive operations. This is the reason why nowadays chemical processes where the plant is expected to operate at high level of production operate with continuous reactors. Our study is focused on stirred tank, heat regulated continuous reactor, although some ideas can be also applied to other types of reactors. The modeling, control and analysis of its highly non-linear behavior, will be considered in the following.

1.1   The Continuous Stirred Tank Reactor

The study of the dynamic behavior of an ideal CSTR with a simple irreversible exothermic reaction $A \rightarrow B$ has been considered in many papers and books [2], [5], [6], [7], [8], [10], [11], [12], [13], [14], [15], [20], [21], [22], [24], [23], [25],

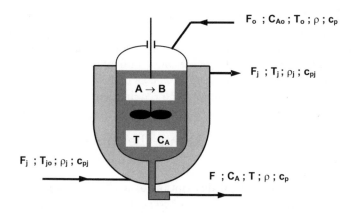

Fig. 1. Perfectly mixed CSTR with reaction $A \rightarrow B$

[26]. The dynamic of this process is highly non-linear mainly due to the heat generation process.

The control of chemical reactors is one of the most challenging problems in control process. Considering that a CSTR is the heart of many processes (i.e. in organic synthesis), its stable and efficient operation is of paramount importance to the success of an entire plant. Many reactors are inherently unstable, so an effective and well-designed control system is necessary in order to assure stable operation. The instability appears when irreversible exothermic reactions are carried out in a CSTR. These reactions tend to produce a large increment in temperature, forcing the rupture of safety and reducing the lifetime of the reactor. The faster the reaction goes, the more heat is generated heating up the reaction mass, and consequently, raising the reactor temperature and increasing the rate of reaction. So, under this positive feedback, the system can reach an undesirable high temperature state. The solution to this problem is a temperature control system capable of detecting the rising of the reactor temperature and quickly removing heat from the reactor.

In this chapter, the modeling and local control around the equilibrium points of perfectly mixed reactors is discussed. From an industrial point of view, there are many suitable configurations for temperature control. Some of them are the following: i) The coolant flows through an external jacket surrounding the reactor, ii) The coolant flows through an internal cooling coil, iii) Reactant liquid is pumped through an external heat exchanger. In this analysis only the case i) will be considered, but the other configurations give similar results. Figure 1 shows a CSTR where the coolant flows through a jacket. The reactor outlet concentration is the main variable of interest and, for the reasons above, the reactor temperature should be also controlled. The usual manipulated variables are the inlet flow rate to the reactor, F_0, and the coolant flow rate, F_j. Usually, the reactor volume, V, is assumed to be constant or perfectly regulated by the outlet reactor flow rate.

This chapter is organized in the following way. First, the general model of the CSTR process, based on first principles, is derived. A linearized approximate model of the reactor around the equilibrium points is then obtained. The analysis of this model will provide some hints about the appropriate control structures. Decentralized control as well as multivariable (MIMO) control systems can be designed according to the requirements.

2 Some Considerations About the Mathematical Modeling of a CSTR

A first principle mathematical description of a CSTR is based on balance equations expressing the general laws of conservation of mass and energy. Assuming that n components are mixed, the material balance of the i-component, taking into account all forms of supply and discharge in the volume V of the reactor, is the following:

$$\frac{dN_{i,acc}}{dt} = F_{i0} - F_i \pm G_{i,r} \tag{1}$$

where

$\frac{dN_{i,acc}}{dt}$ = rate of i-component accumulated in the volume V (mol/time)
F_{i0} = rate of i-component in the inlet flow (mol/time)
F_i = rate of i-component in the outlet flow (mol/time)
$G_{i,r}$ = rate of i-component used or created by the chemical reaction (mol/time).

Similarly, energy balances in the reactor and the jacket can be drawn. However, the energy balance in the reactor can be rather complicated depending on the rate law and the stoichiometry of the reaction [3], [9], [14]. The general equation of the unsteady state of the energy balance is the following:

$$\frac{dE_{sys}}{dt} = \dot{Q} - \dot{W} + \left(\sum_{i=1}^{n} F_i E_i \right)_0 - \left(\sum_{i=1}^{n} F_i E_i \right) \tag{2}$$

where

$\frac{dE_{sys}}{dt}$ = rate of accumulation of energy in the system (kJ/time)
$\dot{Q}$ = rate of the flow heat to the system from the surroundings (kJ/time)
$\dot{W}$ = rate of work done by the system (kJ/time)
$\left(\sum_{i=1}^{n} F_i E_i \right)_0$ = rate of energy added to the system by the mass flow (kJ/time)
$\left(\sum_{i=1}^{n} F_i E_i \right)$ = rate of energy eliminated from the system by the mass flow (kJ/time).

The energy E_i is the sum of the internal energy, the kinetic energy, the potential energy, and other energies in the i-component. In a CSTR, the value of E_i is equal to the internal energy. Also, in a CSTR, the work term $\dot{W}$ is reduced to the flow work and can be written as:

$$\dot{W} = - \left(\sum_{i=1}^{n} F_i P V_i \right)_0 + \left(\sum_{i=1}^{n} F_i P V_i \right) \tag{3}$$

where $V_i = V$, $i = 1, \ldots, n$, is the reactor volume and P is the reactor pressure. Substituting Eq.(3) into Eq.(2) the general equation of energy balance takes the following form:

$$\frac{dE_{sys}}{dt} = \dot{Q} + \left(\sum_{i=1}^{n} F_i(E_i + PV_i) \right)_0 - \left(\sum_{i=1}^{n} F_i(E_i + PV_i) \right) \tag{4}$$

Taking into account that the enthalpy is defined by $H_i = E_i + PV_i$ (kJ/mol-i), Eq.(4) can be written as:

$$\frac{dE_{sys}}{dt} = \dot{Q} + \sum_{i=1}^{n} F_{i0} H_{i0} - \sum_{i=1}^{n} F_i H_i \tag{5}$$

where the subscript " 0 " means inlet conditions.

The total energy of the system E_{sys} is the sum of the products of specific energies E_i of the all reactants in the volume of the reactor and the corresponding moles of each reactant:

$$E_{sys} = \sum_{i=1}^{n} N_i E_i = \sum_{i=1}^{n} N_i(H_i - PV_i) \tag{6}$$

By differentiating Eq.(6) with respect to time, considering that the variations of the volume and total pressure are negligible, and using the enthalpy definition, $H_i = c_{pi}T$, where c_{pi} is the heat capacity of i-reactant (kJ/mol °C), Eq.(5) can be written as follows:

$$\sum_{i=1}^{n} \left(N_i c_{pi} \frac{dT}{dt} + c_{pi} T \frac{dN_i}{dt} \right) = \dot{Q} + \sum_{i=1}^{n} F_{i0} H_{i0} - \sum_{i=1}^{n} F_i H_i \tag{7}$$

In order to obtain the enthalpy of reaction, Eq.(7) can be written in a more adequate form by taking into account the Eq.(1) for the i-reactant:

$$\frac{dN_i}{dt} = F_{i0} - F_i - v_i r V \tag{8}$$

where v_i is the variation of moles of the i-reactant and r is the reaction rate. Substituting Eq.(8) into Eq.(7) and considering that the enthalpy of the reaction is $-\Delta H_R = \sum_{i=1}^{n} v_i H_i$, the energy balance for a CSTR can be written as:

$$\sum_{i=1}^{n} N_i c_{pi} \frac{dT}{dt} = \dot{Q} - \sum_{i=1}^{n} F_{i0} c_{pi} (T - T_{i0}) + (-\Delta H_R)(rV) \qquad (9)$$

Remark 1. Eq.(9) can be used on a CSTR in unsteady state operation or in a semibatch reactor when there is no phase change.

Remark 2. For many liquid phase reactions the variation of the heat capacity is negligible, and so the following approximation can be made:

$$\sum_{i=1}^{n} N_i c_{pi} \approx \sum_{i=1}^{n} N_{i0} c_{pi} = N_{10}(1 + \sum_{i=2}^{n} N_i) c_{pi} = N_{10} c_{ps}$$

$$\sum_{i=1}^{n} F_{i0} c_{pi} = F_{10} c_{ps} \qquad (10)$$

where c_{ps} is the heat capacity of the solution and the units of $F_{10} c_{ps}$ are kJ/s·K. Assuming that all of the reactants enter at the same temperature, T_0, Eq.(9) can be written as:

$$N_{10} c_{ps} \frac{dT}{dt} = \dot{Q} - F_{10} c_{ps}(T - T_0) + (-\Delta H_R)(rV) \qquad (11)$$

2.1   Examples of CSTR Modeling

Eq.(1) and (9) can be utilized for modeling a CSTR considering the material and energy balances as well as the expression for the rate flow of heat removed, $\dot{Q}$. This heat rate is obtained from the overall heat transfer coefficient U and the transmission area A by the equation $\dot{Q} = UA(T - T_j)$ [1], [9], [13], [14], [18], [22].

Example 1. Consider an irreversible exothermic reaction A→B. The reaction is first order respect to A-reactant. Taking into account Eq.(1) and (11) and Figure 1 the reactor model includes the following equations:

• Component A mass balance:

$$V \frac{dC_A}{dt} = F_0 C_{A0} - F_0 C_A - Vr; \quad r = \alpha C_A e^{-E/RT} \qquad (12)$$

where F_0 is the volumetric flow rate for the inlet stream, C_{A0} and C_A are the concentration (kmol/m^3) of component A for the inlet and outlet streams respectively, α is the preexponencial factor from Arrhenius law (h^{-1}) and R is the perfect-gas constant (kJ/kmol·K).

- Energy balance in the reactor:

$$\rho c_p V \frac{dT}{dt} = \rho c_p F_0(T_0 - T) + (-\Delta H_R)(V \cdot r) - UA(T - T_j) \qquad (13)$$

where c_p is the heat capacity for the inlet and outlet streams (kJ/kg·K), ρ is the density for the inlet and outlet streams (kg/m³), U is the overall heat transfer in the jacket (kJ/h·m²·K), A is the heat transfer area (m²), T and T_j are the temperatures of the reactor and the jacket respectively (K).

- Energy balance in the jacked:

$$\rho_j c_{pj} V_j \frac{dT_j}{dt} = \rho_j c_{pj} F_j(T_{j0} - T_j) + UA(T - T_j) \qquad (14)$$

where F_j is the volumetric flow rate for the jacket, c_{pj} is the heat capacity of the coolant stream (kJ/kg·K), ρ_j is its density (kg/m³), and T_{j0} is its temperature (K).

Remark 3. Note that Eq.(12), (13) and (14) are written using a different unit system than the one used in Eq.(1) and (11).

Remark 4. Eq.(13) is valid only for a first order irreversible reaction A→B. So, it is the easiest form to writing the energy balance in the reactor.

A space state model of the reactor can be easily deduced, by appropriate selection of variables. The Eq.(12)-(14) can be rewritten as follows:

$$\frac{dC_A}{dt} = f_1[x(t), u(t)] = \frac{F_0}{V}(C_{A0} - C_A) - \alpha C_A e^{-E/RT}$$

$$\frac{dT}{dt} = f_2[x(t), u(t)] = \frac{F_0}{V}(T_0 - T) + \frac{(-\Delta H_R)\alpha}{\rho c_p} C_A e^{-E/RT} - \frac{UA}{\rho c_p V}(T - T_j) \qquad (15)$$

$$\frac{dT_j}{dt} = f_3[x(t), u(t)] = \frac{F_j}{V_j}(T_{j0} - T_j) + \frac{UA}{\rho_j c_{pj} V_j}(T - T_j)$$

where the state variables are $x(t) = [C_A(t), T(t), T_j(t)]^T$, usually taking $T(t)$ and $C_A(t)$ as outputs, and the control vector is $u(t) = [F_0(t), F_j(t)]^T$.

Example 2. In organic synthesis there are many reactions which are carried out with an excess of inert (i.e., hydrolysis reactions) and consequently can be considered as first-order in reactant A. For example the reaction:

$$A + B + M \longrightarrow C$$

is carried out with an excess of the inert B and M, so the reaction is approximately first-order in reactant A and zero-order with respect to B and M. The reactor model includes the following equations:

- Mass balance of reactants A, B, C and M

$$V\frac{dC_A}{dt} = F_0(C_{A0} - C_A) - (r_A \cdot V)$$

$$V\frac{dC_B}{dt} = F_0(C_{B0} - C_B) - (r_B \cdot V) \qquad (16)$$

$$V\frac{dC_C}{dt} = F_0(-C_C) + (r_C \cdot V)$$

$$V\frac{dC_M}{dt} = F_0(C_{M0} - C_M)$$

where there is no component of C in the inlet stream to the reactor. Taking into account that the inert B and M are in excess, the reaction velocity of Eq.(15) can be approximated by the following relation: $r_A = r_B = -r_C \approx kC_A$; $k = \alpha e^{-E/RT}$. C_{A0}, C_{B0}, C_{M0} are the molar concentrations for the inlet stream, T is the reactor temperature, C_A, C_B, C_C and C_M are the molar concentrations for the outlet stream (kmol/m^3), F_0 is the volumetric flow rate for the inlet stream (m^3/h), V is the volume of the reactor. From the molar flow rates of A, B, M, the inlet volumetric flow is:

$$F_0 = \frac{F_{mA0}}{\rho_{A0}} + \frac{F_{mB0}}{\rho_{B0}} + \frac{F_{mM0}}{\rho_{M0}} \qquad (17)$$

where $F_{mA0}, F_{mB0}, F_{mM0}$ are the molar flow rates for the inlet stream (kmol/h) and $\rho_{A0}, \rho_{B0}, \rho_{M0}$ are the molar densities of pure components A, B, M. The inlet concentrations in Eq.(16) are determined by:

$$C_{A0} = \frac{F_{mA0}}{F_0}; \quad C_{B0} = \frac{F_{mB0}}{F_0}; \quad C_{M0} = \frac{F_{mM0}}{F_0} \qquad (18)$$

- Reactor energy balance from Eq.(9)

$$Vc_{pr}\frac{dT}{dt} = F_{mA0}c_{ps}(T_0 - T) + (-\Delta H_R)\alpha \cdot C_A V e^{-E/RT} - UA(T - T_j) \quad (19)$$

where the parameters c_{pr}, c_{ps} are defined by:

$$c_{pr} = c_{pA}C_A + c_{pB}C_B + c_{pC}C_C + c_{pM}C_M \qquad (20)$$

$$c_{ps} = c_{pA} + \frac{F_{mB0}}{F_{mA0}}c_{pB} + \frac{F_{mM0}}{F_{mA0}}c_{pM} \qquad (21)$$

and $c_{pA}, c_{pB}, c_{pM}, c_{pC}$ are the molar heat capacities for A, B, M and C, respectively.

- The energy balance in the jacket coincides with Eq.(14)

Remark 5. Note that the composition of M can be deduced from Eq.(16), so in this case, even with a simple reaction, the presence of inert components gives a mathematical model with six nonlinear differential equations whose numerical analysis and equilibrium points can be difficult to obtain. An analysis of this problem can be found in [16], [17].

2.2   Linearized CSTR Model

Let us consider the steady-state behavior of the reactor in Example 1 under a constant input, $[F_{0e}, F_{je}]$. The reactor variables will reach an equilibrium point, $[C_{Ae}, T_e, T_{je}]$, vanishing the derivative terms in Eq.(15). In the following, incremental variables are considered:

$$\Delta C_A(t) = C_A(t) - C_{Ae}; \quad \Delta T(t) = T(t) - T_e; \quad \Delta T_j(t) = T_j(t) - T_{je}$$
$$\Delta F_0(t) = F_0(t) - F_{0e} \; ; \quad \Delta F_j(t) = F_j(t) - F_{je} \tag{22}$$

but, if there is no confusion, the same notation will be used. In this way, the equations of the linearized system are, $\dot{x}(t) = Ax(t) + Bu(t), \; y(t) = Cx(t)$. That is:

$$\begin{bmatrix} d\Delta C_A(t)/dt \\ d\Delta T_A(t)/dt \\ d\Delta T_j(t)/dt \end{bmatrix} = \begin{bmatrix} a_{11} & a_{12} & 0 \\ a_{21} & a_{22} & a_{23} \\ 0 & a_{32} & a_{33} \end{bmatrix} \begin{bmatrix} \Delta C_A(t) \\ \Delta T(t) \\ \Delta T_j(t) \end{bmatrix} + \begin{bmatrix} b_{11} & 0 \\ b_{21} & 0 \\ 0 & b_{32} \end{bmatrix} \begin{bmatrix} \Delta F_0(t) \\ \Delta F_j(t) \end{bmatrix}$$
$$\begin{bmatrix} y_1(t) \\ y_2(t) \end{bmatrix} = \begin{bmatrix} 1 & 0 & 0 \\ 0 & 1 & 0 \end{bmatrix} \begin{bmatrix} \Delta C_A(t) \\ \Delta T(t) \\ \Delta T_j(t) \end{bmatrix} \tag{23}$$

where:

$$a_{ij} = \left(\frac{\partial f_i}{\partial x_j} \right)_e ; \quad b_{ij} = \left(\frac{\partial f_i}{\partial u_j} \right)_e \tag{24}$$

Note that if $\Delta T_j = y_3$ is also measured, the C matrix will be the identity. From the first equation of Eq.(15) it is possible to deduce the equilibrium concentration of reactant A:

$$C_{Ae} = \frac{C_{A0}}{1 + \frac{\alpha V}{F_0} e^{-E/RT_e}} \tag{25}$$

where the reactor volume remains constant. From the second Eq. of (15) the energy balance in steady state can be written as follows:

$$\frac{F_0}{V}(T_0 - T_e) + \frac{H\alpha}{\rho c_p} C_{Ae} e^{-E/RT_e} - \frac{UA}{\rho c_p V}(T_e - T_{je}) \tag{26}$$

Substituting Eq.(25) into Eq.(26) the energy balance in steady state is formed by two terms:

• Heat generated by the reaction Q_g

$$Q_g = \frac{V H\alpha \, C_{A0} e^{-E/RT_e}}{1 + \frac{\alpha V}{F_0} e^{-E/RT_e}} \tag{27}$$

showing that the heat generated is proportional to the reactor volume, and

- Heat removed Q_r from the coolant flow rate and the inlet and out flow streams.

$$Q_r = \rho c_p F_0 (T_e - T_0) + UA(T_e - T_{je}) \tag{28}$$

At the equilibrium point, $Q_g = Q_r$. From the third Eq.(15) it is possible to deduce the jacket equilibrium temperature:

$$T_{je} = \frac{UAT_e + \rho_j c_{pj} F_j T_{j0}}{UA + \rho_j c_{pj} F_j} \tag{29}$$

Substituting Eq.(29) into Eq.(28) one obtains:

$$Q_r = \left(\rho c_p F_0 + \frac{\rho_j c_{pj} F_j}{1 + \frac{\rho_j c_{pj} F_j}{UA}} \right) T_e - \left(\rho c_p F_0 T_0 + \frac{\rho_j c_{pj} F_j T_{j0}}{1 + \frac{\rho_j c_{pj} F_j}{UA}} \right) = mT_e - b \tag{30}$$

Eq.(30) is a straight line and Eq.(27) is an S shaped curve as shown in Figure 2. This plot shows the curve of heat generated and the line of heat removed versus temperature in an exothermic CSTR. The three steady-states are the points of intersection P_1, P_2, P_3, of the curve Q_g and the line Q_r.

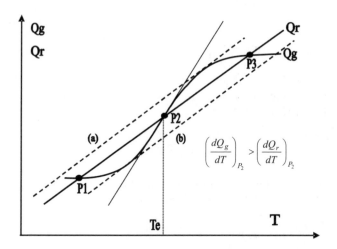

Fig. 2. Three steady states of an exothermic CSTR

It is easy to show that the equilibrium point P2 is unstable, whereas points P1 and P3 are stable (in some cases P3 can be unstable) [3], [13], [14], [22]. If the goal is to operate the reactor at point P2 it is necessary to introduce an external control.

In order to obtain the equilibrium points the linearization steps can be used:

1. Assume fixed values for the parameters of the reactor,
2. From the maximum value of $(dQ_g/dT)_{P_2}$ determine T_e for each reactor,
3. From a fixed value F_0 and the condition:

$$\left(\frac{dQ_g}{dT}\right)_{P_2} > \left(\frac{dQ_r}{dT}\right)_{P_2} \implies \tag{31}$$

$$\frac{V(\Delta H_R)\alpha C_{A0}E}{RT^2} \cdot \frac{e^{E/RT}}{\left(e^{E/RT} + \frac{\alpha V}{F_0}\right)^2} > \rho c_p F_0 + \frac{\rho_j c_{pj} F_j}{1 + \frac{\rho_j c_{pj} F_j}{UA}}$$

calculate a value for F_j,

4. From Eq.(27) and (30) calculate the value of T_0,
5. From Eq.(25) and (29) the equilibrium point $P_2[C_{Ae}, T_e, T_{je}]$ is computed,
6. Linearize the Eq.(15) at the equilibrium point P_2, according to Eq.(24).

From Eq.(23) it is possible to deduce the (2x2) transfer function matrix, $G(s)$, of the linearized MIMO system.

$$y(s) = G(s)u(s); \quad G(s) = C(sI - A)^{-1}B \tag{32}$$

The block diagram is shown in Figure 3, where the variables are incremental ones.

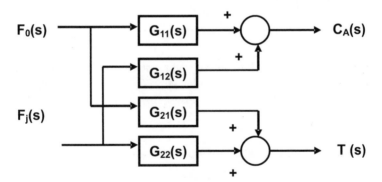

Fig. 3. Block diagram of the CSTR

The transfer functions in the different blocks can be easily computed by selecting the appropriate rows and columns in the matrices C and B, respectively, in Eq.(32). For example, by choosing the first row of C and the second column of B, from Eq.(23), the following transfer function is obtained:

$$G_{12}(s) = \frac{\Delta C_A(s)}{\Delta F_j(s)} = \frac{-b_{32}a_{21}a_{32}}{s^3 + \alpha_2 s^2 + \alpha_1 s + \alpha_0} \tag{33}$$

Similarly, taking the second output, it yields

$$G_{22}(s) = \frac{\Delta T(s)}{\Delta F_j(s)} = \frac{-b_{32}(s - a_{11})a_{32}}{s^3 + \alpha_2 s^2 + \alpha_1 s + \alpha_0} \tag{34}$$

where the parameters α_i are functions of a_{ij}.

3   Analysis of the CSTR's Linear Models

The diagram above shows an interactive MIMO system, where the controlled variables, outlet flow temperature and concentration, both depend on the manipulated variables. In order to design a decentralized control, a pairing of variables should be decided. A look at the state Eq.(23) suggests the assignment of the control of the temperature to the cooling flow and the concentration control to the reactor inlet flow. In this case, the internal variable T_j may be used to implement a cascade control of the reactor temperature. Nevertheless, a detailed study of the elements of the transfer matrix may recommend another option (see, for instance, [1]).

Let us assume some typical data, as summarized in Table 1. From Eq.(23), with the matrix C equal to identity matrix I, $C = I$, the following transfer functions can be easily obtained:

$$G_{11}(s) = \frac{\Delta C_A(s)}{\Delta F_0(s)}; \quad G_{21}(s) = \frac{\Delta T(s)}{\Delta F_0(s)}; \quad G_{31}(s) = \frac{\Delta T_j(s)}{\Delta F_0(s)} \tag{35}$$

$$G_{12}(s) = \frac{\Delta C_A(s)}{\Delta F_j(s)}; \quad G_{22}(s) = \frac{\Delta T(s)}{\Delta F_j(s)}; \quad G_{32}(s) = \frac{\Delta T_j(s)}{\Delta F_j(s)} \tag{36}$$

In order to investigate the controllability of the reactor, different reactor volumes are considered and a fixed inlet flow, F_0, is assumed. For $F_0 = 0.173$ m^3/h and $V = 21.28$ m^3 the eigenvalues of the matrix A, from Eq.(23) are:

$$|\lambda \cdot I - A| = 0 \Longrightarrow \lambda = \{-7.34, 3.41, -135.03\} \tag{37}$$

For the same F_0, but $V = 26.10$ m^3 the eigenvalues of the matrix A, are:

$$|\lambda \cdot I - A| = 0 \Longrightarrow \lambda = \{-9.79, 4.33, -151.91\} \tag{38}$$

In both cases the open-loop system is unstable but the location of the poles makes the second one more difficult to control. Sketching the root locus for the transfer function in Eq.(33), it is easy to verify that the system is conditionally stable. There is only a range of controller gain, $K \in (K_{min}, K_{max})$, leading to a closed-loop stable reactor.

Figure 4 shows the root locus for an example with a typical transfer function of a conditionally stable reactor. Note that from Eq.(35) and (36) different root locus can be deduced. The poles are the same but they have different zeros.

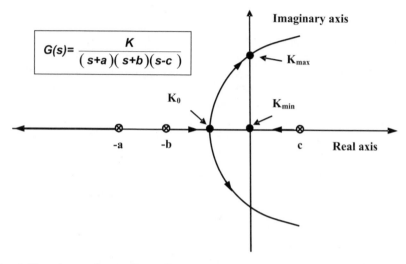

$$G(s)= \frac{K}{(s+a)(s+b)(s-c)}$$

Fig. 4. Root locus of a conditionally stable process. K_{min} means the minimum gain from which the system is stable. For $K > K_0$ the system starts to be underdamped. For $K > K_{max}$ the system is unstable.

Table 1. Parameter values of the reactor

Variable	Description	Value
C_{a0}	Reactant concentration inlet stream (kmol A/m^3)	5
α	Preexponential factor from Arrhenius law (h^{-1})	1.69×10^{13}
E	Activation energy (kJ/kmol)	85000
U	Overall heat transfer in the jacket (kJ/h·m^2·°C)	20124
T_{j0}	Inlet stream cooling water temperature (K)	294.4
R	Perfect-gas constant (kJ/kmol·K)	8.314
$-\Delta H_R$	Enthalpy of reaction (kJ/kmol)	84000
c_p	Heat capacity inlet and out streams (kJ/kg·K)	3.9
c_{pj}	Heat capacity of cooling water (kJ/kg·K)	4.18
ρ	Density of the inlet and out streams (kg/m^3)	1500
ρ_j	Density of cooling water (kg/m^3)	1000

3.1   Scaling Up the Reactor's Volume

It is well known, and it has been seen in the previous example, that to scale-up the size of a CSTR affects the reactor stability, because the ratio of heat transfer area to reactor volume decreases as far as the size of the reactor is increased (they are proportional to the square and power of three, respectively).

Consequently, larger reactors can be more difficult to control than smaller ones.

The study of the effect on the dynamics is carried out varying the volume of the reactor from small to large values. Various reactors (N=31) were considered, ranging from the smallest one (volume 0.0126 m³ numbered by 1 in Table 2) to the largest one (volume 42.41 m³, not shown in this table), their height and diameter being interpolated and indexed by the order i ($i = 1, 2, \ldots N$). Based on the previous results, to take into account that as far as the volume of the reactor increases the inlet flow stream rate must be also increased, the volumetric flow of the inlet stream is calculated from the following empirical relation, for each reactor:

$$F_0(i) = F_0(i-1) \left(\gamma \left(1 + \frac{V(i-1)}{V(i)} \right) \right)^{1/N} \qquad (39)$$

where the exponent of Eq.(39) depends on the total number of the reactors considered, N, is a variable factor between 1000 and 2000, and the subscript denotes a specific reactor.

Using Matlab© the volume and the heat transfer area of the reactors are deduced as follows:

```
>> N = input('Number of reactors. N = ');
>> hi = input('Initial high of reactor 1. hi = ');
>> hf = input('Final height of reactor N. hf = ');
>> di = input('Initial diameter of reactor 1. di = ');
>> df = input('Final diameter of reactor N, df = ');
>> h = [hi:(hf-hi)/N:hf];
>> d = [di:(df-di)/N:df];
>> Nr = length(h);
>> for i = 1:Nr
>>    V(i) = pi*(d(i)/2)^2*h(i);
>>    At(i) = pi*d(i)*h(i) + pi*(d(i)/2)^2;
>> end
```

So, from a reactor number, the volume and heat transfer area can be deduced.

Figure 5 shows the variation of the heat generated for different volume reactors (ranging between 22.82 to 42.41 m³) vs. the line of removed heat. It can be corroborated that as far as the volume of the reactor increases more heat is generated by the reaction. Note that the slope of the heat removed vs. the reactor volume increases more than proportionally, and consequently a high value of the heat removed results in a more difficult control operation.

Figure 6 shows the coolant flow rate and the feed rate curves vs. the reactor volume. Note that the flow rate of cooling increases rapidly as the reactor size increases, whereas the inlet flow rate increases according to Eq.(39).

Table 2. Effect of scale-up on variation of ratio K_{max}/K_0

Number	1	2	3	4	5
Reactor high (m)	0.4	1.1467	2.081	3.0133	4.88
Reactor diameter (m)	0.2	0.5733	1.040	1.5067	2.44
Heat transfer area (m^2)	0.2827	2.3225	7.6454	16.046	42.08
Reactor volume (m^3)	0.0126	0.2960	1.7669	5.3724	22.82
Cooling flow (m^3/min)	9.69×10^{-4}	4.60×10^{-3}	1.94×10^{-2}	7.41×10^{-2}	1.092
Reactor equilibrium temperature (K)	372.1	345.7	340.8	343.9	359.4
Jacket equilibrium temperature (K)	368.5	344.5	339.4	341.3	343.5
Heat transfer (kJ/min)	302	957	3647	1.45×10^4	2.24×10^5
K_{max}	3.73	0.109	0.0417	0.0447	0.322
K_0	0.0854	1.98×10^{-3}	9.38×10^{-4}	1.52×10^{-3}	0.0191
K_{max}/K_0	43.67	55.05	44.45	29.40	16.85

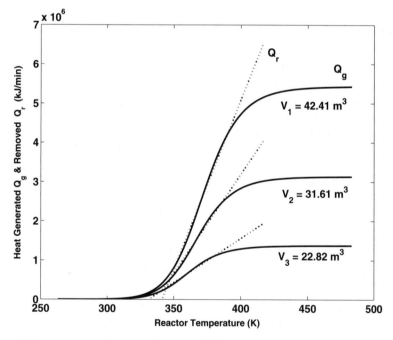

Fig. 5. Heat generated- removed vs. equilibrium reactor temperature at point P_2 of Figure 2

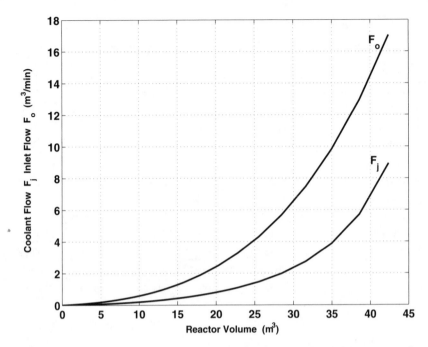

Fig. 6. Curve of flow rate of cooling and the inlet flow rate for N =30 reactors from $V_1 = 0.0126$ m^3 to $V_{30} = 42.41$ m^3

Taking into account these considerations it is possible to obtain a set of transfer functions. Nevertheless, if the range of stable gains K_{max}/K_0 (see Fig. 4) is computed for all the reactors, it appears that as the reactor's volume increases lower jacket temperatures are required, and the range of operation of the feedback controller decreases. Similar results can be obtained using Eq.(34) even, for instance, considering a fixed value of the damping ratio. The transfer function obtained for the small reactor of volume $V = 0.0126$ m^3 is:

$$G_{12} = \frac{\Delta C_A(s)}{\Delta F_j(s)} = \frac{444.92}{(s - 0.0598)(s + 0.514)(s + 2.664)} \qquad (40)$$

From Figure 7, the stabilizing gain range can be obtained giving the ratio $(K_{max}/K_0) = 44.33$. The transfer function for the reactor with volume $V = 42.41$ m^3 is:

$$G_{12} = \frac{\Delta C_A(s)}{\Delta F_j(s)} = \frac{0.0533}{(s - 0.1445)(s + 0.368)(s + 4.441)} \qquad (41)$$

The range of stable gains is: $K_{max} = 4.23$; $K_0 = 0.236$. The ratio between these gains is $(K_{max}/K_0) = 17.92$. So, the effect of a good or poor control can be quantified taking into account the previous considerations.

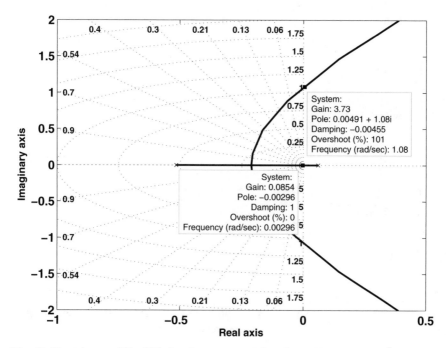

Fig. 7. Root locus of Eq.(36) for the reactor with volume $V = 0.0126$ m^3, $K_{max} = 3.73$, $K_0 = 0.0854$ and $(K_{max}/K_0) = 44.33$

Another important effect that can be analyzed is the relation between the equilibrium reactor temperature and the equilibrium jacket temperature. It is known that temperature difference between cooling jacket and the reactor must be increased as the volume of the reactor increases. Figure 8 shows this effect clearly. When the reactor has a small volume the difference $T_e - T_{je}$ is very small, consequently the heat transfer process is slower and the operation control is easier. Table 2 quantitatively summarizes the effects previously commented for a typical reactor modelled by Eq.(23) with the parameters defined in table 1. As the reactor volume varies from 0.0126 to 42.41 m^3, lower jacket temperatures are required and the operation control is more difficult.

Exercise 1. Compute the range of gains stabilizing $G_{11}(s)$ in the reactors in Table 1 and plot this ratio vs the reactor volume.

Exercise 2. Compute the transfer matrix in Eq.(32) assuming the reactor data corresponding to the smallest one in Table 1. Choose a gain around the value of K_0 in Figure 4 for both, $G_{11}(s)$ and $G_{22}(s)$. Apply this control to the reactor and plot the step responses to changes in the references. Is there something you did not expect? What about the system interaction? See the next Section.

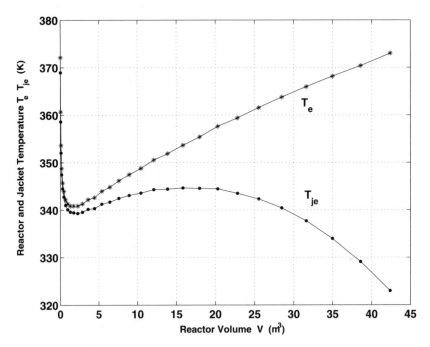

Fig. 8. Reactor and jacket equilibrium temperatures vs. reactor volume

4 Local Control of the CSTR

Based on the linearized models around the equilibrium point, different local controllers can be implemented. In the discussion above a simple proportional controller was assumed (unity feedback and variable gain). To deal with
multivariable systems two basic control strategies are considered: centralized
and decentralized control. In the second case, each manipulated variable is
computed based on one controlled variable or a subset of them. The rest of
manipulated variables are considered as disturbances and can be used in a
feedforward strategy to compensate, at least in steady-state, their effects. For
that purpose, it is typical to use PID controllers. The multi-loop decoupling is
not always the best strategy as an extra control effort is required to decouple
the loops.

4.1 Decentralized Control

In this case, the typical CSTR control structure is as shown in Figure 9,
with a single-loop control of the composition and a cascade control of the
temperature. The block diagram is shown in Figure 10.

Loop 1 is a conventional feedback control where AC shows a PI controller,
G_{r1}, controlling the composition of the outlet stream. The control valve CV1

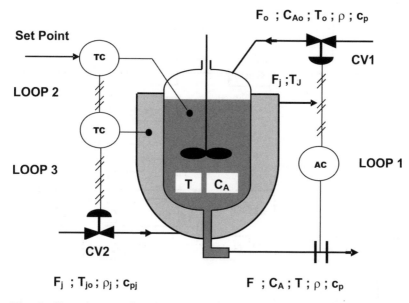

Set Point

F_o ; C_{Ao} ; T_o ; ρ ; c_p

TC

CV1

LOOP 2

F_j ;T_J

TC

AC

LOOP 1

LOOP 3

T C_A

CV2

F_j ; T_{jo} ; ρ_j ; c_{pj}

F ; C_A ; T ; ρ ; c_p

Fig. 9. Cascade control and conventional feedback local control of a CSTR

manipulates the inlet volumetric flow to the reactor in order to maintain the value of C_A at the desired set point. Consequently, the equation of the PI controller will be:

$$F_0(t) = F_{0e} - K_{p1}\left[C_A(t) - C_{Ae} + \frac{1}{\tau_1}\int_0^t (C_A(t) - C_{Ae})dt\right]$$

$$G_{r1}(s) = \frac{F_0}{-C_A} = K_{p1}\left(1 + \frac{1}{\tau_1 s}\right) \tag{42}$$

where K_{p1} and r_1 are the controller proportional and integral action respectively and again, incremental variables are considered. In this loop, the cooling flow acts as a disturbance.

The reactor temperature controller (loop 2) is the primary controller, whereas the jacket temperature controller (loop 3) is the secondary controller. The advantage of the cascade control is that the reactor temperature control quickly reacts by the cascade system to disturbances in cooling fluid inlet conditions. The dynamics of the transfer function G_{32} is faster than that of G_{22}. In the CSTR cascade control there are two control loops using two different measurements temperatures T and T_j, but only one manipulated variable F_j. The transfer function of the primary controller is the following:

$$G_{r2}(s) = \frac{T_i(s)}{T_r(s) - T(s)} = K_{p2}\left(1 + \frac{1}{\tau_2 s}\right) \tag{43}$$

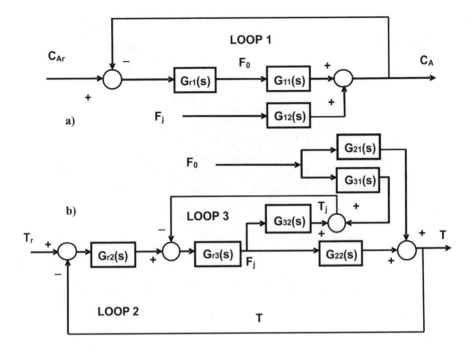

Fig. 10. Block diagrams: a) Single loop concentration control and b) Cascade temperature control

where $T_i(t)$ is a pseudo reference for $T_j(t)$. The transfer function of the secondary controller, which manipulates the coolant flow rate to the reactor, has the following equation:

$$G_{r3}(s) = \frac{F_j(s)}{T_i(s) - T_j(s)} = K_{p3}\left(1 + \frac{1}{\tau_3 s}\right) \tag{44}$$

where the input signal is the difference between an auxiliary temperature $T_i(t)$ and the (incremental) jacket temperature $T_j(t)$.

Remark 6. Note that in order to achieve a good control the loop 3 must be faster than the loop 2. The values of the PI controllers can be obtained from the basic methods of engineering process control. As an example, the following data have been obtained for the reactor number 4 in Table 2:

$$G_{11}(s) = \frac{\Delta C_A(s)}{\Delta F_0(s)} = \frac{0.43743(s + 0.4767)}{(s - 1.354)(s + 2.189)}$$

$$G_{12}(s) = \frac{\Delta C_A(s)}{\Delta F_j(s)} = \frac{287.4774}{(s - 1.354)(s + 2.189)(s + 111.2)}$$

$$G_{21}(s) = \frac{\Delta T(s)}{\Delta F_0(s)} = \frac{-4.0932(s + 101.8)(s + 1.241)}{(s - 1.354)(s + 2.189)(s + 111.2)}$$

$$G_{22}(s) = \frac{\Delta T(s)}{\Delta F_j(s)} = \frac{-600.113(s + 4.451)}{(s - 1.354)(s + 2.189)(s + 111.2)}$$

$$G_{31}(s) = \frac{\Delta T_j(s)}{\Delta F_0(s)} = \frac{-394.126(s + 1.241)}{(s - 1.354)(s + 2.189)(s + 111.2)}$$

$$G_{32}(s) = \frac{\Delta T_j(s)}{\Delta F_j(s)} = \frac{-58.4(s^2 + 10.21s + 40)}{(s - 1.354)(s + 2.189)(s + 111.2)}$$

According to the diagram in Figure 10, by using, for instance the root locus approach, the following controllers have been graphically tuned:

$$G_{r1}(s) = \frac{\Delta F_0(s)}{\Delta E_{C_A}(s)} = 20\left(1 + \frac{1.5}{s}\right) \ ; \ G_{r2}(s) = \frac{\Delta T_i(s)}{\Delta E_T(s)} = -40\left(1 + \frac{2}{s}\right)$$

$$G_{r3}(s) = \frac{\Delta F_j(s)}{\Delta E_{T_i}(s)} = -20\left(1 + \frac{1.5}{s}\right)$$

It can be easily checked that the step response of the two controlled variables, Figure 10, is appropriated under single changes in the respective reference, but the interaction is rather strong and any change in one reference acts as a disturbance in the other control loop.

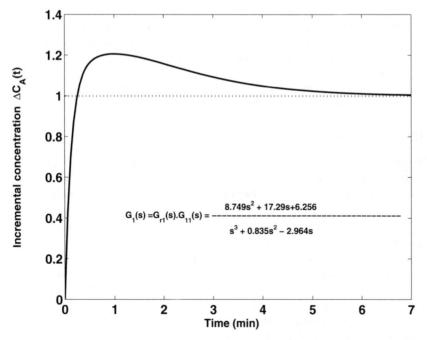

Fig. 11. Step response to changes in the reference: a) Concentration

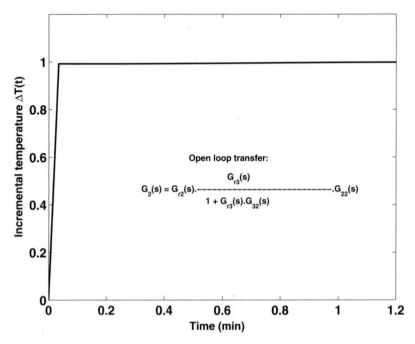

Fig. 12. Step response to changes in the reference: b) Temperature

Exercise 3. From table 2 repeat the previous procedure for the reactor number three and number five and show that if the reactor volume increases the response of Fig. 11 and 12 become more oscillatory.

Exercise 4. From the loop 1 and loop 2 of the block diagram in Fig. 10 determine the response to step disturbances in $F_j(t)$ and $F_0(t)$

4.2 Decoupling

In order to better tune both control subsystems, a feedforward prefiltering to decouple the two loops can be implemented. Nevertheless, this extra effort to decouple both processes may be a drawback in the solution, as both loops could work together to achieve the global goals. Anyway, let us assume a two input two output process (forgetting the access to the jacket temperature), corresponding to the model represented in Figure 3.

The decoupling structure is depicted in Figure 11. The global transfer matrix is diagonal. It is clear that the decoupling, as stated, is not always possible, as the transfer function of the different blocks should be stable and physically feasible. If it is possible, the new input variable u_1 will control the concentration and u_2 will control the temperature, and both control loops could be tuned independently. Sometimes, a static decoupling is more than enough.

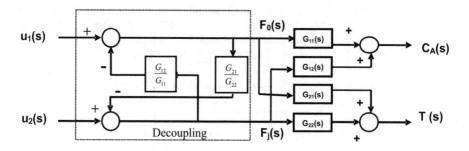

Fig. 13. Dynamic decoupling

4.3   Centralized Control

In this setting, all the measured variables are used to compute all the control actions (see, for instance, [1]). A special case is possible if all the state variables are measured. Using state feedback control law, that is:

$$u(t) = Kx(t) \qquad (45)$$

the feedback gain, K, can be computed based on different approaches:

- *Pole placement.* K is determined to assign the closed loop poles given by $eig(A + BK)$,
- *Optimal control.* K is determined to minimize a cost index such as:

$$J = \int_0^\infty (x^r Qx + u^r Ru)dt \qquad (46)$$

- *Decoupling.* K is determined to decouple some blocks of process/ manipulated variables. This is usually complemented by a decentralized control or a pole placement strategy.

In any case, the general control structure is as depicted in Figure 14, where integral actions have been introduced to avoid the static errors. For pole placement, there are many options to implement the feedback law and additional requirements (such as decoupling, as already mentioned) can be settled.

If all the state variables are not measured, an observer should be implemented. In the Figure 14, the jacket temperature is assumed as not measured, but it can be easily estimated by the rest of inputs and outputs and based on the separation principle, the observer and the control can be calculated independently. In this structure, the observer block will provide the missing output, the integrators block will integrate the concentration and temperature errors and, these three variables, together with the directly measured, will input the state feedback (static) control law, K. Details about the design of these blocks can be found in the cited references.

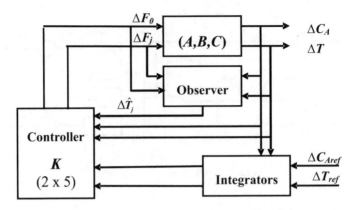

Fig. 14. State feedback control

Centralized control can be also designed based on disturbance rejection or robustness requirements. In this case, the controller is not a static linear feedback law, as (45), but a dynamic feedback controller is obtained. Additionally, two degree of freedom controllers allow for a better control behavior in tracking and regulation. All these alternatives are beyond the scope of this introductory local control design treatment and are the subject of specialized references (see, for instance, [19]).

Example 3. A MIMO design of the CSTR will illustrate the many possibilities of these approaches. Assume again the same reactor parameters (number 4 in Table 2). The system equations (23) are:

$$
\begin{bmatrix} d\Delta C_A(t)/dt \\ d\Delta T(t)/dt \\ d\Delta T_j(t)/dt \end{bmatrix} = \begin{bmatrix} -4.45 & -0.479 & 0 \\ 30.0376 & -5.5747 & 10.2743 \\ 0 & 96.287 & -101.83 \end{bmatrix} \begin{bmatrix} \Delta C_A(t) \\ \Delta T(t) \\ \Delta T_j(t) \end{bmatrix}
$$

$$
+ \begin{bmatrix} 0.4374 & 0 \\ -4.0932 & 0 \\ 0 & -58.41 \end{bmatrix} \begin{bmatrix} \Delta F_0(t) \\ \Delta F_j(t) \end{bmatrix} \tag{47}
$$

$$
\begin{bmatrix} y_1(t) \\ y_2(t) \end{bmatrix} = \begin{bmatrix} 1 & 0 & 0 \\ 0 & 1 & 0 \end{bmatrix} \begin{bmatrix} \Delta C_A(t) \\ \Delta T(t) \\ \Delta T_j(t) \end{bmatrix}
$$

It is also assumed that all the state variables are measured, so it is not necessary to estimate the jacket temperature $T_j(t)$. Then, the block diagram of the Figure 14 can be simplified as shown in Figure 15.

The purpose of this example is to show how to design the control system by using the pole-placement technique and the use of integrators. The integrators can be represented by introducing a new set of state variables $\nu(t)$, so the equations of the global system are the following:

$$\dot{x}(t) = Ax(t) + Bu(t)$$
$$y(t) = Cx(t) \qquad (48)$$
$$\dot{\nu}(t) = r(t) - y(t)$$

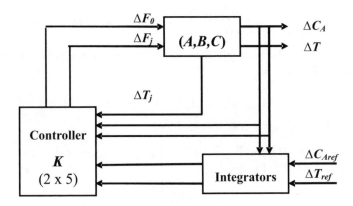

Fig. 15. State feedback control without estimation of ΔT_j

Now, the vector and matrix dimensions are: $A(3 \times 3)$; $B(3 \times 2)$; $C(2 \times 3)$; $x(3 \times 1)$; $u(2 \times 1)$; $y(2 \times 1)$; $\nu(2 \times 1)$; $r(2 \times 1)$. Eqs.(48) can be written as follows

$$\begin{bmatrix} \dot{x}(t) \\ \dot{\nu}(t) \end{bmatrix} = \begin{bmatrix} A & 0 \\ -C & 0 \end{bmatrix} \begin{bmatrix} x(t) \\ \nu(t) \end{bmatrix} + \begin{bmatrix} 0 \\ I_\nu \end{bmatrix} r(t) + \begin{bmatrix} B \\ 0 \end{bmatrix} u(t)$$
$$\begin{bmatrix} y(t) \\ \nu(t) \end{bmatrix} = \begin{bmatrix} C & 0 \\ 0 & I_\nu \end{bmatrix} \begin{bmatrix} x(t) \\ \nu(t) \end{bmatrix} \qquad (49)$$

In accordance with the pole-placement technique, the linear time-invariant feedback control is defined as:

$$u(t) = -\begin{bmatrix} K & K_\nu \end{bmatrix} \begin{bmatrix} x(t) \\ \nu(t) \end{bmatrix} = -K_r \begin{bmatrix} x(t) \\ \nu(t) \end{bmatrix} \qquad (50)$$

Substituting Eq.(50) into Eq.(49) the closed loop equations of the system can be written as:

$$\begin{bmatrix} \dot{x}(t) \\ \dot{\nu}(t) \end{bmatrix} = (A_a - B_a K_r) \begin{bmatrix} x(t) \\ \nu(t) \end{bmatrix} + \begin{bmatrix} 0 \\ I_\nu \end{bmatrix} r(t) \qquad (51)$$

where

$$A_a = \begin{bmatrix} A & 0 \\ -C & 0 \end{bmatrix}; \quad B_a = \begin{bmatrix} B \\ 0 \end{bmatrix} \qquad (52)$$

The input vector $r(t)$ are references (step functions). Taking into account an state error vector $e(t)$, $u(t)$ and $r_r(t)$ defined by:

$$e(t) = \begin{bmatrix} x(t) \\ \nu(t) \end{bmatrix} - \begin{bmatrix} x_e \\ \nu_e \end{bmatrix}; \quad u_r(t) = u(t) - u_e; \quad r_r(t) = r(t) - r_e \qquad (53)$$

where the sub-index e means steady state, from Eq.(49) the following error equation can be deduced:

$$\left. \begin{array}{l} \dot{e}(t) = A_a e(t) + B_a u_r(t) \\ u_r(t) = -K_r e(t) \end{array} \right\} \Longrightarrow \dot{e}(t) = (A_a - B_a K_r)e(t) \qquad (54)$$

The matrix K_r can be uniquely determined if the system (54) is completely state controllable, i.e., if:

$$C = [B_a \vdots A_a B_a \vdots A_a^2 B_a \vdots A_a^3 B_a \vdots A_a^4 B_a]; \quad rank(C) = 5 \qquad (55)$$

From Eq.(49), in steady state, it is deduced

$$\begin{bmatrix} 0 \\ 0 \end{bmatrix} = \begin{bmatrix} A & B \\ -C & 0 \end{bmatrix} \begin{bmatrix} x_e \\ u_e \end{bmatrix} + \begin{bmatrix} 0 \\ r_e \end{bmatrix} \Longrightarrow \begin{bmatrix} x_e \\ u_e \end{bmatrix} = - \begin{bmatrix} A & B \\ -C & 0 \end{bmatrix}^{-1} \begin{bmatrix} 0 \\ r_e \end{bmatrix}$$

$$u_e = -K_r \begin{bmatrix} x_e \\ \nu_e \end{bmatrix} \qquad (56)$$

Eqs.(56) can be used to verify the values of steady state from the simulation of Eqs. (49).

The simulations are carried out by using Matlab© throughout the following steps:

1. Check that the rank of the controllability matrix is 5:

   >> Aa = [A zeros(3,2); -C zeros(2)];Ba=[B; zeros(2)];
   >> Co = ctrb(Aa,Ba); rank(Co)

2. 2. Choose a set of desired closed-loop poles, for example:

   >> P = [-1+j   -1-j   -6   -7   -8];

3. 3. From the command place determine the value of K_r:

   >> Kr = place(Aa,Ba,p)

   $$Kr = \begin{bmatrix} -11.2142 & -1.5508 & -3.1753 & 3.0584 & 9.6902 \\ -0.4784 & -1.7551 & 1.5459 & 0.5480 & 0.8971 \end{bmatrix}$$

4. From Eq.(54) determine the matrices:

   >> Aaa = Aa - Ba*Kr;
   >> Baa = [zeros(3,2); eye(2);
   >> Caa = eye(5);
   >> Daa = zeros(5,2);

5. Simulate Eq.(51) with the command *lsim*:

```
>> t = [0:0.005:15]; N = length(t);
>> u = [-0.1*ones(1,N); ones(1,N); % Input reference
>> z = lsim(Aaa,Baa,Caa,Daa,u,t);
>> plot(t,z);
```

The results are shown in Figure 16

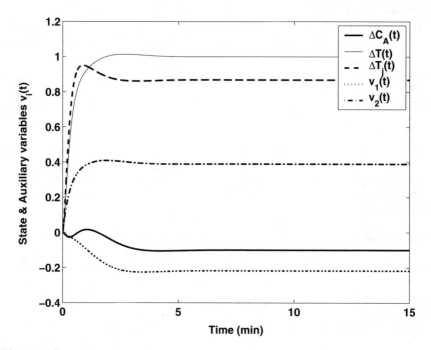

Fig. 16. Step response by using auxiliary state variables $\nu(t)$. Set point values: $\Delta C_{Aref} = -0.1$ kmolA/m^3; $\Delta T_{ref} = 1$ °C.

From Figure 16 it follows that the response of the reactor has no offset due to the integrators block. Another interesting aspect is that the incremental jacket temperature is lower than the incremental reactor temperature, so the heat transmission goes from the reactant to the coolant flow rate. The values of the control signals, i.e. the outlet and inlet flow rate deduced from Eq.(50), are shown in Figure 17.

From Eqs.(56) is possible to obtain the steady state values of the state variables $\Delta C_A(t)$, $\Delta T(t)$, $\Delta T_j(t)$, the control signals $\Delta F_0(t)$, $\Delta F_j(t)$ and the auxiliary variables $\nu(t)$. The commands by using Matlab$^{©}$ are the following:

```
>> P = [A B; -C =]; rank(P)
>> % rank(P) = 5
>> xuinf = -inv(P)*[0 0 0 -0.1 1]';
>> % xuinf = [-0.1 1 0.8682 0.0777 0.1349]';
```

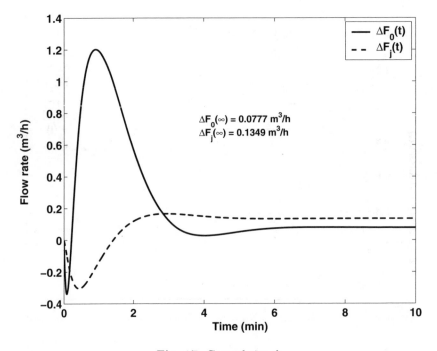

Fig. 17. Control signals

Note that the values xuinf(1:3) = [-0.1 1 0.8682] coincide with the values obtained from the steady state of the solution vector: z(1:3,N)= [-0.1 1 0.8682]. The values xuinf(3:4) = [0.0777 0.134] are the same of those in Figure 17. The steady state auxiliary variables are deduced as follows:

$$
\begin{bmatrix} u_{1e} \\ u_{2e} \end{bmatrix} = -Kr(1:2,1:3) * \begin{bmatrix} -0.1 \\ 1 \\ 0.8682 \end{bmatrix} - Kr(1:2,4:5) * \begin{bmatrix} \nu_{1e} \\ \nu_{2e} \end{bmatrix}
$$

$$
\begin{bmatrix} \nu_{1e} \\ \nu_{2e} \end{bmatrix} = -inv(Kr(1:2,4:5)) * \left\{ \begin{bmatrix} 0.0777 \\ 0.1349 \end{bmatrix} + Kr(1:2,1:3) * \begin{bmatrix} -0.1 \\ 1 \\ 0.8682 \end{bmatrix} \right\}
$$

$$
= \begin{bmatrix} -0.2178 \\ 0.3896 \end{bmatrix}
$$

Consequently, the calculation process is correct.

5   Conclusion

From the results presented in this chapter, it has been shown that the first step in the control problem of a CSTR should be the use of an appropriate mathematical model of the reactor. The analysis of the stability condition at the steady states is a previous consideration to obtain a linearised model for control purposes. The analysis of a CSTR linear model is carried out trough a scaling up reactor's volume in order to investigate the difference between the reactor and jacket equilibrium temperatures as the volume of the reactor changes from small to high value.

From these models and these data, a set of open-loop transfer functions and the corresponding root-locus are deduced, showing that these transfer functions are conditionally stable, i.e. there is a range of the controller gain which gives a close-loop stable reactor. In this setting, it is interesting to obtain the relation between the maximum and minimum controller gain, when the reactor volume varies. This relation can be used as an intuitive measure of the reactor's controllability.

The local control of the CSTR by using centralized and decentralized control has been also analyzed. The decentralized control is studied with a cascade control with two PI primary and secondary controllers. From the block diagram, the step response to change in the concentration and temperature references are deduced. A short reference to the decoupling problem is also discussed.

The centralized control can be approached using different techniques: pole-placement, optimal control and loop decoupling. When the whole state is not accessible, a motivation to introduce a state observer is discussed. A detailed example when all state variables are accessible, i.e. when the state observer it is not necessary, has been explained. It is important to remark that the previously cited techniques are not widely used in CSTR control. This is due to the fact that these procedures require non-intuitive matrix tuning and computations, which are not familiar in the process industry. Nevertheless, for complex processes, these procedures can be the only solution to the control problem, when a limited set of sensors are available.

References

[1] P. Albertos and A. Sala. *Multivariable Control Systems: An Engineering Approach*. Springer Verlag, 2004.

[2] J. Alvarez-Ramírez, J. Suarez, and R. Femat. Robust stabilization of temperature in continuous-stirred tank reactors. *Chem. Eng. Sci.*, 52(14):2223–2230, 1997.

[3] R. Aris. *Elementary Chemical Reactor Analysis*. Prentice-Hall, Inc. Englewood Cliffs, New Jersey, 1980.

[4] R. Aris and N.R. Amundson. An analysis of chemical reactor stability and control. *Chem. Eng. Sci.*, 7:121–130, 1958.

[5] B.W. Bequette. *Process Control: Modeling, Design and Simulation*. Prentice Hall, New York, 2003.

[6] D.R. Coughanowr. *Process Systems Analysis and Control*. McGraw-Hill, New York, 2nd. edition, 1991.

[7] M. Dolnik, A.S. Banks, and I.R. Epstein. Oscillatory chemical reaction in a CSTR with feedback control of flow rate. *Journal Phys. Chem. A*, 101:5148–5154, 1997.

[8] R. Femat. Chaos in a class of reacting systems induced by robust asymptotic feedback. *Physica D*, 136:193–204, 2000.

[9] H.S. Fogler. *Elements of Chemical Reaction Engineering*. Prentice Hall, 1999.

[10] G. Kevrekidis, R. Aris, and L.D. Schmidt. The stirred tank forced. *Chem. Eng. Sci.*, 41(6):1549–1560, 1986.

[11] Z. Kubickova, M. Kubicek, and M. Marek. Feed-batch operation of stirred reactors. *Chem. Eng. Sci.*, 42(2):327–333, 1987.

[12] M.J. Kurtz, G. Yan Zhu, and M.A. Henson. Constrained output feedback control of a multivariable polymerization reactor. *IEEE Trans. Contr. Syst. Technol.*, 8:87–97, 2000.

[13] W.L. Luyben. *Process Modeling, Simulation and Control for Chemical Engineering*. McGraw-Hill, New York, 2nd. edition, 1990.

[14] T. Marlin. *Process Control*. McGraw-Hill, New York, 2000.

[15] B. Ogunnaike and W.H. Ray. *Chemical Control Process*. Oxford University, 1994.

[16] M. Pérez and P. Albertos. Self-oscillating and chaotic behaviour of a PI-controlled CSTR with control valve saturation. *J. Process Control*, 14:51–57, 2004.

[17] M. Pérez, R. Font, and M.A. Montava. Regular self-oscillating and chaotic dynamics of a continuos stirred tank reactor. *Comput. Chem. Eng.*, 26:889–901, 2002.

[18] W.H. Ray. *Advanced process control*. McGraw-Hill, New York, 1981.

[19] S. Skogestad and I. Postlewhite. *Multivariable Feedback Control*. John Wiley & Sons, New York, 1996.

[20] C.A. Smith and A.B. Corripio. *Principles and Practice of Automatic Process Control*. John Wiley & Sons, New York, 1997.

[21] M. Soroush. Nonlinear state-observer design with applications to reactors. *Chem. Eng. Sci.*, 52(3):387–404, 1987.

[22] G. Stephanopoulos. *Chemical Process Control: An introduction to theory and practice*. Prentice Hall, New Jersey, 1984.

[23] F. Teymour. Dynamics of semibatch polymerization reactors: I. Theoretical analysis. *A.I.Ch.E. Journal*, 43(1):145–156, 1997.

[24] F. Teymour and W.H. Ray. The dynamic behavior of continuous solution polymerization reactors-IV. Dynamic stability and bifurcation analysis of an experimental reactor. *Chem. Eng. Sci.*, 44(9):1967–1982, 1989.

[25] Y. Uppal, W.H. Ray, and A.B. Poore. On the dynamic behavior of continuous stirred tank reactors. *Chem. Eng. Sci.*, 29:967–985, 1974.

[26] D.A. Vaganov, N.G.V. Samoilenko, and V.G. Abranov. Periodic regimes of continuous stirred tank reactors. *Chem. Eng. Sci.*, 33:1133–1140, 1978.

Temperature Control Via Robust Compensation of Heat Generation: Isoparaffin/Olefin Alkylation

G. Carrizales-Martínez[1], R. Femat[2], and V. González-Álvarez[3]

[1] Division of Postgraduate Studies and Research, I.T.C.M.
carrizales@itcm.edu.mx
[2] Department of Applied Mathematics and Computational Systems, IPICyT
rfemat@ipicyt.edu.mx
[3] Department of Chemical Engineering, UdG
victorga@ccip.udg.mx

Summary. The temperature control of the heterogeneous alkylation, where H_2SO_4 is used as catalyst, is discussed in this paper. The problem is to design a control function to stabilize temperature in face of uncertain kinetic model. The proposed controller, whose structure resembles a PI-controller, is based on energy balance. The resulting feedback controller is robust, leads to an acceptable performance, and computes the temperature of the coolant from the refrigeration section to the reactor. The effectiveness and robustness of the designed controller for computing the coolant temperature is tested by means of simulations in such a manner that we study the effects of: (i) load disturbances, (ii) model uncertainties and (iii) initial conditions for estimation values.

1 Introduction

The catalytic alkylation involves the addition of an isoparaffin containing a tertiary hydrogen to an olefin. Basically, alkylation process combines olefines (propylene, butylenes and/or amylenes) with paraffins, as isobutane, in presence of strong acid to produce high octane branched chain hydrocarbons. Thus, mixtures of isoparaffin, $C_5 - C_{16}$, are alkylate products and can be divided in five families: (i) Trimethylpentanes (TMP's), (ii) light ends (less than $C_5 - C_7$ isoparaffin), (iii) dymethylhexanes (DMH's), (iv) heavy ends (higher than C_{10}) and (v) acid-soluble hydrocarbons, as conjunct polymers, ester, red oil, etc. The process is used by petroleum industry to prepare highly branched paraffins. In last 15 years, new alkylation plants were built. About 70% of this alkylate product is or will be produced using H_2SO_4 as catalyst (these new plants announcement agree well with information reported along 90's in Hydrocarbon Processing journal). There are some remainder plants where HF is

H.O. Méndez-Acosta et al. (Eds.): Dyn. & Ctrl. of Chem. & Bio. Proc., LNCIS 361, pp. 33–51, 2007.
springerlink.com

used as catalyst, however HF is a very hazard catalyst. Thus, although HF unit operation is well understood and there has been a large effort to ensure safe handing of HF catalyst [11], however H_2SO_4 units are preferable due to secure operation. Moreover, alkylation with sulfuric acid as catalyst has been studied early 70's (see, for instance, [16]). On the other hand, large efforts are being performed to develop acid-solid catalyst (see, for instance, [7], [19], [24]); however results in this direction are not complete yet. Although much progress has been made during past 50 years for understanding the complicated chemistry of alkylation, control of alkylation process is not an easy task if we think that alkylation chemistry is significantly different when distinct catalyst is used or when operating conditions change [1], [8].

Alkylation is a quite complex chemical process in which control can result in economic and environmental benefits. There are different control problems for an alkylation plant; for instance, regulation of isoparaffin/olefin ratio, reactor temperature stabilization, etc. Alkylation involves complex kinetics and mass transfer and one of the difficulties of this class of complex chemical systems is that the kinetic terms and/or mass transfer phenomena can be required by the feedback function to compute control action (e.g., heat generation terms can be required for stabilizing temperature [12]). A refrigeration section is required to regulate the reactor temperature because alkylation reaction is highly exothermic. Indeed, reactor should be maintained at uniform temperature; i.e., temperature fluctuation less than one degree. Several operations can be comprised into the refrigeration system [22]. The reaction section can be constructed as contactor (STRATCO) or as a reactor chain (EXXON). However refrigeration system basically consists of compressor and depropanizer (see Figure 1). That is, in the *refrigeration system*, the hydrocarbons that are vaporized due to the heat reaction are routed into the refrigeration compressor and, once compressed and depropanized, are returned to the reaction section. Due to very complex chemistry of alkylation, high computational effort can be required to compute the heat generation terms in such a manner that the refrigeration system reach the computed temperature for stabilizing the plant. Although high computational technology is on hand, high computational effort is undesirable. Then, a function with simple computational procedures but robustness is required to control the alkylation reactor. On the other hand, alkylation process involves mass and heat recycle streams (see Figure 1). Recycle streams are not feedback in the control sense. Recycle streams increases the order of the characteristic polynomial with subsequent complexity. In fact, process with recycle becomes unclear at present. Few efforts has been recently devoted to understand this dynamics and control of such kind of process (see [18] and references therein; [3]; [4]; [5]; [15]; [21]). Luyben in [18] showed that a recycle system requires a large time constant (related to residence time) because as recycling-rate increases as instability can be induced. That is, the time response of the recycle systems is needed large in order to open-loop dynamics remains stable. Hence, plants for recycle are designed with large tanks for secure operation. Large tanks isolates sequences

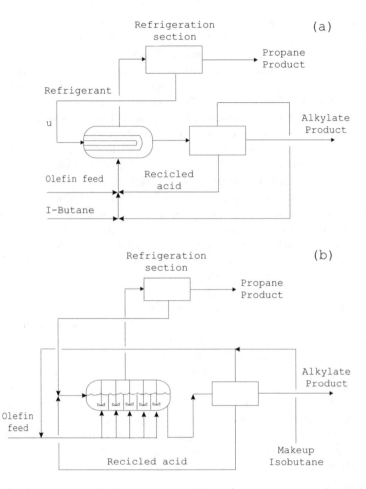

Fig. 1. Configuration or alkylation reactors. The refrigeration system basically consists in compressor and depropanizer. The main contribution is the design of an approach to the robust control temperature via heat reaction compensation. Thus, dynamical behavior of the refrigeration system is not considered.

of units and permits the use of cascade control configuration; i.e., each downstream unit (inner loop) simply sees disturbances coming from its upstream loop (external loop). If such disturbances are attenuated, the reactor can be stabilized. Results in regard to dynamics and control of recycle systems are promissory; however, up today, cascade scheme is, by secure operation, best option for controlling "wide plant".

In this work, the goal is to design a control function in such a manner that neither the reaction heat nor kinetic nor mass transfer terms are required for stabilizing temperature. The scheme provides an estimated value of the heat generation from energy balance. Alkylation isobutane/propylene using sulfuric acid as catalyst [16] was chosen to illustrate the design of a control function for complex systems. An estimated value of the heat generation is obtained from on line measurements (output feedback) and the proposed controller does not require full information. Indeed, the proposed feedback comprises an estimation algorithm based on closed-loop heat balance. The proposed controller exploits similarities among the alkylation processes for obtaining the estimated value of heat generation by reaction. The key similarities are: (i) alkylation reaction is exothermic. Therefore the control function represents the heat removal required for isothermic operation, i.e., the control action is *nonnegative*; (ii) mass transfer from hydrocarbon to acid solution restricts the concentration in acid solution and reaction occurs in acid phase, i.e., the inlet concentration in reaction phase is a bounded time function (see below); and (iii) since wide-plant process comprises very large tanks, the process is characterized by very slow dynamics, i.e., the time constant is large. Thus, the control function in this paper computes the coolant temperature such that the alkylation reactor is stable and can be used in cascade control. The idea behind the proposed controller is that the computed coolant temperature enters to the refrigeration section as reference. The main contribution of the paper is to perform the reactor temperature regulation with least prior knowledge about the heat generation by reaction (i.e., neither kinetic nor thermodynamics terms are required) for a very complex chemical system and recycle process. The controller is a bounded PI-like feedback with dynamic estimation of uncertainties. In addition for stabilizing temperature under uncertain kinetics, the disturbance attenuation is carried out by the proposed controller.

The text is organized as follows. Next section contains a brief review on robust temperature control of complex chemical systems. Alkylation system is presented in third section. Problem statement is discussed in fourth section. Temperature control is analyzed in fifth section. Controller performance and tuning rules are shown in sixth section. The paper is closed with some concluding remarks.

2   Brief Review on Reactor Temperature Regulation

There are several control problems in chemical reactors. One of the most commonly studied is the temperature stabilization in exothermic monomolecular irreversible reaction A→B in a cooled continuous-stirred tank reactor, CSTR. Main theoretical questions in control of chemical reactors address the design of control functions such that, for instance: (i) feedback compensates the nonlinear nature of the chemical process to induce linear stable behavior; (ii) stabilization is attained in spite of constrains in input control (e.g., bounded control or anti-reset windup); (iii) temperature is regulated in spite of uncertain kinetic model (parametric or kinetics type); or (iv) stabilization is achieved in presence of recycle streams. In addition, reactor stabilization should be achieved for set of physically realizable initial conditions, (i.e., global stabilization is required). Some questions have been solved and among solutions the following proposals are remarkable.

i) *Compensation of nonlinear terms via state feedback.* Henson and Seborg discussed in [12] the input-output linearization for temperature stabilization via nonlinear control theory. Their results were a timely contribution in the sense that geometrical control framework has potential application in chemical process for global stabilization. Such a procedure is based on nonlinear coordinate transformation from Lie derivatives, which, when applied to a given chemical process, can provide an input-output closed-loop linear dynamical behavior.

ii) *Stabilization under constrained control input.* Alvarez et al. reported in [2] that nonlinear bounded control is capable to stabilize exothermic CSTR. Authors departed from a global stabilization under control without saturation, namely $u = \mu(x)$, toward a bounded (constrained) control. The first one (unbounded control) is based on directional derivatives along vector fields (as in [12]). Alvarez et al. proved in [2] that under certain conditions (e.g., uniqueness of the open-loop CSTR behavior) the bounded control input yields asymptotic stabilization for a given set of the initial conditions, which is the physically realizable. On the other hand, their results implies that the stabilization of exothermic CSTR cannot necessarily reach the prescribed point. From practical point of view, these results make sense since bounded input implies that required heat removal could not be reach. Hence, under bounded feedback control, a given exothermic CSTR can be stabilized at an undesired critical point.

iii) *Stabilization against uncertain kinetic model.* a) An extension for global stabilization of chemical reactors was proposed by Viel et al. in [28]. That is, they proposed a feedback control such that stabilization can be achieved "whatever the initial conditions are", which means that initial conditions are located inside the physical domain of the system CSTR. Viel et al. have shown in [28] that if the sign of the reaction heat is constant the stabilization of chemical reactors can be attained in spite of unknown kinetic terms. First, the authors assumed that all states are available for feedback. Such an assumption was partially relaxed to develop an observed-based controller via temperature and concentration measurements. b) Femat et al. reported in [9] that temperature regulation in CSTR's can be achieved against unknown kinetics only from temperature measurements and for initial conditions belonging to the physical domain of the reactor even under bounded control actions. Triangle reaction was used to show the controller performance. Promissory results were found. However, results in [9] were developed for a fluidized bed reactor. This kind of reactors involves a small time constant (on the order of seconds) which, according to above discussion, does not necessarily imply cascade feedback control. In addition, one complex problem in process with small time constant is the noisy measurements.

Above results have been timely contributions for temperature regulation and reactor stabilization. However, the mechanism of the uncertainties compensation from measurements is not clear for systems with mass transfer. That is, from the control theory point of view, the design algorithms are proved and clearly established (for example the geometrical control theory). However, from the process point of view, there is an unclear meaning of the resulting controllers. Thus, there remain some questions. For example, a) can temperature regulation via robust feedback be interpreted from heat balance? and b) which is the physical meaning of the temperature stabilization against unknown kinetics via observed-based control?. Questions a) and b) have been solved for classical control (see for example [17]). However, robustness via geometrical control holds obscure in this sense. Here, the temperature regulation against uncertainties from measurements in terms of the heat balance for a class of rector with complex reaction and mass transfer are discussed. It is show that a robust observed-based controller can be interpreted from the heat balance point of view. To this end, an alkylation reactor allows us to present the robust temperature regulation via measurements and its physical interpretation from heat balance.

3   Alkylation Reactor Model: Kinetics, Mass Transfer and Dynamics

The goal of this section is to show the main features in reactor that affect its dynamics.

Kinetic model: First complexity in alkylation consists in the kinetic model. The alkylation of isobutane with propylene with sulfuric acid as catalyst has been chosen. A 17 reaction mechanism model was postulated in [16] to describe such an alkylation. Such a mechanism comprises 20 chemical species and including saturated hydrocarbon species. The kinetic model is based on Schmerling carbonium ion mechanism with modifications for accounting the formation of iC_9 and iC_{10} (see Table 1). This model was experimentally validated in a pilot scale CSTR under isothermal operation. Several experiments were carried out over a temperature range from 290 to 330K. There are differences in opinion concerning to the complexity of alkylation mechanism, then the model proposed in [16] is not the only one. However, their model is good enough for illustrating the objectives of this chapter. The results here discussed concern only for model including the kinetics proposed in [16] for alkylation. Experimental implementation is expected to corroborate simulation results and possibly extend to other alkylation mechanisms. Thus, the model for the kinetics taken in this contribution is given by

$$R_1(x_1, x_2^*) = x_{1,1}(k_2 x_{1,13} + k_3 x_{1,15} + k_4 x_{1,16} + k_5 x_{1,17}$$
$$+ k_6 x_{1,18} + k_7 x_{1,19} + k_8 x_{1,20})$$
$$R_2(x_1, x_2^*) = x_{1,2}(k_1[HX] + k_{11} x_{1,14} + k_{15} x_{1,17})$$
$$R_3(x_1, x_2^*) = k_2 x_{1,1} x_{1,13}$$
$$R_4(x_1, x_2^*) = k_3 x_{1,1} x_{1,15}$$
$$R_5(x_1, x_2^*) = k_4 x_{1,1} x_{1,16}$$
$$R_6(x_1, x_2^*) = k_5 x_{1,1} x_{1,17}$$
$$R_7(x_1, x_2^*) = k_6 x_{1,1} x_{1,18}$$
$$R_8(x_1, x_2^*) = k_7 x_{1,1} x_{1,19}$$
$$R_9(x_1, x_2^*) = k_8 x_{1,1} x_{1,20} \tag{1}$$
$$R_{10}(x_1, x_2^*) = k_9 x_{1,14} - k_{10} x_{1,10} x_{1,14}$$
$$R_{11}(x_1, x_2^*) = k_{13} x_{1,12} x_{1,14} + k_{17} x_{1,20} - k_{14} x_{1,11}[HX]$$
$$- k_{16} x_{1,11} x_{1,14}$$
$$R_{12}(x_1, x_2^*) = k_{12} x_{1,17} - k_{13} x_{1,12} x_{1,14}$$
$$R_{13}(x_1, x_2^*) = k_1 x_{1,2}[HX] - k_2 x_{1,13} x_{1,1}$$
$$R_{14}(x_1, x_2^*) = -x_{1,14}(k_9 + k_{10} x_{1,10} + k_{11} x_{1,2} + k_{13} x_{1,12} + k_{16} x_{1,11}$$
$$R_{15}(x_1, x_2^*) = k_{14} x_{1,11}[HX] + k_{17} x_{1,20} - k_3 x_{1,15} x_{1,1}$$
$$R_{16}(x_1, x_2^*) = k_{13} x_{1,12} x_{1,14} - k_4 x_{1,16} x_{1,1}$$
$$R_{17}(x_1, x_2^*) = k_{11} x_{1,14} x_{1,2} - k_5 x_{1,17} x_{1,1} - k_{15} x_{1,17} x_{1,2} - k_{12} x_{1,17}$$
$$R_{18}(x_1, x_2^*) = k_{10} x_{1,14} x_{1,10} - k_6 x_{1,18} x_{1,1}$$
$$R_{19}(x_1, x_2^*) = k_{16} x_{1,11} x_{1,14} - k_7 x_{1,19} x_{1,1}$$
$$R_{20}(x_1, x_2^*) = k_{15} x_{1,17} x_{1,2} - k_{17} x_{1,20} - k_8 x_{1,20} x_{1,1}$$

where $x_1 \in \mathbb{R}^{20}$ denotes the concentration vector of the chemical species (see below for notation), $[HX]$ denotes the sulfuric acid concentration, $k_i = k_i(x_2^*)$, $i = 1, 2, \ldots, 20$ are the constant rate for any fixed value of the temperature x_2^* in $[290, 330]$ and $R_i(x_1, x_2^*)$ denotes the i-th reaction rate corresponding to each chemical specie. In model (1) the parameters k's are temperature functions; in this sense the kinetics can be denoted as $R(x_{1,i}; k_j(x_2^*))$, indexes $i = 1, 2, \ldots, 20$ and $j = 1, 2, \ldots, 17$ stand for chemical specie and reaction in mechanisms, respectively.

Mass transfer: One more difficulty arises from the fact that there are two phases in the reactor: (i) hydrocarbon and (ii) acid. The reaction occurs in the acid phase while reactants are feed in hydrocarbon phase. This implies that, in order to reaction occurs, there is mass transfer from hydrocarbon to acid phase. The mass transfer is a very complex phenomenon which can involve the reaction-diffusion equation. However, such a phenomenon is beyond of the goal of this chapter. Both isobutane and propilene are feed in hydrocarbon phase. Solubility of propylene in acid phase is very fast whereas mass transfer of isobutane is dictated by the smaller mass transfer rate. That is the reactor is mass-transfer-limited by isobutane. Based on the following assumptions, a simplistic model can be found for mass transfer [27]: (MT.1) The propylene concentration at the hydrocarbon phase is constant. (MT.2) The steady-state of the mass transfer of iC_4 is equal to its consumption rate in the acid phase. (MT.3) The consumption rate of the isobutane in acid phase involves the intermediate and carbonium ion rate equations. In this manner, under the above assumptions, the mass transfer results in the isobutane concentration, $C_{ic,a}$, which can be expressed by

$$C_{iC4,a} = \frac{K_d x_{1,1}^h}{\left(1 + \frac{zH_a}{k_L a_V}\right)} \tag{2}$$

where $C_{iC4,a}$ is the isobutane concentration at acid phase, $x_{1,1}^h$ denotes the concentration of the isobutane at hydrocarbon phase, $z = k_2 x_{1,13} + k_3 x_{1,15} + k_4 x_{1,16} + k_5 x_{1,17} + k_6 x_{1,18} + k_7 x_{1,19} + k_8 x_{1,20}$ involves the rate of the isobutane consumption and subscript a means acid phase and the constant K_d (distribution coefficient for iC_4), $k_L a_V$ (term of mass transfer), H_a (the acid/hydrocarbon fraction) are positive.

In this manner the dynamical model can be obtained from mass and heat balance for the open-loop system [17]. STRATCO reactor (see Figure 1.a) was chosen for modeling under the following assumptions. (D.1) reactor is perfect mixed and volume in the reactor is constant, (D.2) reaction is carried out in the acid phase and kinetics is given by Eq.(1) and (D.3) mass transfer is given by: i) expression (2) for isobutane while propylene remains constant. Parameters of Eq.(2) are constant. It should be noted that the kinetic model (1) was obtained for isothermic process. Then, according to Langley and Pike in [16], the temperature dependence of the isobutane/propylene alkylation is given by $k_i(x_2) = k_{0,i} exp(-\alpha_i/x_2)$, where x_2 denotes the temperature, $k_{0,i}$ is the pre-exponential factor and $\alpha = Ea/R_g$, (Ea and R_g are, respectively, the activation energy and the constant of the ideal gases, $R_g = 8.314$ J mol^{-1} K^{-1}). The values of the these parameters are shown in Table A.1. Note that neither the products nor olefinic intermediate nor carbonium ions are feed into the reactor. In this way, the dynamical model of the alkylation reactor is given by

$$\dot{x}_{1,1} = \Theta(x_{1,1}^{IN} - x_{1,1}) - R_1(x_1, x_2)$$
$$\dot{x}_{1,2} = \Theta(x_{1,2}^{IN} - x_{1,2}) - R_2(x_1, x_2) \tag{3a}$$
$$\dot{x}_{1,i} = -\Theta x_{1,i} + R_i(x_1, x_2); \quad i = 3, 4, \ldots, 20$$

$$\dot{x}_2 = \Theta(x_2^{IN} - x_2) + \beta \left[\sum_{i=1}^{17} R_i'(x_1; k(x_2))\right] - \gamma(x_2 - u) \tag{3b}$$

Table 1. Reaction mechanism and rate constants at 330K with 95% of H_2SO_4 as catalyst

Reactions	Rate law $R'_i(x_1, x_2)$	Pre-exponential Factor (cc/gmol-s)	Activation Energy (Kcal/gmol)
Initiations			
$C_3^= + HX \xrightarrow{k_1} C_3^+X^-$	$k_1 AB$	1.01×10^7	2.35
$C_3^+X^- + iC_4 \xrightarrow{k_2} C_3 + iC_4^+X^-$	$k_2 AB$	2.07×10^{11}	0.0
Primary			
$iC_4^+X^- + C_3^= \xrightarrow{k_{11}} iC_7^+X^-$	$k_{11} AB$	1.99×10^{17}	2.36
$iC_7^+X^- + iC_4 \xrightarrow{k_5} iC_7 + iC_4^+X^-$	$k_5 AB$	4.20×10^{10}	0.0
Self-alkylation			
$iC_4^+X^- \xrightarrow{k_9} iC_4^= + HX$	$k_9 A$	3.92×10^4	0.40
$iC_4^+X^- + iC_4^= \xrightarrow{k_{10}} iC_8^+X^-$	$k_{10} AB$	5.63×10^{19}	4.10
$iC_8^+X^- + iC_4 \xrightarrow{k_6} iC_8 + iC_4^+X^-$	$k_6 AB$	5.35×10^{10}	0.0
Destructive Alkylation			
$iC_7^+X^- \xrightarrow{k_{12}} iC_7^= + HX$	$k_{12} A$	7.49×10^5	1.08
$iC_7^= + iC_4^+X^- \xrightarrow{k_{13}} iC_5^= + iC_8^+X^-$	$k_{13} AB$	3.64×10^{21}	5.73
$iC_5^= + HX \xrightarrow{k_{14}} iC_5^+X^-$	$k_{14} AB$	1.62×10^{11}	3.69
$iC_5^+X^- + iC_4 \xrightarrow{k_3} iC_5 + iC_4^+X^-$	$k_3 AB$	3.29×10^{10}	0.0
$iC_6^+X^- + iC_4 \xrightarrow{k_4} iC_6 + iC_4^+X^-$	$k_4 AB$	4.04×10^{10}	0.0
$iC_7^+X^- + C_3^= \xrightarrow{k_{15}} iC_{10}^+X^-$	$k_{15} AB$	3.72×10^{17}	2.59
$iC_{10}^+X^- + iC_4 \xrightarrow{k_8} iC_{10} + iC_4^+X^-$	$k_8 AB$	6.68×10^{10}	0.0
$iC_5^= + iC_4^+X^- \xrightarrow{k_1} iC_9X^-$	$k_{16} AB$	4.26×10^{19}	2.65
$iC_9X^- + iC_4 \xrightarrow{k_7} iC_9 + iC_4^+X^-$	$k_7 AB$	6.02×10^{10}	0.0
$iC_{10}^+X^- \xrightarrow{k_{17}} iC_5^= + iC_5^+X^-$	$k_{17} A$	4.45×10^{11}	8.40

where $x := (x_1, x_2) \in \mathbb{R}^{21}$, $x_1 \in \mathbb{R}^{20}$ is the concentration of the chemical species, x_2 is the temperature in the reactor, $(x_{1,1}^{IN}, x_{1,2}^{IN}) \in \mathbb{R}^2$ are the concentration feeding of isobutane and propylene, respectively. $R'_i(x_1, x_2)$, $i = 1, 2, \ldots, 17$ is the i-th rate laws in Table 1 with Arrenhuis-like temperature dependence. γ stands for the heat transfer coefficient of coolant system and β the reaction enthalpy, whose value has been experimentally obtained [20]. Θ is the inverse of the residence time. One should note that the concentration feed

$x_{1,1}^{IN}$ and $x_{1,2}^{IN}$ are restricted by the mass transfer. That is, $x_{1,1}^{IN}$ corresponds to the isobutane concentration at acid phase $C_{iC4,a}$, which is computed from Eq.(2) whereas $x_{1,2}^{IN}$ is constant due to mass transfer of propylene is fast.

4   Temperature Regulation Problem

In order to state the control problem, the following property of the alkylation reactor is presented.

Property 1. Consider an exothermic continuous stirred-tank reactor with temperature dependence Arrenhius-type, there is a stable equilibrium point such that, under the isothermic operation (i.e., as reactor temperature x_2^* is constant).

The above property was theoretically studied for CSTR's by Gavalas in [10] and guarantees the existence of, at least, one stable equilibrium point. Of course, multiple equilibrium points can be found in chemical reactors. Such an assumption also implies that reactor stabilization can be achieved via temperature regulation (see [2], [9], [12], [28]). Thus, according to Property 1, it is enough to design a feedback controller from subsystem (3b) in order to attain the reactor temperature stabilization at any equilibrium point. This makes sense in terms of the heat balance. That is, in order to control the reactor temperature, the most important term to be compensated in equation (3b) is the heat generation due to reaction, namely $\beta \sum_i R_i'(x_1)$, which, according to discussion in previous sections, can be often uncertain. Also the inlet heat term, Θx_2^{IN}, contributes to the reactor temperature; however, small contribution is expected for highly exothermic reactions. In addition, inlet heat term is not hard to compute or estimate because it does not comprise neither kinetic nor mass transfer terms.

From heat balance, the heat removal by jacket along time such that reactor operates in constant temperature x_2^* (isothermic operation) can be computed. Thus, from heat balance, the heat removal to induce steady-state in reactor temperature is as follows

$$\begin{pmatrix} \text{Heat} \\ \text{removal} \end{pmatrix} = \begin{pmatrix} \text{Heat generation} \\ \text{by reaction} \end{pmatrix} - \begin{pmatrix} \text{Heat entering} \\ \text{by inlet flow} \end{pmatrix} + \begin{pmatrix} \text{Heat leaving} \\ \text{by outlet flow} \end{pmatrix}$$

or   (4)

$$-\gamma(x_2 - u) = -\beta \sum_{i=1}^{m} R(x_1, x_2) - \Theta x_2^{IN} + \Theta x_2$$

where γ, Θ and β are defined in equation (3). Here, the reaction rate $R(x_1, x-2)$ corresponds to kinetics in Table 1 including temperature dependence (for example, Arrhenuis type).

As Property 1 holds, the heat removal (4) implies that the jacket temperature $u = (\gamma x_2 - \beta \sum_i R_i'(x_1, x_2) - \Theta x_2^{IN} + \Theta x_2)/\gamma$ induces the steady-state in subsystem (3b) from measurements of reactor temperature and assuming that model of reaction rate has no errors. This implies that the heat accumulation and generation are required in order to compute the heat removal. Thus, regarding (4), there are two drawbacks: (i) heat generation involves the reaction laws which, according to discussion in previous sections, are uncertain and (ii) although the jacket temperature computed from heat balance (4) induces temperature steady-state, the reactor dynamics is not regulated in a desired value (that is the isothermic operation does not assure uniqueness of the stable equilibrium points in chemical systems, e.g., autocatalysis, see Chapter 5 in [25]). On contrary, controller from heat balance (4) has the following advantages: i) a linear closed-loop behavior can be induced and the convergence rate can be directly related to the control gain, ii) if a gradient term $Kc(x_2 - x_2^r)$ is added to the feedback, then the coordinate of the (unique) equilibrium point x_2^* are located at the constant x_2^r (which is stable for suitable control parameter value). In fact, the jacket temperature $u = (\gamma x_2 - \beta \sum_i R_i'(x_1, x_2) - \Theta x_2^{IN} + \Theta x_2 + Kc(x_2 - x_2^r))/\gamma$ can also be derived from Lie derivative of the measured temperature along vector field of (3); i.e., as $L_g L_f^0 x_2 = \gamma$ and $L_f x_2 = \gamma x_2 - \beta \sum_i R_i'(x_1, x_2) - \Theta x_2^{IN} + \Theta x_2$. The gradient terms corresponds to the desired dynamics [12], [13]. Thus, one has that the jacket temperature to stabilize the reactor becomes: $u = (\gamma x_2 - \beta \sum_i R_i'(x_1, x_2) - \Theta x_2^{IN} + \Theta x_2 + Kc(x_2 - x_2^r))/\gamma$. Such control law can be interpreted in terms of the classical control theory (indeed the term $Kc(x_2 - x_2^r)$ is a proportional control action [17]) and exactly corresponds to the nonlinear state-feedback designed by geometric control theory (compare above equation with results in [12]). Nevertheless, the term $Kc(x_2 - x_2^r)$ is known as high-gain control and undesirable behavior can be induced; as, for example, scattering.

All above advantages and drawbacks should be considered to design the control function for the alkylation reactor. In this way the temperature control problem in alkylation can be worded *for finding the jacket temperature such that the heat balance* (4) *is held to reach the desired temperature value despite kinetic uncertainties.* This implies that reaction laws should be estimated in order to obtain a feedback control against errors in kinetic model. In what follows a feedback controller is proposed for accounting the heat generation with least prior information about the kinetic and chemical mechanism but by exploiting structure from heat balance. The main idea behind our proposal is to obtain an estimated value of temperature dynamics due to an estimated value of the heat generation by reaction.

5   Estimation of the Heat Generation

Let us define the time variable $\eta(t) = \beta \sum_i R_i'(x_1(t), x_2(t))$ toward obtainment of estimated value of heat generation term. Such a variable describes the generation heat due to reaction along the solution of the dynamical model (3).

That is, at time $t \in \mathbb{R}$, the heat generation is evolving according with rule $\beta \sum_i R_i'(x_1(t), x_2(t))$ along time and such an time-evolution can be seem as the variable $\eta(t)$. In other words, the heat reaction is a continuous-time assignation $\eta : \mathbb{R}^{21} \to \mathbb{R}$ whose time derivative $\dot{\eta}$ is defined for all $x \in \mathbb{R}^{21}$. In this sense, subsystem (3b) can be rewritten as follows

$$\dot{x}_2 = \Theta(x_2^{IN} - x_2) + \eta + \gamma(x_2 - u) \tag{5a}$$

$$\dot{\eta} = \beta \sum_{j=1}^{21} \frac{\partial \left(\sum_{i=1}^{21} R_i'(x_1(t)) \right)}{\partial x_j} f_j(x) \tag{5b}$$

where $f_j(x)$, for $j = 1, 2, \ldots, 21$ stands the right side terms in dynamical model (3). Note that time-derivative of the augmented state η can be derived by Lie derivative as follows $\dot{\eta} = L_f R_i'(x_1(t), x_2(t))$. Thus, the heat balance (4) can be interpreted in terms of the extended temperature system (5). One should note that, if the reaction law terms are uncertain, then the time derivative of the augmented state η is also uncertain. Thus, although $\dot{\eta}$ describes the dynamics of the heat generation term, the augmented state cannot be available for feedback. That is, the augmented state η cannot be measured from system (5). As matter of fact, equation (5) stands an intermediate dynamical system toward final controller such that final controller remains the main features of the heat balance (4), i.e., $u = (\gamma x_2 - \eta - \Theta x_2^{IN} + \Theta x_2 + Kc(x_2 - x_2^r))/\gamma$. However, system (5) has the property that it provides information about the directional derivative of the heat generation terms along the reactor trajectory. That is, the augmented state indicates when an uncertain term increases or decreases at any time $t \in [t, t + \Delta t]$. Now, since η is not available for feedback an estimation procedure is needed. Thus, following the ideas reported by Teel and Praly in [26], a state observer can be designed for system (5) from the output $y = x_2$ to obtain

$$\dot{\hat{x}}_2 = \Theta(x_2^{IN} - y) + \hat{\eta} - \gamma(y - u) + g_1(y - \hat{x}_2) \tag{6}$$
$$\dot{\hat{\eta}} = g_2(y - \hat{x}_2)$$

where $(\hat{x}_2, \hat{\eta})$ are estimated values of the reactor temperature x_2 and the augmented state η, which, by definition is equal to heat generation term, $\beta \sum_i R_i'(x_1(t))$. The parameters g_1 and g_2 are such that the polynomial $P_2(\lambda) = \lambda^2 + g_1 \lambda + g_2$ is Hurwitz. In this manner, the jacket temperature can be computed from the output y and the estimated value of generation heat term as follows

$$u = \frac{1}{\gamma} \left[\gamma y - \hat{\eta} - \Theta x_2^{IN} + \Theta y + Kc(y - x_2^r) \right] \tag{7}$$

where parameters are defined as above.

The aim of system (6) is to estimate the motion of the heat generation, which is the uncertain term. Fortunately, estimator (6) and feedback function

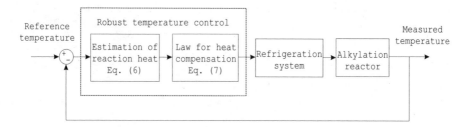

Fig. 2. Block diagram for the complete closed-loop behavior. Here, the block corresponding to the refrigeration system is neglected.

(7) leads to feedback control with PI-like structure. To prove this, we can take the Laplace transformation to obtain the transfer function of the system (6),(7). Thus, we have that the transfer function of the (heat compensation) control law becomes: $C(s) = u(s)/y(s) = Kc\left[1 + (1/\tau_I s) + (K_e/(s^2 + k_1 s))\right]$. Moreover, note that as $\hat{\eta}(t) \to \eta(t)$ as the controller (7) behaves like the heat compensation (4). That is, $\hat{\eta}(t) \to \eta(t)$ for all $t > 0$ then controller (7) is the linearizing feedback control. In addition, note that estimator (6) is a linear dynamical system with constant parameters. Hence, the transfer function of the estimator can be computed to get $\hat{\eta}(s) = G_E(s)y(s) + G_d(s)x_2^{IN}(s)$ where

$$G_E(s) = \frac{K_E}{s(\tau_E s + 1)}; \qquad G_d(s) = \frac{K_{IN}}{\tau_{IN} s + 1} \qquad (8)$$

and the parameters becomes $K_E = g_2(\Theta + \gamma - g_1 + Kc)/(\gamma + \Theta - g_1)$, $\tau_E = 1/(\gamma + \Theta - g_1)$, $K_{IN} = g_2 Kc$ and $\tau_{IN} = 1/(\gamma + \Theta - g_1)$. Transfer function $G_E(s)$ depends on the control and estimator gain. That is, the estimator and controller should be designed together (i.e., there is no separation principle).

Figure 2 shows the block diagram of the closed-loop process under above feedback. In regard the transfer functions $G_d(s)$ and $G_E(s)$, one should note that: a) $G_d(s)$ has low-pass structure. Indeed, the cut frequency is $\omega_c = 1/\tau_{IN}$; and increases as the estimation constant g_1. Then, high frequency disturbances from inlet temperature $x_2^{IN}(t)$ can be filtered by means of parameters tuning of the estimator (6). Nevertheless, as the control constant Kc increases the gain K_{IN} increases. These facts can involve an estimation/regulation tradeoff. However, under appropriate tuning, for suitable values of the control parameters the noisy measurements cannot affect the uncertainties estimation. b) Figure 3 shows the bode diagram for several values of the parameters Kc, g_1 and g_2. It is easy to see that there is a frequency ω^*, which depends on the control and estimation parameters, such that for any $\underline{\omega} < \omega^*$ the estimated value $\hat{\eta}(s)$ is more sensitive to output $y(s)$ than disturbance $x_2^{IN}(s)$, but there is any $\overline{\omega} > \omega^*(s)$ such that $\hat{\eta}(s)$ is more sensitive to disturbances $x_2^{IN}(s)$ than output $y(s)$. That is, $|G_E(\omega j)| > |G_d(\omega j)|$ for any $\omega < \omega^*$. Therefore, since an accurate estimated value $\hat{\eta}(t)$ is needed in order to proposed controller behaves like linearizing feedback, a tuning procedure is required in order to

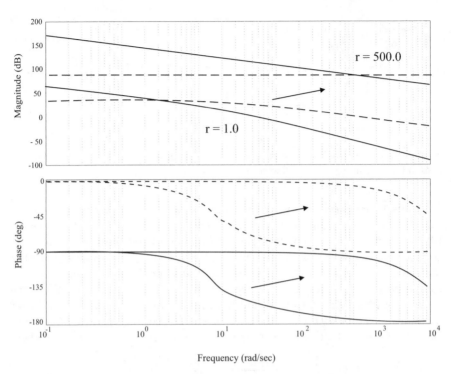

Fig. 3. Bode diagram of the closed-loop under the estimation/compensation control approach. Closed-loop system is expected to be non-sensitive to high frequency signals. Arrows indicate how the frequency response of the closed-loop as Kc is increased.

estimated value of uncertain term $\hat{\eta}(s)$ be not sensitive to external disturbance $x_2^{IN}(s)$.

The objective of the tuning procedure is to find the value of the parameters Kc, g_1 and g_2 such that the magnitude of $G_d(s)$ be smaller than magnitude of $G_E(s)$ for any ω less than the given frequency ω^*. Classical tuning methods include criteria based on trial and error and procedures which involve to determine process models and use heuristic basis [6], [14]. Here, since process model is available, we use heuristic basis to derive a tuning procedure via frequency response. Let $r > 1$ be an integer number such that $g_1 = -2rKc$ and $g_2 = (rKc)^2$. The number r denotes the estimation/regulation ratio; that is, for a given control gain Kc (which stands the regulation rate), as r increases as the roots of the polynomial $P_2(\lambda) = \lambda^2 + g_1\lambda + g_2$ are shifted to left within open left-hand complex plane. Figure 3 shows Bode plots for several estimation/regulation ratio r and control gains Kc. Then, since $|G_E(\omega j)| > |G_d(\omega j)|$ for any $\omega < \omega^*$.

6   Numerical Simulation

Here, the performance of the heat compensation controller is tested. To this end, the model parameters were taken as follows [16]: $x_{1,1}^{IN} = 862.5$ mol m^{-3}, $x_{1,2}^{IN} = 145.4$ mol m^{-3}, [HX] $=1776$ mol m^{-3}, $x_2^{IN} = 283$ K, $\Theta = 8.33 \times 10^{-4}$, 1.49×10^{-3}, $\beta = 0.3266$ K m^3 mol^{-1}. The controller (6),(7) was interconnected with nonlinear model (3) to numerically simulate its performance.

The idea behind the proposed feedback is the estimation of the generated heat by reaction, then simulations are aimed to show how the reactor temperature stabilization is affected by the initial value of the estimated heat, $\hat{\eta}(0)$. Figure 4 shows the reactor temperature and the computed coolant temperature for several initial values $\hat{\eta}(0)$. Here the estimation parameter was arbitrarily chosen $L = 0.5$. Note that as the estimated value decreases the convergence to the reference temperature, 283 K, is reached. Figure 5 shows the same effect for the value of the estimation parameter $L = 5.0$. By comparing both Figures 4 and 5, we can observe that as the value of L increases

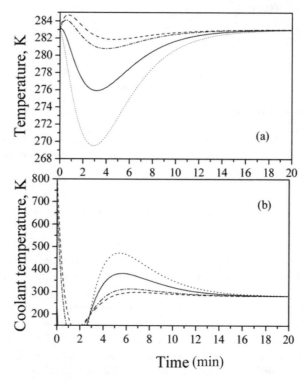

Fig. 4. Effects of the estimated initial value of the reaction heat for $L = 0.5$. Dotted-line, $\hat{\eta}(0) = 10.0$; Solid-line, $\hat{\eta}(0) = 5.0$; Dash-dotted line, $\hat{\eta}(0) = 1.0$; Dashed-line, $(0) = 0.0$. Smaller initial values of the heat reaction lead to better performance than large values. a) Reactor temperature and b) coolant temperature.

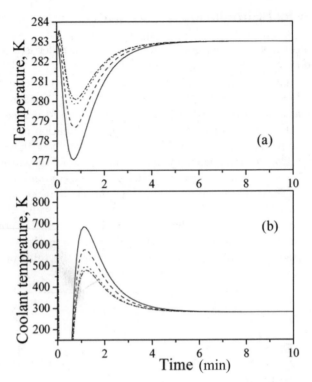

Fig. 5. Effects of the initial value of the reaction heat for $L = 5.0$. Note that as L increases as faster reactor temperature convergence (compare with Figure 2). Solid-line, $\hat{\eta}(0) = 10.0$; Dashed-line, $\hat{\eta}(0) = 5.0$; Dotted-line, $\hat{\eta}(0) = 1.0$; Dash-dotted line, $\hat{\eta}(0) = 0.0$. In regard initial value of estimated value $\hat{\eta}(0)$, the effect is similar than for $L = 0.5$. a) Reactor temperature and b) coolant temperature.

as the faster convergence; however, the control cost for faster convergence is the saturation of the coolant temperature. Thus, a trade off between fast convergence and control performance is found. The effect of increasing the estimation parameter L is shown in Figure 6. After extensive simulations, we found that the performance satisfies the Integral Square Errors and Control (ISEC) performance index (see Chapter 7 in [17]) corresponds to $L = 1.0$.

Note that, for large value of estimation parameter L, the control input (coolant temperature) can be saturated due to finite-time escape of the control signal (peaking phenomenon). Saturation is a nonlinear function that induces discrepancy between the computed and actual control input. Such a discrepancy is often related to actuator constraints [23]. Thus, the feedback is broken and the process behaves as an open-loop plant with the possible performance degradation. Indeed, the degradation can include windup phenomenon in presence of integration actions. Since the estimation of the heat generation is computed from (6) (i.e., it includes integration parts), an antireset-windup

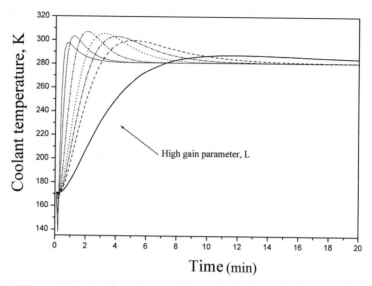

Fig. 6. Effect of increasing the high gain parameter L on the performance of the control action. The response of the coolant temperature is faster as L increases. However, a very large value of L can induce overshot and saturation. On contrary, a very low value on L can induce off set at steady state. The arrow in picture indicates the direction of the effect; that is, L was increased as the arrow direction shows.

strategy can be needed. We do not find reset windup in the extensive simulations. This can be associated to the slow dynamics of the process and low value of the estimation gain; however, the possibility of appearing the reset windup phenomenon should be considered as well.

7   Concluding Remarks

Here, a control law for chemical reactors had been proposed. The controller was designed from compensation/estimation of the heat reaction in exothermic reactor. In particular, the paper is focused on the isoparaffin/olefin alkylation in STRATCO reactors. It should be noted that control design from heat compensation leads to controllers with same structure than nonlinear feedback. This fact can allow to exploit formal mathematical tools from nonlinear control theory. Moreover, the estimation scheme yields in a linear controller. Thus, the interpretation for heat compensation/estimation is simple in the context of process control.

On the other hand, the proposed approach has structure of low pass filter; see Eq.(8). Thus, we can expect that the closed-loop response is not sensitive to high frequency signals (as, for example, by noisy measurements or fluctuations in fluids mechanics by agitation). Although Figure 3 depicts the frequency response, the robustness of the closed-loop should be guaranteed by weighting functions. To this end, H_∞ theory can be exploited. One more issue about robustness is related to delay by lag transport. In this direction, the question is how the delay by lag transport affects the closed-loop performance?. These issues are beyond the goals of the present contribution, and are being analyzed to be reported elsewhere.

Acknowledgments. G. Carrizales-Martínez wishes to tank to CONACyT for financial support under scholarship grant 130350 and Instituto Mexicano del Petróleo by partial support under grant FIES-96-31-III.

References

[1] L.F. Albright. Updating alkylate gasoline technology. *CHEMTECH*, 28(1):40–46, 1998.
[2] J. Alvarez, J. Alvarez, and R. Suárez. Nonlinear bounded control for a class of continuous agitated tank reactors. *Chem. Eng. Sci.*, 46(12):3341–3354, 1991.
[3] K.Y. Chang and W.L. Luyben. Design and control of coupled reactor/column systems- Part 1. A binary coupled reactor/rectifier systems. *Comput. Chem. Eng.*, 21(1):25–46, 1997.
[4] K.Y. Chang and W.L. Luyben. Design and control of coupled reactor/column systems- Part 2. More complex coupled reactor/column systems. *Comput. Chem. Eng.*, 21(1):47–67, 1997.
[5] K.Y. Chang and W.L. Luyben. Design and control of coupled reactor/column systems- Part 3. A reactor/stripper with two columns and recycle. *Comput. Chem. Eng.*, 21(1):69–86, 1997.
[6] Y. Cheng-Ching. *Autotuning of PID Controllers: Relay Feedback Approach (Advances in Industrial Control)*. Springer-Verlag, London, 1999.
[7] M.C. Clark and B. Subramaniam. Extended alkylate production activity during fixed-bed supercritical 1-butene/isobutane alkylation on solid-acid catalyst using carbon dioxide as a diluent. *Ind. Eng. Chem. Res.*, 37(4):1243–1250, 1998.
[8] A. Corma and A. Martínez. Chemistry, catalysts and processes for isoparaffin-olefin alkylation: Actual situation and future trends. *Catalysis Review: Sci. and Eng.*, 35:485–523, 1993.
[9] R. Femat, J. Alvarez-Ramírez, and M. Rosales-Torres. Robust asymptotic linearization via uncertainty estimation: Regulation of temperature in a fluidized bed reactor. *Comput. Chem. Eng.*, 23:697–708, 1999.
[10] G. Gavalas. *Nonlinear differential equations of chemical reacting systems*. Springer, New York, 1968.
[11] R.G. Gonzalez. Alkylation beyond 1995 will play a key role in clean fuels. *Fuel Reformulation*, pages 37–, 1995.
[12] M.A. Henson and D.E. Seborg. Input-output linearization of general non-linear process. *A.I.Ch.E. Journal*, 36:1753–, 1990.
[13] A. Isidori. *Nonlinear Control Systems*. Springer Verlag, third edition, 1995.
[14] T. Kok-Kiong and H. Chang-Chieh. *Advances in PID control*. Springer-Verlag, London.

[15] N. Kunimatsu. Stabilization of nonlinear tubular reactor dynamics with recycle. In *1st Int. Conference on Control of Oscillations and Chaos*, volume 2, pages 291–295, 1997.

[16] J.R. Langley and R.W. Pike. The kinetics of alkylation of isobutane and propylene. *A.I.Ch.E. Journal*, 18:698–705, 1972.

[17] W.L. Luyben. *Process Modeling, Simulation and Control for Chemical Engineering*. McGraw-Hill, New York, 2nd. edition, 1990.

[18] W.L. Luyben. Dynamics and control of recycle systems 1. Simple open-loop and closed-loop systems. *Ind. Eng. Chem. Res.*, 32:466–475, 1993.

[19] A. Mantilla-Ramírez, G. Ferrat-Torres, J.M. Domínguez, C. Aldana-Rivero, and M. Bernal. Influence of reaction parameters and comparison of fluorinated alumina and silica supports in the heterogeneous alkylation of isobutane with olefins. *Appl. Cat. A*, 143:203–214, 1996.

[20] W.L. Nelson. *Petroleum Refiney Engeening*. McGraw-Hill, Tokyo, 1984.

[21] B.O. Recke and S.B. Jorgensen. Nonlinear dynamics and control of a recycle fixed bed reactor. In *American Control Conference (ACC)*, volume 4.

[22] A.K. Rhodes. New process schemes, retrofits, fine tune alkylation capabilities. *Oil and Gas J.*, August:56–59, 1994.

[23] S. Rönnbäck. *Linear Control of Systems with Actuator Constraints*. PhD thesis, Division of Automatic Control, Lulea University of Technology, May 1993.

[24] M.F. Simpson, J. Wei, and S. Sudaresan. Kinetic analysis of isobutane/butene alkylation over ultrastable H-Y zeolite. *Ind. Eng. Chem. Res.*, 35:3861–3873, 1996.

[25] J.I. Steinfeld, J.S. Francisco, and W.L. Hase. *Chemical Kinetics and Dynamics*. Prentice-Hall, USA, 1989.

[26] A. Teel and L. Praly. Tools for semiglobal stabilization by partial state and output feedback. *SIAM J. Control Optim.*, 33(5):1443–1488, 1995.

[27] M. Vichalaik, J.R. Hooper, C.L. Yaws, and R.W. Pike. Alkylation of mixed olefin with isobutane is a STRATCO chemical reactor. In *5th World Congress of Chem. Eng.*, pages 238–243, San Diego, Cal., July 1996.

[28] F. Viel, F. Jadot, and G. Bastin. Robust feedback stabilization of chemical reactors. *IEEE Trans. Automat. Contr.*, 42:473–481, 1997.

Control Performance of Thermally Coupled Distillation Sequences with Unidirectional Flows

J.G. Segovia-Hernández[1], S. Hernández[2], and A. Jiménez[1]

[1] Departamento de Ingeniería Química, Instituto Tecnológico de Celaya
  {gabriel,arturo}@iqcelaya.itc.mx
[2] Facultad de Química, Universidad de Guanajuato
  hernasa@quijote.ugto.mx

Summary. The Petlyuk distillation system has been considered with special interest in the field of separation processes because of the high energy savings that it can provide with respect to the operation of sequences based on conventional columns. The original design of the Petlyuk structure, however, shows two interconnections that seem to affect its operational and controllability properties. For one thing, the arrangement is more complex than the common structure of the conventional distillation systems of one feed and two products with no recycle streams. Furthermore, the interconnections show a bidirectional flow of the vapor streams between the two columns of the Petlyuk arrangement, posing a major control challenge for its operation. To overcome this problem, two new structures have been suggested that use unidirectional flows of the vapor or liquid interconnecting streams. The new options have been conceptually developed to improve the controllability properties of the Petlyuk system, but no formal analysis has been conducted on this matter. In this work, a comparative analysis of the control properties of the Petlyuk column and the new arrangements with unidirectional interconnecting flows is presented. Through a singular value decomposition analysis, it is shown that the new schemes provide better theoretical controllability properties than the Petlyuk system. Closed loop tests using proportional-integral controllers were also carried out, and the results showed that, in most of the cases considered, the new arrangements improved the dynamic responses of the Petlyuk column. Such arrangements, therefore, show promising perspectives for its practical consideration.

1 Introduction

A chemical process is based on the transformation of raw material into more valuable chemical products, and two of the steps commonly encountered in such processes are the chemical reaction equipment and the section for purification of product species. A common task in the chemical industry, therefore,

H.O. Méndez-Acosta et al. (Eds.): Dyn. & Ctrl. of Chem. & Bio. Proc., LNCIS 361, pp. 53–72, 2007.
springerlink.com

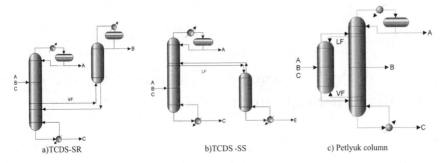

Fig. 1. Three thermally coupled distillation schemes for the separation of ternary mixtures

is the separation of multicomponent mixtures into specified product streams. By far, the dominant choice to carry out such separation tasks is provided by the use of distillation processes; the main disadvantage of the distillation units is that they consume significant amounts of energy, which many times constitute a major portion of the operating costs of the overall process. To improve the energy efficiency of separation processes based on distillation, several strategies have been proposed; one of them calls for the integration of the distillation columns with the rest of the process [23], [32]. The layout of the chemical plant or the energetic characteristics of the streams involved in the process may make such integration limited or not possible, in which case other options can be considered, for example the use of heat pumps or the implementation of thermally coupled arrangements [6], [?], [24]. In particular, the use of columns with thermal coupling has received considerable attention in recent years, with a special development reported for the case of separation problems of ternary mixtures. Thermally coupled distillation systems (TCDS) are obtained through the implementation of interconnecting streams (one in the vapor phase and the other one in the liquid phase) between two columns; each interconnection typically replaces one condenser or one reboiler from one of the columns, thus providing potential savings in capital investment. Furthermore, through a proper selection of the flow values for the interconnecting streams of TCDS, one can obtain significant energy savings with respect to the energy consumption of conventional distillation sequences.

Three thermally coupled schemes have been particularly analyzed. Two of them are fairly similar and make use of a main column and a side column. One can use a side extraction in the vapor phase from the first column and feed it to a side rectifier that purifies the intermediate component. The reboiler of the side column is eliminated by recycling the bottom stream, in the liquid phase, to the first column. The arrangement is known as a thermally coupled distillation system with a side rectifier (TCDS-SR), and its structure is shown in Figure 1a. If the side extraction from the first column is carried out in the liquid phase, a side stripper is used to obtain the intermediate component as a bottom product. In such case, the condenser of the side column is eliminated if the top vapor stream is sent as a recycle stream to the first column. The resulting structure, a thermally coupled distillation system with a side stripper, TCDS-SS, is shown in Figure 1b. The third option that has been considered with special interest uses two interconnections in an arrangement that consists of a first column (or prefractionator) interconnected through thermal couplings with a main column from which the three components are produced. Through the use of the interconnections, both the condenser and the reboiler of the prefractionator are eliminated. Such option that uses the maximum number of interconnections for the separation of a ternary mixture is referred to as the fully thermally coupled system, or Petlyuk column (Figure 1c).

One of the first studies in which TCDS schemes were taken into account was reported in [34]. The thermally coupled systems with side columns were included as part of a set of eight separation sequences, and they were shown to provide the lowest yearly costs for the separation of ternary mixtures over a wide range of feed compositions. Other theoretical studies followed for the analysis of the structures of Figure 1, most of them under the assumption of minimum reflux conditions. In general, the studies showed that expected savings of minimum internal vapor flows, and therefore of minimum energy requirements, of up to 30 percent could be obtained with respect to the operation of conventional direct and indirect distillation sequences [7], [8], [11], [12], [13]. The savings depend on the amount of the intermediate component in the feed mixture, and are more noticeable for feeds with low or high contents of the intermediate component. More recent works have reported the use of optimization strategies for TCDS to detect designs with minimum energy consumption under un-pinched (finite reflux) conditions [4], [16], [18], [25], [37]. When the performance of the three integrated schemes has been compared, it has been found that in general the Petlyuk system offers better energy savings than the systems with side columns. The energy savings of TCDS can be explained in terms of the internal composition profiles of the intermediate component. The implicit objective of a proper design for a thermally coupled system is to reduce or eliminate the remixing of the intermediate component of the ternary mixture that is typically observed during the operation of a conventional sequence [19], [35]. Thus, if the resulting design of a TCDS scheme is such that the extraction of the side stream is performed at the tray with the maximum internal composition of the intermediate component, then the external energy required for the separation will reach a minimum value for such scheme.

Despite the well-established potential of TCDS in terms of their energy savings, these structures had not been considered for industrial implementation until recent times [22], largely because of the lack of operational experience and to the expectations that they might be rather difficult to control [9]. One may notice the more complex structure of each of the sequences shown in

Figure 1, which in general pose higher control challenges than the case of the conventional sequences, in which no recycle streams are used.

Typically, the control task in distillation processes consists of the production of the main products meeting specified purities. In a conventional distillation column, with one feed and two products, the control of the top product is generally implemented through the adjustment of the reflux flowrate, and the control of the bottom product is tied to the reboiler heat duty. These simple control loops have been used extensively in practice. The TCDS structures, on the other hand, introduce new interconnections for which no operational experience has been gained. Also, a product does not necessarily appear as a top or as a bottom stream. New control loops must be tried, for which no sufficient industrial experience is available. Therefore, a special effort is needed to gain a proper understanding on the control behavior of the more complex TCDS configurations. Some initial works dealing with some of the control aspects of TCDS have recently been reported [1], [17], [20], [28], [29], [31], [36], and in general they have shown that some of these integrated options are controllable, so that the predicted savings in energy would probably not be obtained at the expense of operational and control problems. Further development in this area is still needed.

Another line of research has been observed in recent times, in which efforts to generate alternate structures to the systems of Figure 1, but with more promising expectations as far as their control perspectives, have been suggested. Some newer configurations for TCDS that appear to have some operational advantages over the expected dynamic properties of the designs of Figure 1 have been reported [2], [3], [5], [10], [26], [27]. In particular, the Petlyuk column is the most energy-efficient option, but its structure poses two interesting features. One, it has the highest number of interconnections, and two, the streams in the vapor phase (and therefore also in the liquid phase) flow in the two directions, back and forth between the columns. These aspects affect not only the design but also the control properties of the Petlyuk system. Alternative options can be developed through the correction of each of these two items. Thus, through a reduction on the number of interconnections, or through the use of unidirectional interconnecting flows, one can conceptually generate simpler schemes, in principle easier to control. Six alternative schemes to the Petlyuk column that arise from such modifications have been recently analyzed and reported to show similar energy savings as the original arrangement [21]. These results open a wide interest for the analysis of the dynamic properties of such new schemes; if the options with similar energy efficiencies are proved to provide better controllability properties than the Petlyuk system, then their potential for industrial implementation would be clearly enhanced.

In this work a set of two modifications of the Petlyuk system that use unidirectional flows of both interconnecting streams are considered to analyze their control behavior. The aim of the study is to show if such design modifications may lead to an improvement on the control properties of the original Petlyuk arrangement.

2   Design and Model of the Petlyuk Column

A model was first developed for the Petlyuk system. Consider a generic equilibrium stage as shown in Figure 2. The model for the distillation system is based on the following set of equation (with all symbols defined in the Nomenclature section).

Total mass balance:

$$\frac{dM_j}{dt} = L_{j-1} + V_{j+1} + F_j^L + F_j^V - (L_j + U_j) - (V_j + W_j) \qquad (1)$$

Component mass balance:

$$\frac{d(M_j X_{i,j})}{dt} = L_{j-1} X_{i,j-1} + V_{j+1} Y_{i,j+1} + F_j^L Z_{i,j}^L + F_j^V Z_{i,j}^V$$
$$- (L_j + U_j) X_{i,j} - (V_j + W_j) Y_{i,j} \qquad (2)$$

Energy balance:

$$\frac{d(M_j \bar{U}_{i,j})}{dt} = L_{j-1} \bar{h}_{j-1} + V_{j+1} \bar{H}_{j+1} + F_j^L \bar{h}_j^L + F_j^V \bar{H}_j^V$$
$$- (L_j + U_j) \bar{h}_j - (V_j + W_j) \bar{H}_j + Q_j \qquad (3)$$

Equilibrium relationships:

$$Y_{i,j} = K_{i,j} X_{i,j} \qquad (4)$$

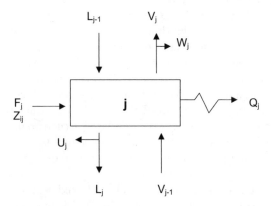

Fig. 2. A generic equilibrium stage for the model of the Petlyuk system

Summation constraints:

$$\sum_{i=1}^{NC} K_{i,j} X_{i,j} - 1.0 = 0 \tag{5}$$

Stage hydraulics (Francis weir formula):

$$L_j = k_j L_{Wj} (h_{OW})_j^{3/2} \tag{6}$$

As written, this set of equations provide the dynamic model for the Petlyuk column. Since a design for the column must first be obtained, the same set of equations but written for steady state conditions provides the basis for such a design problem. In any case, the model shows a coupled structure because of the recycle streams between the two columns such that the full set of equations must be solved simultaneously.

The complete design of the Petlyuk system should provide the tray structure of each column, the tray positions for the interconnecting streams, and the operating conditions (such as pressure and reflux ratios or values of the interconnecting streams) that minimize a given objective function, for instance the total yearly cost. This case yields a complicated optimization problem, which would require a mixed-integer nonlinear programming formulation. To overcome such situation, a sequential solution on the tray structure and operating conditions was implemented. For the tray structure, a base design is obtained from the information provided by a conventional distillation system (design methods for conventional distillation systems are well known) consisting of a prefractionator followed by two binary separation columns. Figure 3 shows how the six tray sections of the conventional sequence are related to the structure of the Petlyuk system; such sections perform similar separation tasks in both arrangements. Therefore, the design of the conventional sequence, readily obtained through the use of shortcut methods, provides the tray distribution for the integrated system. The design depends on the operating pressure, which can be set such that the use of refrigerants in the condenser is avoided. Once the design (tray structure) is obtained, it needs to be validated. Steady state rigorous simulations are then conducted to test the preliminary design. If the design specifications (product compositions) are met with the tray structure obtained with the section analogy procedure, the preliminary design was successful; otherwise, proper arrangements in the tray structure are implemented until the specified product compositions are obtained. To complete the design, the validated structure of the Petlyuk system is finally subjected to an optimization procedure to obtain the operating conditions that minimize the energy consumption required for the separation task. Two degrees of freedom remain at this stage, which are used as search variables for the optimization process. The selected search variables are the interconnecting streams of the Petlyuk column (LF and VF, see Figure 1c). Further details on this design procedure are available in [18].

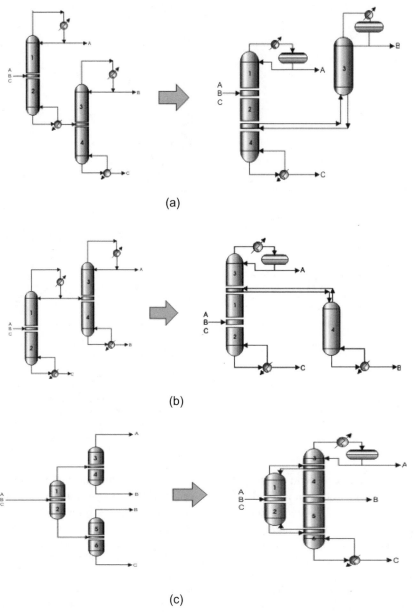

(a)

(b)

(c)

Fig. 3. Rearrangement of a conventional distillation sequence into the Petlyuk column

3   More Operable Thermally Coupled Distillation Sequences

As mentioned earlier, the Petlyuk column (Figure 1c) poses potential operational problems because of the two directions of the interconnecting vapor

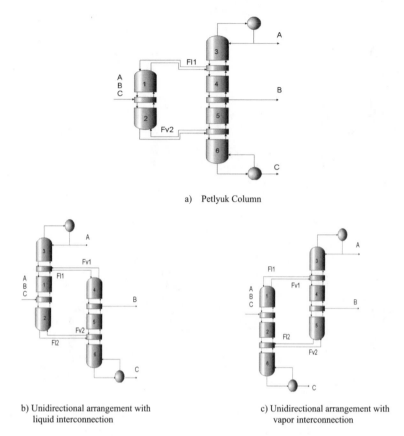

a) Petlyuk Column

b) Unidirectional arrangement with
   liquid interconnection

c) Unidirectional arrangement with
   vapor interconnection

Fig. 4. Alternative schemes to the Petlyuk column

streams. The overhead interconnecting stream of the prefractionator, in the vapor phase, goes from a high pressure point of the prefractionator to a low pressure point of the main column, but the recycle vapor stream from the second column requires a higher pressure of the main column with respect to the prefractionator. This creates an operational conflict since the pressure of neither column can be set higher or lower than the other one. This problem could be avoided, in principle, by implementing unidirectional flows in the interconnecting streams in an alternate arrangement. In [2], [3] have been proposed some of such modifications; two of them are taken in this work for their analysis. Figure 4b shows a modified Petlyuk arrangement with unidirectional liquid interconnecting streams from the first column to the second one, while Figure 4c shows the second option of a modified arrangement with unidirectional vapor interconnecting streams. In the remaining of this work, an analysis on the energy requirements and control properties of the unidirectional alternative schemes of Figures 4b and 4c is conducted, with a comparison shown with respect to those required by the Petlyuk column.

The design of the alternative arrangements is carried out through a section analogy procedure with respect to the design of the Petlyuk system. The system shown in the Figure 4b is built by moving tray section 3 from the Petlyuk column (Figure 4a) along with the condenser to the top of the first column. This modification provides an arrangement that is connected at the top of the second column (with a liquid stream) and at the bottom of the first column (with another liquid stream). Both thermal couplings are preserved, but both liquid flows go from the first column to the second column. We identify this configuration as a fully coupled arrangement with liquid flows (PUL). Similarly, one can remove the bottom tray section of the Petlyuk column (section 6 of Figure 4a) with the reboiler and implement them in the bottom part of the first column to produce a fully coupled arrangement with vapor streams (PUV, Figure 4c). The resulting structures with unidirectional interconnecting flows also show two degrees of freedom each, which can be used to find out the operating conditions under which minimum energy consumptions are obtained.

4   Energy Consumption

To carry out the analysis, a case study of the separation of a ternary mixture of n-pentane (A), n-hexane (B) and n-heptane (C) with a feed composition low in the intermediate component (A, B, C equal to 0.4, 0.2, 0.4) was considered. The selected composition reflects typical values for which significant energy savings of TCDS have been reported. Design specifications included 98.7% purity in A, 98% purity in B and 98.6% purity in C. A feed flowrate to the sequence of 100 $lbmol/hr$, available as a saturated liquid, was taken. To avoid the use of refrigerants that would have a high impact on the economics of the separation sequence, the design pressure for each column was chosen such that all condensers could be operated with cooling water. The thermodynamic properties of the mixture were estimated with the Chao-Seader correlation [15].

The first part of the analysis was conducted to detect the designs with minimum energy consumption for the integrated sequences. Once a validated design (tray structure) was obtained, an optimization procedure was carried out on the recycle streams for each of the three coupled sequences to detect the operating conditions under which each design was more energy efficient.

Table 1 shows the values of the minimum energy consumption obtained for the two systems with unidirectional flows and compares them to the one

Table 1. Minimum energy requirements for each arrangement, Btu/h

Petlyuk Column	PUL	PUV
1,772,479	1,777,537	1,834,960

obtained for the Petlyuk system. It can be noted that similar energy requirements are obtained for the three options. The energy consumption of the
structure with interconnecting vapor streams was slightly higher, but probably because of numerical convergence aspects. The results are consistent with
those presented in [21], who conducted a more exhaustive study on the energy
consumption of these types of sequences and reported that the new structures
have similar energy consumption levels as the Petlyuk scheme. The new sequences are therefore suitable candidates for a complementary analysis on
their dynamic properties.

5   Control Behavior

The controllability analysis was conducted in two parts. The theoretical control properties of the three schemes were first predicted through the use of the
singular value decomposition (SVD) technique, and then closed-loop dynamic
simulations were conducted to analyze the control behavior of each system and
to compare those results with the theoretical predictions provided by SVD.

5.1   Singular Value Decomposition

The application of the SVD technique provides a measure of the controllability properties of a given dynamic system. More than a quantitative measure,
SVD should provide a suitable basis for the comparison of the theoretical control properties among the thermally coupled sequences under consideration.
To prepare the information needed for such test, each of the product streams
of each of the thermally coupled systems was disturbed with a step change in
product composition and the corresponding dynamic responses were obtained.
A transfer function matrix relating the product compositions to the intended
manipulated variables was then constructed for each case. The transfer function matrix can be subjected to SVD:

$$G = V \Sigma W^H \tag{7}$$

where $\Sigma = diag(\sigma_1 \ldots \sigma_n)$, $\sigma_1 =$ singular value of $G = \lambda_i^{1/2}(GG^H)$; $V = (V_1, V_2, \ldots)$ matrix of left singular vectors, and $W = (w_1, w_2, \ldots)$ matrix of
right singular vectors. Two parameters of interest are the minimum singular
value, σ_*, and the ratio of maximum to minimum singular values, or condition
number:

$$\gamma^* = \sigma^*/\sigma_* \tag{8}$$

The minimum singular value is a measure of the invertibility of the system
and therefore represents a measure of the potential problems of the system
under feedback control. The condition number reflects the sensitivity of the
system under uncertainties in process parameters and modelling errors. These
parameters provide a qualitative assessment of the dynamic properties of a

Table 2. Minimum singular value and condition number for each arrangement

Sequence	σ_*	γ^*
Petlyuk Column	0.0055	1260.47
PUL	0.4383	14.33
PUV	0.0363	252.68

given design. The objective is to identify the systems with higher minimum singular values and lower condition numbers; those systems are expected to show the best dynamic performance under feedback control. A nice feature of the SVD test is that it is independent of the type of controller to be implemented; the basic idea is that the controllability properties of the system are instead limited or imposed by its inner dynamic structure.

The full application of the SVD method should consider a range of frequencies sufficiently high to give a complete coverage on the behavior of minimum singular values and conditions numbers, as shown in [17], [20] for their controllability analysis of some conventional and nonconventional distillation systems. As a preliminary attempt, however, the SVD analysis was conducted in this work only at zero frequency. Such analysis requires only the steady-state gain matrix, and gives the theoretical controllability properties of each system around its nominal operating point, in this case around the point for which a minimum energy consumption was obtained. This information should be sufficient at this point for the intended comparative analysis on the controllability properties of the three thermally coupled systems.

Table 2 shows the results obtained from the application of the SVD at zero frequency for each sequence. The PUL column has the lowest value of the condition number, which implies that this sequence is better conditioned to the effect of disturbances than the other two integrated systems; it also presents the highest value of σ_*, which means that the sequence is expected to show the lowest control effort. Therefore, from this initial analysis, it is apparent that the PUL scheme will show the best operational properties of the three options under consideration. On the other hand, it is interesting to notice that the Petlyuk column offers the worst theoretical control properties (lowest minimum singular value and highest condition number), which supports the notion that the new arrangements may indeed improve the control properties of the Petlyuk system through the correction of the bidirectionality of the vapor interconnecting streams observed in the original design.

5.2 Closed-Loop Simulations

To supplement the SVD analysis, rigorous dynamic simulations under closed loop operation were carried out. For the closed-loop analysis, several issues must be defined first, such as the control loops for each system, the type of process controller to be used, and the values of the controller parameters.

Several techniques, such the relative gain array method, can be used to fix the loops for a control system. In the case of distillation columns, however, such loops are fairly well established and used successfully in practice, at least for conventional columns. A well-known structure is based on energy balance considerations, which yields to so-called LV control structure in which the reflux flowrate L and the vapor boilup rate V (affected directly by the heat duty supplied to the reboiler) are used to control the distillate and bottom outputs compositions (see for instance [14]). The control loops for the integrated systems were chosen from extensions of the practical considerations observed for conventional distillation columns. The control objective was to preserve the output streams at their design purity specifications. Two control loops arise naturally from the experience on the operation of conventional columns. For the control of product stream A, that is obtained as an overhead product, the reflux flowrate was used, whereas for the control of product stream C, that is obtained as a bottom product, the reboiler heat duty was chosen. Product stream B is obtained as a side stream for the integrated arrangements, and in the lack of operational experience, its control was simply set through the manipulation of the side stream flow rate. It should be mentioned that such control loops have been used with satisfactory results in previous studies we have conducted on thermally coupled systems [20], [28], [29], [30]. The choice of the type of controller was based on the ample use that the Proportional-Integral (PI) mode has for distillation systems in industrial practice. Also, since this is the first reported analysis on the controllability properties of the alternative schemes of figures 4b and 4c, the choice may also provide a basis upon which the use of more elaborated control laws can be compared. As for the selection of the parameters of the PI controllers, care was taken to provide a common method for each of the sequences under comparison. A tuning procedure that involved the minimization of the integral of the absolute value (IAE) for each loop of each scheme was used [33]. Therefore, for each loop, an initial value of the proportional gain was set; a search over the values of the integral reset time was conducted until a local optimum value of the IAE was obtained. The process was repeated for other values of the proportional gain. The selected set of controller parameters was then the one that provided a global minimum value of the IAE. Although the tuning procedure is fairly elaborated, the control analysis is conducted based on a common tuning method for the controller parameters, Table 3 shows the parameters obtained for the control of each product stream.

The simulations involve the solution of the rigorous tray-by-tray model of each sequence, given by equations 1 to 6, together with the standard equations for the PI controllers for each control loop (with the parameters obtained through the minimization of the IAE criterion). The objective of the simulations is to find out how the dynamic behavior of the systems compare under feedback control mode. To carry out the closed-loop analysis, two types of cases were considered: i) servo control, in which a step change was induced in the set point for each product composition under SISO feedback control,

Table 3. Parameters of the PI controllers for each control loop

Sequence	$\tau_l(min)$	$K_c(\%/\%)$
Petlyuk Column	A=30	A=20
	B=5	B=130
	C=10	C=20
PUL	A=60	A=80
	B=150	B=140
	C=10	C=100
PUV	A=70	A=20
	B=40	B=40
	C=30	C=180

and ii) regulatory control, in which a feed disturbance with a 5% change in the composition of one component (with a proportional adjustment in the composition of the other components to keep the same total feed flowrate) was implemented.

Dynamic Behavior of the Heavy Component (Set Point Change)

When a set point change from 0.986 to 0.99 for the heavy component (C) was considered, the responses shown in Figure 5 were obtained. The three options show good dynamic responses, with relatively low values of settling times, lower than 0.1 hours. The systems with unidirectional flows show some overshooting, while the Petlyuk system offers a more sluggish response.

In addition to the visual observations of the dynamic responses, a quantitative measure is needed to provide a better comparison. With such an objective, IAE values were evaluated for each closed-loop response. The PUL option shows the lowest IAE value of 5.607×10^{-5}, while the value for the Petlyuk column turns out to be 2.35×10^{-4}. Therefore, the results of the test indicate that, for the SISO control of the heaviest component of the ternary mixture, the PUL option provides the best dynamic behavior and improves the performance of the Petlyuk column. Such result is consistent with the prediction provided by the SVD analysis.

Dynamic Behavior of the Light Component (Set Point Change)

The results of the dynamic test for a positive change in the set point of the light component (A) are displayed in Figure 6. For a change in the set point from 0.987 to 0.991, the three systems are shown to be controllable and reach the new value of product composition, although the PUL scheme shows a quicker adjustment. IAE values were also calculated for each response; the two best IAE values correspond to the new arrangements: 4.20×10^{-5} for the PUL system, and 6.10×10^{-5} for the PUV system. The IAE value for the Petlyuk column was 2.35×10^{-4}, which again shows a case in which the dynamic

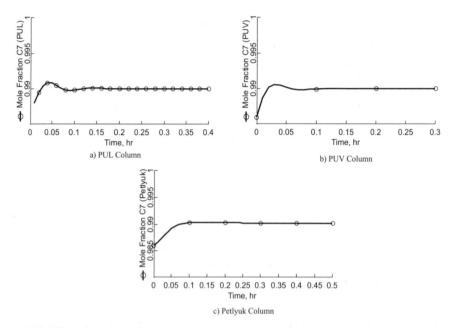

Fig. 5. Closed loop responses for a set point change in the composition of the heavy component

properties of the Petlyuk system are improved through the modifications that provide unidirectional flows of the interconnecting streams.

Dynamic Behavior of the Intermediate Component (Set Point Change)

Figure 7 shows the dynamic responses obtained when the set point for the intermediate component was changed from 0.98 to 0.984. One may notice the better response provided by the Petlyuk column in this case, which is faster than the other two systems and without oscillations. When the IAE values were calculated, a remarkable difference in favor of the Petlyuk system was observed: 2.87×10^{-4} for the Petlyuk column, compared to 0.0011 for the PUL system and 0.0017 for the PUV system. The results from this test may seem unexpected, since the new arrangements have been proposed to improve the operation capabilities of the Petlyuk column. The SISO control of the intermediate component, interestingly, seems to conflict with that of the other two components in terms of the preferred choice from dynamic considerations.

Feed Disturbance Analysis

The last set of tests conducted in this work had to do with the implementation of feed disturbances. The regulatory control tests consisted of a change in the feed composition with the same nominal feed flowrate. Under such scenario, it is clear that the feed composition of all components must change, but the

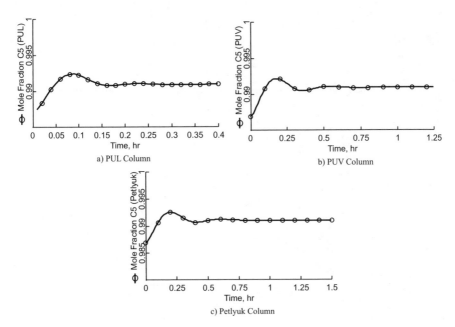

Fig. 6. Closed loop responses for a set point change in the composition of the light component

component with the highest change (of 5 percent) was used to identify the tests. Figure 8 shows the results obtained when a feed disturbance in the composition of component A was implemented. The PI controllers successfully rejected the feed disturbance for each of the separation systems, although the settling times are higher than those obtained for the servo tests. It can be observed that, consistent with the SVD results, the best response of the three alternatives is provided by the PUL system, which also implies a lower control effort. When the systems were compared according to their IAE values, the Petlyuk column showed the highest value of 0.0018.

The results for this test and for the responses to disturbances in feed compositions of B and C are summarized in Table 4. A consistent trend with the servo tests was observed, in the sense that one option provides the best common choice for the control of the system under feed disturbances on the extreme components of the mixture, but a different arrangement yields a superior dynamic performance for the control task under a feed disturbance on the intermediate component. From Table 4, the IAE values indicate that the PUL system shows the best behavior for feed disturbances in the light and heavy component. However, the PUL arrangement shows the worst response when the feed disturbance in the intermediate component was considered, in which case both the Petlyuk and the PUV systems show fairly similar rejection capabilities.

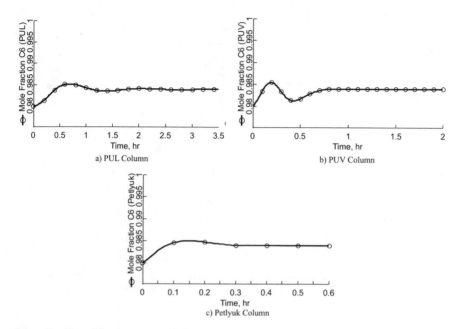

Fig. 7. Closed loop responses for a set point change in the composition of the intermediate component

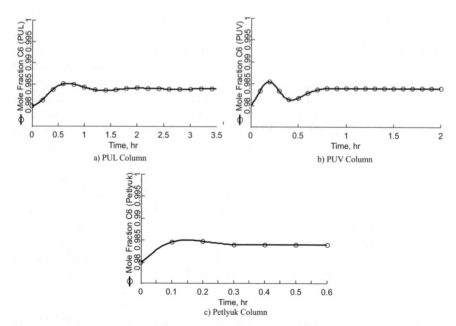

Fig. 8. Closed loop responses for feed disturbance in the composition of the light component

Table 4. IAE values for the feed disturbance test

Sequence	A	B	C
Petlyuk Column	0.0018	1.66×10^{-4}	5.83×10^{-4}
PUL	3.63×10^{-4}	0.0059	9.60×10^{-5}
PUV	0.0011	1.50×10^{-4}	6.20×10^{-4}

6 Conclusion

To improve the low energy efficiency observed during the operation of conventional distillation sequences, some integrated arrangements have been suggested. Distillation sequences with thermal coupling have been shown to provide promising options because of their potential energy savings. Among these options, the Petlyuk configuration stands as the most efficient arrangement in terms of its energy requirements. Originally developed for the separation of ternary mixtures, the Petyuk system shows two thermal couplings whose structure seems to conflict in terms of its operational perspectives. The thermal couplings show a bidirectional flow of the vapor interconnecting streams, which implies that it is not possible to design one of the columns with a pressure uniformly higher than the other one. Among the possible revisions that have been suggested to improve the operational perspectives of the Petlyuk system, two modifications that keep the same number of thermal couplings but use a unidirectional flow of the interconnecting (liquid or vapor) streams from the first column to the second one have been suggested. An analysis on the dynamic properties of the two new schemes with unidirectional flows of the interconnecting streams has been conducted in this work and compared to the behavior of the original Petlyuk arrangement. A simplified analysis via singular value decomposition at zero frequency indicates that the new arrangements may improve the controllability properties around the nominal operating point with respect to the Petlyuk system. A set of rigorous dynamic simulations under closed loop operation with PI controllers was also conducted. The closed loop simulations were based on a SISO control strategy, and the results showed an interesting trend. The control behavior of the Petlyuk column was improved by the new options when the control tasks called for the control of the extreme components (A or C) of the ternary mixture. However, when the control of the intermediate component was considered, the Petlyuk column provided a better dynamic performance. Overall, the results here obtained show that the new arrangements with unidirectional interconnections provide interesting choices for further consideration. For one thing, they seem to preserve the energy efficiency of the original Petlyuk system; for the other, they were shown to offer proper controllability properties. In particular, the structure with unidirectional liquid interconnections provided the best theoretical control properties, and yielded the best overall dynamic behavior under the operation scenarios considered in this work.

List of Abbreviations and Symbols

Related to the process:
F	feed flow-rate
$\bar{H}$	enthalpy of vapor
$\bar{h}$	enthalpy of liquid
h_{OW}	liquid height on weir
K	equilibrium constant
L	liquid flow-rate
L_W	weir length
M	moles of liquid retained
Q	heat added or extracted
t	time
U	liquid side stream
$\bar{U}$	internal energy of liquid retained
u	manipulated input
V	vapor flow-rate
W	vapor side stream
X	mole fraction in liquid phase
Y	mole fraction in vapor phase

Superscripts:
L	liquid
V	vapor

Subscripts:
i	component
j	stage

References

[1] M.I. Abdul-Mutalib and R. Smith. Operation and control of dividing wall distillation columns. Part I: Degrees of freedom and dynamic simulation. *Trans. IchemE*, 76(Part A):308–318, 1998.

[2] R. Agrawal and Z. Fidkowski. More operable arrangements of fully thermally coupled distillation columns. *A.I.Ch.E. Journal*, 44(11):2565–2568, 1998.

[3] R. Agrawal and Z. Fidkowski. New thermally coupled schemes for ternary distillation. *A.I.Ch.E. Journal*, 45(3):485–496, 1999.

[4] K.A. Amminudin, R. Smith, D.Y.C. Thong, and G.P. Towler. Design and optimization of fully thermally coupled distillation columns. Part I: Preliminary design and optimization methodology. *Trans. IchemE*, 79(Part A):701–715, 2001.

[5] O. Annakou, A. Meszaros, Z. Fonyo, and P. Mizsey. Operability investigation of energy integrated distillation schemes. *Hung. J. Ind. Chem.*, 24:155–160, 1996.

[6] O. Annakou and P. Mizsey. Rigorous investigation of heat pump assisted distillation. *Heat Recovery Syst. CHP*, 15(3):241–247, 1995.

[7] O. Annakou and P. Mizsey. Rigorous comparative study of energy – integrated distillation schemes. *Ind. Eng. Chem. Res.*, 35:1877–1885, 1996.

[8] J. Cerda and A.W. Westerberg. Shortcut methods for complex distillation columns. I. Minimum reflux. *Ind. Eng. Chem. Process Des. Dev.*, 20:546–557, 1981.

[9] G. Dünnebier and C. Pantelides. Optimal design of thermally coupled distillation columns. *Ind. Eng. Chem. Res.*, 38:162–176, 1999.

[10] M. Emtir, P. Mizsey, E. Rev, and Z. Fonyo. Economic and controllability investigation and comparison of energy – integrated distillation schemes. *Chem. Biochem. Eng. Q.*, 17(1):31–40, 2003.

[11] Z. Fidkowski and L. Krolikowski. Energy requirements of nonconventional distillation systems. *A.I.Ch.E. Journal*, 36(8):1275–1278, 1991.

[12] A. Finn. Consider thermally coupled distillation. *Chem. Eng. Prog.*, 10:41–45, 1993.

[13] K. Glinos and M.F. Malone. Optimality regions for complex column alternatives in distillation systems. *Chem. Eng. Res. Des.*, 66:229–240, 1988.

[14] K.E. Häggblom and K.V. Waller. *Practical Distillation Control*, chapter Control Structures, Consistency, and Transformations. Van Nostrand Reinhold, New York, 1992.

[15] E.J. Henley and J.D. Seader. *Equilibrium–Stage Separation Operations in Chemical Engineering*. Wiley, New York, 1981.

[16] S. Hernández and A. Jiménez. Design of optimal thermally – coupled distillation systems using a dynamic model. *Trans. IchemE*, 74(Part A):357–362, 1996.

[17] S. Hernández and A. Jiménez. Controllability analysis of thermally coupled distillation systems. *Ind. Eng. Chem. Res.*, 38(10):3957–3963, 1999.

[18] S. Hernández and A. Jiménez. Design of energy - efficient petlyuk systems. 23(8):1005–1010, 1999.

[19] S. Hernández, S. Pereira-Pech, and V. Rico-Ramírez. Energy efficiency of an indirect thermally coupled distillation sequence. *Can. J. Chem. Eng.*, 81(5):1087–1091, 2003.

[20] A. Jiménez, S. Hernández, F.A. Montoy, and M. Zavala-García. Analysis of control properties of conventional and nonconventional distillation sequences. *Ind. Eng. Chem. Res.*, 40(17):3757–3761, 2001.

[21] A. Jiménez, N. Ramírez, A. Castro, and S. Hernández. Design and energy performance of alternative schemes to the petlyuk distillation system. *Trans. IchemE*, 81(Part A):518–524, 2003.

[22] G. Kaibel and H. Schoenmakers. Process synthesis and design in industrial practice. In J. Grievink and J.V. Schijndel, editors, *European Symposium on Computer Aided Process Engineering-12*, pages 9–22. Elsevier, Holland, 2002.

[23] P. Mizsey and Z. Fonyo. Toward a more realistic process synthesis – The combined approach. *Comput. Chem. Eng.*, 14(11):1303–1310, 1990.

[24] P. Mizsey and Z. Fonyo. Energy integrated distillation system design enhanced by heat pumping distillation and absorption. *Inst. Chem. Eng.*, pages B69–B75, 1992.

[25] K. Muralikrishna, K.P. Madhavan, and S.S. Shah. Development of dividing wall distillation column design space for a specified separation. *Trans. IchemE*, 80(Part A):155–165, 2002.

[26] B.G. Rong and A. Kraslawski. Optimal design of distillation flowsheets with a lower number of thermal couplings for multicomponent separations. *Ind. Eng. Chem. Res.*, 41:5716–5726, 2002.

[27] B.G. Rong and A. Kraslawski. Partially thermally coupled distillation systems for multicomponent separations. *A.I.Ch.E. Journal*, 49:1340–1347, 2003.

[28] J.G. Segovia-Hernández, S. Hernández, and A. Jiménez. Análisis dinámico de secuencias de destilación térmicamente acopladas (in spanish). *Información Tecnológica*, 13:103–108, 2002.

[29] J.G. Segovia-Hernández, S. Hernández, and A. Jiménez. Control behaviour of thermally coupled distillation sequences. *Trans. IchemE*, 80(Part A):783–789, 2002.

[30] J.G. Segovia-Hernández, S. Hernández, V. Rico-Ramírez, and A. Jiménez. A comparison of the feedback control behavior between thermally coupled and conventional distillation schemes. *Comput. Chem. Eng.*, 28:811–819, 2004.

[31] M. Serra, A. Espuña, and L. Puigjaner. Control and optimization of the divided wall column. *Chem. Eng. Process.*, 38:549–562, 1999.

[32] R. Smith and B. Linnhoff. The design of separators in the context of overall process. *Chem. Eng. Res. Des.*, 66:195–210, 1988.

[33] G. Stephanopoulos. *Chemical Process Control: An introduction to theory and practice*. Prentice Hall, New Jersey, 1984.

[34] D.W. Tedder and D.F. Rudd. Parametric studies in industrial distillation: Part I. Design comparisons. *A.I.Ch.E. Journal*, 24:303–315, 1978.

[35] C. Triantafyllou and R. Smith. The design and optimization of fully thermally coupled distillation columns. *Trans. Inst. Chem. Eng.*, 70(Part A):118–132, 1992.

[36] E. Wolff and S. Skogestad. Operation of integrated three - products (petlyuk) distillation columns. *Ind. Eng. Chem. Res.*, 34(6):2094–2103, 1995.

[37] H. Yeomans and I. Grossmann. Optimal design of complex distillation columns using rigorous tray-by-tray disjunctive programming models. *Ind. Eng. Chem. Res.*, 39(11):4326–4335, 2000.

Robust Tracking for Oscillatory Chemical Reactors

J.P. García-Sandoval[1], B. Castillo-Toledo[1], and V. González-Álvarez[2]

[1] Centro de Investigación y de Estudios Avanzados del IPN, Unidad Guadalajara
{pgarcía,toledo}@gdl.cinvestav.mx
[2] Departamento de Ingeniería Química, Universidad de Guadalajara
victorga@ccip.udg.mx

Summary. In this chapter the control problem of output tracking with disturbance rejection of chemical reactors operating under forced oscillations subjected to load disturbances and parameter uncertainty is addressed. An error feedback nonlinear control law which relies on the existence of an internal model of the exosystem that generates all the possible steady state inputs for all the admissible values of the system parameters is proposed, to guarantee that the output tracking error is maintained within predefined bounds and ensures at the same time the stability of the closed-loop system. Key theoretical concepts and results are first reviewed with particular emphasis on the development of continuous and discrete control structures for the proposed robust regulator. The role of disturbances and model uncertainty is also discussed. Several numerical examples are presented to illustrate the results.

1 Introduction

Periodically time-varying (PTV) systems are often used to model natural or forced periodic phenomena occurring in various engineering applications. Many processes experience periodic disturbances due to natural cycle times of upstream processes or other cyclical environmental influences such as diurnal temperature fluctuations. In wastewater treatment plants, for example, the feed flow rate and its composition can exhibit strong diurnal variations [7]. In some applications, forced periodic operations can be used either to improve selectivity and yields [2] or to make a continuous operation feasible [42],[49]. A potential drawback of adopting such an operation mode is that operation becomes more complex and difficult to control. Also, processes with periodic characteristics may show strong non-stationary or cyclo-stationary behavior. Additionally, because of the wide envelope of operating space covered by a periodic operation, the process dynamics can be strongly nonlinear. These

H.O. Méndez-Acosta et al. (Eds.): Dyn. & Ctrl. of Chem. & Bio. Proc., LNCIS 361, pp. 73–115, 2007.
springerlink.com   © Springer-Verlag Berlin Heidelberg 2007

factors make the control of periodic processes more complicated than the designed for constant operation processes.

This chapter is concerned with the control of continuous chemical reactors in order to keep some process state variables (or some function of the state variables) close to a prespecified periodic reference value, in the face of disturbances as load changes and variations in process parameters. Oscillations in chemical reactions, which have been the intense focus of recent investigations, are often deemed undesirable in practice because they may cause various operational complications. However, recent works groups have shown that spontaneous or forced oscillatory chemical reactions may enhance the performance of continuous stirred-tank reactors (CSTR's). For instance, periodic operation of polymerization reactors under sinusoidal perturbations in monomer and initiator feed concentrations produces a material that may not be generated by other ways while improving the performance of the operation [26],[39]. The benefits of oscillatory operation of catalytic reactors have been shown in [29],[30],[35] while the conditions for advantageous oscillatory operation of a CSTR have been simulated in [47],[48] and have demonstrated that oscillatory operation enhances the performance of CSTR's in series for competitive reactions but not for successive reactions. It has been also demonstrated that the average concentration of the product through time from a biochemical reaction occurring in a CSTR is magnified substantially by forced oscillations [12],[27]. In [45] has been theoretically demonstrated that low frequency periodic perturbations in the reaction temperature of the CSTR enhance the yield of a certain set of chemical reactions and that the maximum yield can be thrice that of the steady state operation. A common characteristic of these studies relies on studying the effect of applying those periodic changes rather than trying to find the appropriate periodic operating conditions to meet the stringent quality requirements and the suitable control system to track the nominal oscillating trajectories.

On the other hand, a number of authors have also shown that certain chemical reactors exhibit sustained oscillations as a result of applying different control schemes near an unstable steady state [5],[6],[11]. Fast zero-average oscillations has been introduced in the parameter of the reactor to achieve stabilized behavior with small swings in the controlled variables [13]. This control scheme, known as vibrational control, has been shown to yield higher conversions than the conversions corresponding to the stabilized reactor temperature [14]. A control scheme of a forced periodic fixed bed reactor has been devised to avoid the reactor extinction [22]. They have found that using the cycle time as an actuator gives rise to a slow non-minimum phase behavior, which severely limits the design of a single loop control. It has been demonstrated that both, linear and nonlinear feedback mechanisms, substantially increase the performance of a CSTR where a spontaneous oscillating autocatalytic reaction takes place [33],[34]. A homoclinic chaotic behavior can be induced in a controlled reacting system by continuous robust asymptotic feedback [16] while in [36] was analyzed the time response of a PI-controlled CSTR, where a first order irreversible reaction occurs, finding that self-oscillations and chaotic dynamics may appear. Although all the aforementioned studies have demonstrated that periodic operation render better results than those obtained by operating the CSTR at steady state, the resulting oscillating behavior was induced by some type of feedback control action and not by following a predetermined periodic profile to deliberately satisfy certain product quality properties.

For a periodic system, the standard control objective is to track periodic reference trajectories and/or to reject disturbances having periodic effects on the controlled variables. The effectiveness of conventional feedback controllers such as the proportional-integral-derivative (PID) type or dynamic matrix control (DMC) type on such problems can be quite limited in general because the system might not settle down to a constant condition. For servomechanical systems that are designed to track periodic reference trajectories, the technique called repetitive control (RC) has been widely used [23]. The idea is based on the internal model principle, which allow embedding into the controller an element that has poles on all the harmonic and subharmonic frequencies of the fundamental frequency. However, the assumption of unconstrained linear-time invariant system inherent in the development of RC is rather limiting for chemical process applications. Aside from RC, extensions of optimal control methods such as LQR and LQG to periodic systems have been developed [4]. However, most optimal controllers, when formulated as in the literature, would lead to periodic off-sets. This issue has been addressed in [28] by proposing an approach based on the model predictive control (MPC). Such an approach is referred to as R-MPC (R for repetitive) to embed into the controller the integral action over period index and thereby to ensure convergence to a periodic set-point trajectory and offset-free rejection of periodic disturbances. A number of control methods have also addressed the control problem of periodic processes in a fundamental context. These range from open-loop approaches for modeling analysis and design, to linear time-varying (continuous and discrete time) techniques [3],[20]. However, the extension to nonlinear systems has been limited to use linear time-varying models valid along some nominal periodic trajectory. Moreover, all the aforementioned control approaches were not tested under model/plant mismatches and therefore, the robustness was not guaranteed.

Broadly speaking, one of the solutions to the tracking problem in dynamical systems which are subjected to external disturbances and reference signals (both generated by a dynamic system better known as the exosystem) can be formulated as the problem of finding a solution to the so called regulation problem. Under such a formulation, the control problem consists, in fact, in finding a feedback of the state or the error such that the equilibrium point of the closed loop system is asymptotically stable and the tracking error approaches zero even under the influence of the exosystem. Notice that in this context, regulation stands in the more general sense of tracking a possibly non-constant trajectory. The regulation problem has been extensively studied in linear systems and recently has been extended to nonlinear systems. The solution to the linear case has been established in terms of the solution of certain matrix equations (the Francis equations), which describe the existence of a steady state subspace, on which the output tracking error is zeroed [17]. In order to guarantee the asymptotic tracking of the output in the face of parametric disturbances, it has been demonstrated that the solution of the robust regulation problem may be provided by a dynamic controller containing an internal model [17]; that is, a model that generates all the necessary steady state inputs for any admissible value of the system parameters. These ideas were then extended to the nonlinear case in [25], where has been demonstrated that the corresponding solution depends on the solution of a pair of matrix partial differential equations, known henceforth as the Francis-Isidori-Byrnes equations. Recently, a robust regulator based on the inclusion of an internal model of the exosystem has been developed [37]. Two interesting features of this regulator are that it is constructed from a linear approximation of the system and that if the internal model is linear, the regulator is also fully linear. It is worth mentioning that even if the controller is linear, it guarantees the tracking requirement for the original nonlinear system. This property will be fully detailed in the following sections.

The aim of this chapter is to discuss issues relevant to the development and application of a class of robust regulators to solve the tracking problem in chemical reactors. This class of robust regulators is an error feedback controller which relies on the existence of an internal model, obtained by finding if possible, an immersion of the exosystem dynamics into an observable one, which allows to generate all the possible steady state inputs for the admissible values of the system parameters [9]. First, it will be given a general overview for the case of the linear systems and will then extend the theory to nonlinear systems and derive both the continuous-time and the discrete-time versions of the error feedback controller. Also, it will be demonstrated the feasibility of the proposed robust regulator by applying the different versions developed previously to track a periodic trajectory in several models of chemical reactors. Finally, conclusions from this work will be drawn.

2   Basic Facts for the Regulation Problem

In the control literature and control applications, regulation is often addressed as forcing the output of a dynamical system to reach a desirable constant value. While for many physical systems this is the case due to the proper nature of the system, for other interesting systems, time varying reference signals are imposed to obtain a suitable behavior of the system. In this section, a review of some results relative to the regulator problem, for the linear and non linear case is presented . Extension of these results to the case of discrete-time systems will be also introduced.

2.1   Regulation Problem for Linear Systems

Consider a linear invariant system represented in Figure 1, described by

$$\dot{x}(t) = Ax(t) + Bu(t)$$
$$y(t) = Cx(t), \tag{1}$$

where $x \in \mathbb{R}^n, u \in \mathbb{R}^m$ and $y \in \mathbb{R}^p$ are the state, input and output signals, respectively. For this system, an output tracking error is chosen as the difference between the system output and a reference signal; namely

$$e(t) = y(t) - y_r(t), \tag{2}$$

where the reference signal $y_r(t) \in \mathbb{R}^p$ is supposed to be generated by an external generator, the exosystem, given by

$$\dot{\omega}(t) = S\omega(t)$$
$$y_r(t) = R\omega(t), \tag{3}$$

with $\omega \in \mathbb{R}^q$. Since we are interested in tracking or rejecting signals that do not decay asymptotically to zero, it is assumed that in general S is a neutral stable matrix, i.e., all its eigenvalues lie on the imaginary axis [24]. In this case, the regulation problem is defined as the problem of tracking the reference signal, maintaining at the same time the stability of the closed-loop structure. This goal can be achieved either by a static, or state-feedback controller of the form

$$u = K_1 x + K_2 \omega, \tag{4}$$

or by a dynamic, or error-feedback controller given by

$$\dot{z} = Fz + Ge$$
$$u = Hz. \tag{5}$$

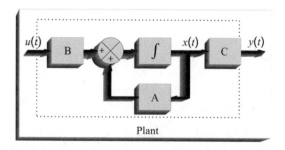

Fig. 1. Block diagram representation of a linear system

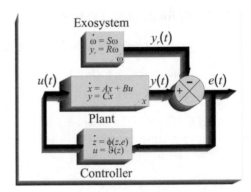

Fig. 2. Control scheme for error feedback linear regulation

To be more precise, the state-feedback regulation problem consists of finding a controller of the form (4) such that the following conditions hold

SS **Stability:** The equilibrium point $x = 0$, of the closed-loop system
$$\dot{x}(t) = Ax(t) + BK_1x$$
is asymptotically stable.

SR **Regulation:** For each initial condition $(x(0), z(0), \omega(0))$ in a neighborhood of the origin, the solution of the closed-loop system
$$\dot{x}(t) = Ax(t) + BK_1x + BK_2\omega$$
$$\dot{\omega}(t) = S\omega(t)$$
guarantees that $\lim_{t \to \infty} e(t) = 0$.

Likewise, the error-feedback regulation problem consists of finding a controller of the form (5) such that the following conditions hold

ES **Stability:** The equilibrium point $(x, z) = (0, 0)$ of the closed-loop system
$$\dot{x}(t) = Ax(t) + BHz(t)$$
$$\dot{z}(t) = Fz(t) + GCx(t)$$
is asymptotically stable.

ER **Regulation:** For each initial condition $(x(0), z(0), w(0))$ in a neighborhood of the origin, the solution of the closed-loop system
$$\dot{x}(t) = Ax(t) + BHz(t)$$
$$\dot{z}(t) = Fz(t) + G(Cx(t) - Rw(t))$$
$$\dot{\omega}(t) = S\omega(t)$$
guarantees that $\lim_{t \to \infty} e(t) = 0$ (Figure 2 shows the block diagram of the aforementioned closed loop case).

In other words, the regulation problem is equivalent to the problem of finding a subset Z of the Cartesian product $\mathbb{R}^n \times \mathbb{R}^q$ on which the output tracking error $e(t)$ is zeroed, and an input signal which makes attractive and invariant this subset (Figure 3). The trajectories described by the state and the input on the invariant subset Z, are thereafter referred as the *steady state*

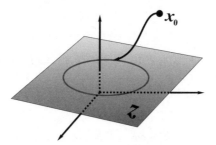

Fig. 3. Invariance subspace

and *steady state input*, and are denoted by $x_{ss}(t)$ and $u_{ss}(t)$, respectively. Assume that these steady-state and input behavior are given by

$$x_{ss}(t) = \Pi\omega(t)$$
$$u_{ss}(t) = \Gamma\omega(t), \tag{6}$$

where Π and Γ are constants matrices to be defined later. If we define the state and input deviation as

$$\delta x = x - \Pi\omega$$
$$\delta u = u - \Gamma\omega, \tag{7}$$

by taking their derivative with respect to time, we obtain

$$\delta\dot{x} = Ax + Bu + -\Pi S\omega$$
$$= A\delta x + B\delta u + (A\Pi + B\Gamma - \Pi S)\omega, \tag{8}$$

while the output tracking error becomes

$$e(t) = Cx - R\omega$$
$$= C\delta x + (C\Pi - R)\omega. \tag{9}$$

At this point, notice that if we can separate the effect of the exosystem states on the previous equations, we may handle the problem of regulation by simply finding a feedback that stabilize system (8). This situation may be then obtained if and only if there exist solutions Π, and Γ to the following Sylvester-like conditions:

$$\Pi S = A\Pi + B\Gamma$$
$$0 = C\Pi - R. \tag{10}$$

These equations are known in the literature as the Francis equations [17]. Equations (8)-(9) take now the form

$$\delta \dot{x} = A\delta x + B\delta u$$
$$e = C\delta x, \tag{11}$$

which can be stabilized by a state feedback $\delta u = K\delta x$, such that K makes the matrix $(A + BK)$ stable, provided that the pair (A, B) is controllable. In this case, regulation can be achieved by a static state-feedback controller of the form (4) given by

$$u(t) = \Gamma\omega(t) + K\delta x = \Gamma\omega(t) + K(x(t) - \Pi\omega(t)). \tag{12}$$

Figure 4 shows the control scheme for state feedback. In the event that only the output measurements are available, the controller (12) cannot be directly implemented. In this case, an observer of the states may be used. To this end, we may rewrite the first expression in (11) and the second one of (3) as follows

$$\delta \dot{x} = A\delta x - B\Gamma\omega + Bu$$
$$\dot{\omega}(t) = S\omega(t).$$

An observer for the states $(\delta x, \omega)$ is then constructed as a Luenberger observer of the form

$$\dot{z}_1(t) = Az_1(t) - B\Gamma z_2 + Bu(t) + G_1(C\delta x - Cz_1(t))$$
$$\dot{z}_2(t) = Sz_2(t) + G_2(C\delta x - Cz_1(t)) \tag{13}$$
$$u = Kz_1 + \Gamma z_2$$

where $z_1(t)$ is the estimation of $\delta x(t)$ given by the first expression in equation (7), $z_2(t)$ is the estimation of $\omega(t)$, while matrix $G = (G_1 \ G_2)^T$ is such that, matrix $\left[\begin{pmatrix} A & -B\Gamma \\ 0 & S \end{pmatrix} - G(C \ 0) \right]$ is stable, from which the detectability of the pair $\left[\begin{pmatrix} A & -B\Gamma \\ 0 & S \end{pmatrix}, (C \ 0) \right]$ must be satisfied.

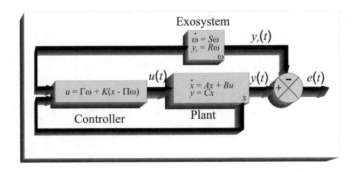

Fig. 4. The controller scheme for state feedback regulation

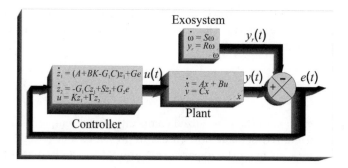

Fig. 5. The controller scheme for error feedback regulation

Substituting the expression for $u(t)$, the controller becomes

$$\dot{z}_1(t) = (A + BK_1 - G_1 C) z_1(t) + G_1 e(t)$$
$$\dot{z}_2(t) = -G_2 C z_1(t) + S z_2(t) + G_2 e(t) \tag{14}$$
$$u(t) = K z_1(t) + \Gamma z_2(t) = H z(t)$$

The implementation of this controller is depicted in Figure 5. The previous discussion is summarized in the following theorem.

Theorem 1. *Let us assume that for the linear system (1) there exists a matrix K such that $A + BK$ is stable and S is neutral stable. Then, the state feedback regulator problem is solvable if and only if the Francis equations*

$$\Pi S = A\Pi + B\Gamma,$$
$$0 = C\Pi - R,$$

have solutions given by Π, and Γ. The controller that solves the problem is then given by $u(t) = \Gamma\omega + K(x - \Pi\omega)$. Moreover, if there exist matrices G_1, G_2 such that the matrix $\begin{pmatrix} A & -B\Gamma \\ 0 & S \end{pmatrix} - \begin{pmatrix} G_1 \\ G_2 \end{pmatrix} (C \ 0)$ is stable, then the error feedback regulator problem is also solvable by the controller

$$\dot{z} = \begin{pmatrix} A + BK - G_1 C & 0 \\ -G_2 C & S \end{pmatrix} z + \begin{pmatrix} G_1 \\ G_2 \end{pmatrix} e$$
$$u = (K \ \Gamma) z.$$

∎

The following example illustrate the calculations involved in the construction of both the state and error feedback regulators.

Example 1. Periodic Control of an Open Loop Unstable CSTR

Consider the problem of synthesizing a controller in order to track a periodic temperature profile in the decomposition reaction of hydrogen peroxide,

$$H_2O_2 \rightarrow H_2O(\ell) + \frac{1}{2}O_2(g)$$

conducted in a CSTR [38]. Tracking control is achieved by manipulating the coolant flowrate. A model for the reacting system is derived from the mass and energy balances and it is given by

$$\dot{x} = A_0 x + B_0 u$$
$$y = C_0 x$$

where x_1 and x_2 denote the normalized composition and temperature, respectively. The normalized coolant flow rate is represented by u while matrices A_0, B_0 and C_0 result from the linearization around the desired steady state. These are given by

$$A_0 = \begin{pmatrix} -0.03318 & -0.002511 \\ 0.05237 & 0.010216 \end{pmatrix}, \quad B_0 = \begin{pmatrix} 0 \\ -0.0988 \end{pmatrix}, \quad C_0 = (0 \ 1)$$

It is straightforward to show that the desired steady-state (i.e., the origin) is unstable. A robust tracking control law can be constructed to stabilize the CSTR under forced oscillatory operation. That is, we can derive a controller to track an oscillatory temperature profile (say, $y_r(t) = a + b\sin(4\pi t)$), which can be generated by the exosystem (3) where

$$S = \begin{pmatrix} 0 & \frac{2\pi}{5} & 0 \\ -\frac{2\pi}{5} & 0 & 0 \\ 0 & 0 & 0 \end{pmatrix} \quad \text{and} \quad R = (1 \ 0 \ 1).$$

For this example, the invariant subset Z is found by solving the invariant equations (10). Here

$$\Pi = \begin{pmatrix} -5.2723 \times 10^{-5} & 1.9968 \times 10^{-3} & -0.075678 \\ 1 & 0 & 1 \end{pmatrix} \quad \text{and}$$
$$\Gamma = (0.10337 \ -12.718 \ 0.063287).$$

For this system the pairs $[A_0, B_0]$ and $\left[\begin{pmatrix} A_0 & -B\Gamma \\ 0 & S \end{pmatrix}, (C_0 \ 0) \right]$ are controllable and observable, respectively.

On the other hand, let us choose

$$K = (-1442.4 \ \ 12.925)$$
$$G = (0.70706 \ \ 2.9671 \ \ 1.5033 \ -2.6415 \ \ 2.2361)^T,$$

such that $(A_0 + B_0 K)$ and $\left[\begin{pmatrix} A_0 & -B_0\Gamma \\ 0 & S \end{pmatrix} - G(C_0 \ 0) \right]$ are Hurwitz (the correspondent eigenvalues are $(-0.5, -0.8)$ and $(-0.614 \pm 1.6i, -1.7, -0.047, -0.0032)$, respectively). Notice that in this case, K and G were calculated using simple pole assignments but they can be calculated by any other method,

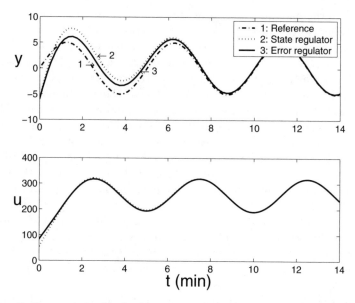

Fig. 6. System behavior for Example 1 where y and u are dimensionless

such as the LQR for optimal control. Thus, the static controller (12) takes the form

$$u(t) = \begin{pmatrix} -1442.4 & 12.925 \end{pmatrix} x(t) + \begin{pmatrix} -12.898 & -9.8378 & -122.02 \end{pmatrix} \omega(t)$$

while the dynamic controller (14) becomes

$$\dot{z}_1(t) = \begin{pmatrix} -0.03318 & -0.70957 \\ 142.56 & -4.234 \end{pmatrix} z_1(t) + \begin{pmatrix} 0.70706 \\ 2.9671 \end{pmatrix} e(t)$$

$$\dot{z}_2(t) = \begin{pmatrix} 0 & -1.5033 \\ 0 & 2.6415 \\ 0 & -2.2361 \end{pmatrix} z_1(t) + \begin{pmatrix} 0 & \frac{2\pi}{5} & 0 \\ -\frac{2\pi}{5} & 0 & 0 \\ 0 & 0 & 0 \end{pmatrix} z_2(t) + \begin{pmatrix} 1.5033 \\ -2.6415 \\ 2.2361 \end{pmatrix} e(t)$$

$$u(t) = \begin{pmatrix} -1442.4 & 12.925 \end{pmatrix} z_1(t) + \begin{pmatrix} 0.10337 & -12.718 & 0.063287 \end{pmatrix} z_2(t).$$

The results of applying these strategies under the influence of initial error conditions are shown in Figure 6. As can be observed, both linear regulators ensure the asymptotic tracking of the desired oscillatory temperature profile. Both regulators performance is different because the error feedback regulator behavior depends on the initial states of z_1 and z_2.

2.2   Robust Regulation of Linear Systems

A desirable feature of a controller is the ability of maintaining the properties for which it has been designed, even in the presence of unmodelled dynamics or variation on the parameters. This feature gives a measure of the robustness of the controller. In the particular case of regulation, the state or error feedback controllers are in general not robust in the presence of parametric variations. In this section, we show how to design a robust regulator. To this end, let us denote $\mu \in \mathbb{R}^p$ as the parameters vector of the system and $A(\mu)$, $B(\mu)$, $C(\mu)$ and $R(\mu)$ the values of the matrices A, B, C and R depending on the values of μ. Assume, without lost of generality, that $\mu = 0$ represents the nominal values of these parameters; thus, $A(0)$, $B(0)$, $C(0)$ and $R(0)$ are the nominal values of these matrices. From classical results of linear systems, it is well known that the property of stabilization is robust; namely, if for a pair $(A(0), B(0))$ there exists a matrix K such that $A(0) + B(0)K$ is a stable matrix, then $A(\mu) + B(\mu)K$ is also stable in a neighborhood P of the nominal parameter vector ($\mu = 0$). This property will be exploited to construct the robust regulator.

However, when the system parameters change, the steady-state input also changes. If the actual values of the parameters are unknown, then it will be necessary to adjust the steady-state input in order to maintain the invariance of the zero output tracking subset. A way to attain this goal is presented in the following.

Let us assume that for a linear system, equation (10) has solution for all values of the parameters in the neighborhood P; namely, there exist solutions $\Pi(\mu), \Gamma(\mu)$ to the equations

$$\Pi(\mu) S = A(\mu) \Pi(\mu) + B(\mu) \Gamma(\mu), \tag{15}$$

$$0 = C(\mu) \Pi(\mu) - R(\mu). \tag{16}$$

From these equations, as in the previous section, $u_{ss}(t) = \Gamma(\mu)\omega(t)$ represents the steady-state input. The main difference here is that this steady state input is unknown since the exact value of the parameter μ is also unknown. A way to generate all possible values of the steady state inputs is given as follows.

Define the $m \times 1$ vector

$$z_1(t) = \Gamma(\mu)\omega(t) \tag{17}$$

and take the derivative with respect to time q times to find,

$$\begin{aligned}
\dot{z}_1(t) &= \Gamma(\mu)S\omega(t) =: \dot{z}_2(t) \\
\dot{z}_2(t) &= \Gamma(\mu)S^2\omega(t) =: \dot{z}_3(t) \\
&\quad\vdots \qquad\quad \vdots \qquad\quad \vdots \\
\dot{z}_{q-1}(t) &= \Gamma(\mu)S^{q-1}\omega(t) =: \dot{z}_q(t) \\
\dot{z}_q(t) &= \Gamma(\mu)S^q\omega(t).
\end{aligned} \tag{18}$$

At this point, since S is a $q \times q$ matrix, according to the Cayley-Hamilton Theorem, S must satisfy its own characteristic polynomial, i.e.

$$S^q + a_{q-1}S^{q-1} + \ldots + a_2 S^2 + a_1 S + a_0 I = 0$$

where a_i, $i = 1, 2, \ldots, q-1$ are the coefficients of the characteristic equation $\det(\lambda I - S) = \lambda^q + a_{q-1}\lambda^{q-1} + \ldots + a_2\lambda^2 + a_1\lambda + a_0 = 0$. From this, and the last equation of (18), we can write

$$\dot{z}_q(t) = -\left(a_0\Gamma(\mu) + a_1\Gamma(\mu)S + a_2\Gamma(\mu)S^2 + \ldots + a_{q-1}\Gamma(\mu)S^{q-1}\right)w(t)$$
$$= -a_0z_1(t) - a_1z_2(t) - a_3z_2(t) - \ldots - a_{q-1}z_q(t).$$

Equation (18) may be then rewritten as

$$\dot{z}(t) = \widehat{\Phi}z(t)$$
$$u_{ss}(t) = \widehat{H}z(t) \tag{19}$$

where

$$Z(t) = \begin{pmatrix} z_1(t) \\ z_2(t) \\ \vdots \\ z_q(t) \end{pmatrix}, \quad \widehat{\Phi} = \begin{pmatrix} 0 & I_m & 0 & \cdots & 0 \\ 0 & 0 & I_m & \cdots & 0 \\ \vdots & \vdots & \vdots & \ddots & \vdots \\ 0 & 0 & 0 & \cdots & I_m \\ -a_0I_m & -a_1I_m & -a_2I_m & \cdots & -a_{q-1}I_m \end{pmatrix},$$

$$\widehat{H} = \begin{pmatrix} I_m & 0 & 0 & \cdots & 0 \end{pmatrix}_{m \times qm}.$$

Here, $z_1, z_2, \ldots, z_q$, are vectors of m dimension. By reordering the vector $Z(t)$, matrices $\widehat{\Phi}$ and $\widehat{H}$ take the block forms,

$$\Phi = \begin{pmatrix} \widetilde{S} & \cdots & 0 \\ \vdots & \ddots & \vdots \\ 0 & \cdots & \widetilde{S} \end{pmatrix}, \quad H = \begin{pmatrix} \widetilde{H} & \cdots & 0 \\ \vdots & \ddots & \vdots \\ 0 & \cdots & \widetilde{H} \end{pmatrix} \tag{20}$$

where

$$\widetilde{S} = \begin{pmatrix} 0 & 1 & 0 & \cdots & 0 \\ 0 & 0 & 1 & \cdots & 0 \\ \vdots & \vdots & \vdots & \ddots & \vdots \\ 0 & 0 & 0 & \cdots & 1 \\ -a_0 & -a_1 & -a_2 & \cdots & -a_{q-1} \end{pmatrix} \quad \text{and} \quad \widetilde{H} = \begin{pmatrix} 1 & 0 & 0 & \cdots & 0 \end{pmatrix}_{1 \times q}.$$

and system (19) becomes

$$\dot{z}(t) = \Phi z(t)$$
$$u_{ss}(t) = Hz(t) \tag{21}$$

Notice that in this case, all the terms $\widetilde{S}$ have the same known dimension. This feature will not be maintained in the nonlinear case, but it will be treated under a slightly different approach.

The procedure outlined above represents an immersion of the exosystem into an observable system (19), which can generate, for some appropriate initial conditions, the exact steady-state input for all the values of the parameter in a suitable neighborhood. Since these initial conditions are also unknown, the structure of the immersion will be used in the controller to estimate asymptotically the required exact steady-state input required. This feature will allow the controller to incorporate the desirable robustness property.

Finally, the controller solving the robust regulation problem for linear system (1) takes the form

$$
\begin{aligned}
\dot{z}_1(t) &= (A_0 + B_0 K_1 - G_1 C_0)\, z_1(t) + G_1 e(t) \\
\dot{z}_2(t) &= -G_2 C_0 z_1(t) + \Phi z_2(t) + G_2 e(t) \\
u(t) &= K z_1(t) + H z_2(t),
\end{aligned}
\tag{22}
$$

where K is such that $(A_0 + B_0 K)$ is Hurwitz, while $G = \begin{pmatrix} G_1 & G_2 \end{pmatrix}^T$ makes stable $\left(\overline{A} - G\overline{C}\right)$ where

$$
\overline{A} = \begin{pmatrix} A_0 & -B_0 H \\ 0 & \Phi \end{pmatrix}, \quad \overline{C} = \begin{pmatrix} C_0 & 0 \end{pmatrix}.
$$

Obviously, the pairs (A_0, B_0) and $\left(\overline{A}, \overline{C}\right)$ must be stabilizable and detectable, respectively. As we can see, controller (22) has the form of (5) and does not contain the mappings $\Pi(\mu)$ and $\Gamma(\mu)$; thus, although the initial condition for $z_2(t)$ is not exactly known, the immersion observer (second expression in (22)) estimates the correct steady-state input and as a result, the controller is capable to drive the system towards the correct zero-error submanifold in spite of parametric variations. It can be seen from the first equation in (22) that as $e(t)$ approaches asymptotically zero, so does z_1. Notice also that the dynamics of z_2 is similar to immersion (21). It is important to point out that this design procedure does not require the exact calculation of mappings $\Pi(\mu)$ and $\Gamma(\mu)$, but it suffices only to know the dimension of matrix S.

Example 2. Let us now test the robust linear regulator under model mismatch. For this application let us consider again the CSTR of Example 1. Here, we assumed that the CSTR model dynamics is given by

$$
A = \begin{pmatrix} -0.029862 & -0.0027621 \\ 0.05237 & 0.0097052 \end{pmatrix}, \quad B = \begin{pmatrix} 0 \\ -0.08892 \end{pmatrix}, \quad C = \begin{pmatrix} 0 & 1 \end{pmatrix}.
$$

Notice that A and B contain coefficients that change up to 10% with respect to the nominal values used in the previous example. In this example, the robust controller design is based on nominal matrices defined in Example 1.

The exosystem defined in Example 1 produces the immersion

$$
\Phi = \begin{pmatrix} 0 & 1 & 0 \\ 0 & 0 & 1 \\ 0 & -\frac{4}{25}\pi^2 & 0 \end{pmatrix} \quad \text{and} \quad H = \begin{pmatrix} 1 & 0 & 0 \end{pmatrix},
$$

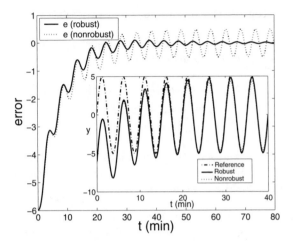

Fig. 7. Tracking output error for the robust and the nonrobust controllers where error and y are dimensionless

which makes the pair $\left[\begin{pmatrix} A_0 & -B_0 H \\ 0 & \Phi \end{pmatrix}, \begin{pmatrix} C_0 & 0 \end{pmatrix}\right]$ observable.

The gains

$$K = \begin{pmatrix} 1.0124 & 1.5031 \end{pmatrix}$$

$$G = \begin{pmatrix} 0.60243 & 1.7625 & 5.0968 & -0.071920 & -0.029351 \end{pmatrix}^T$$

are chosen such that $(A_0 + B_0 K)$ and $\left[\begin{pmatrix} A_0 & -B_0 H \\ 0 & \Phi \end{pmatrix} - G \begin{pmatrix} C_0 & 0 \end{pmatrix}\right]$ are Hurwitz.

Thus, the resulting robust regulator (22) is given by

$$\dot{z}_1(t) = \begin{pmatrix} -0.03318 & -0.60494 \\ -0.047655 & -1.9007 \end{pmatrix} z_1(t) + \begin{pmatrix} 0.60243 \\ 1.7625 \end{pmatrix} e(t)$$

$$\dot{z}_2(t) = \begin{pmatrix} 0 & -5.0968 \\ 0 & 0.071920 \\ 0 & 0.029351 \end{pmatrix} z_1(t) + \begin{pmatrix} 0 & 1 & 0 \\ 0 & 0 & 1 \\ 0 & -\frac{4}{25}\pi^2 & 0 \end{pmatrix} z_2(t) + \begin{pmatrix} 5.0968 \\ -0.071920 \\ -0.029351 \end{pmatrix} e(t)$$

$$u(t) = \begin{pmatrix} 1.0124 & 1.5031 \end{pmatrix} z_1(t) + \begin{pmatrix} 1 & 0 & 0 \end{pmatrix} z_2(t).$$

In Figure 7 we compare the performance of the robust regulator with that obtained from applying a nonrobust controller. From inspecting this figure, it is clear that the robust controller yields excellent response in the face of initial error conditions and model mismatch. In contrast, the nonrobust controller performance is influenced by the periodic reference signal and as a result, it does not settle down around a constant value (in fact, the off-set is periodic).

2.3 Robust Regulation of Linear Discrete Systems

All the concepts previously presented for a continuous linear system, can be formulated *vis a vis* for the case of a linear discrete time system described by

$$
\begin{aligned}
x_d\,(k+1) &= A_d x_d\,(k) + B_d u_d\,(k) + P_d w(k) \\
e_d\,(k) &= C_d x_d\,(k) - R_d w_d\,(k) \\
w_d\,(k+1) &= S_d w_d\,(k)
\end{aligned}
\tag{23}
$$

where $k = 0, 1, 2, \ldots$, and A_d, B_d, C_d, S_d and R_d, are matrices representing the discrete system. In this case, the following theorem, which mimics Theorem 1, gives conditions for the existence of a solution.

Theorem 2. *Let us assume that for the discrete linear system (23) there exists a matrix K_d such that $A_d + B_d K_d$ has all its eigenvalues inside the unit circle in the complex plane and S_d has all its eigenvalues in the unit circle. Then the state feedback regulator problem is solvable if and only if the Francis equations*

$$
\Pi_d S = A_d \Pi_d + B_d \Gamma_d,
\tag{24}
$$

$$
0 = C_d \Pi_d - R_d.
\tag{25}
$$

have solutions Π_d and Γ_d. The controller that solves the problem is then given by $u_d(k) = \Gamma_d w(k) + K_d(x(k) - \Pi_d w(k))$. Moreover, if there exist matrices G_{d1}, G_{d2} such that the matrix $\begin{pmatrix} A_d & -B_d\Gamma_d \\ 0 & S_d \end{pmatrix} - \begin{pmatrix} G_{d1} \\ G_{d2} \end{pmatrix} \begin{pmatrix} C_d & 0 \end{pmatrix}$ has its eigenvalues inside the unit circle, then the error feedback regulator problem is also solvable by the controller

$$
\begin{aligned}
z(k+1) &= \begin{pmatrix} A_d + B_d K_d - G_{d1} C_d & 0 \\ -G_2 C & S_d \end{pmatrix} z + \begin{pmatrix} G_{d1} \\ G_{d2} \end{pmatrix} e_d \\
u(k) &= \begin{pmatrix} K_d & \Gamma_d \end{pmatrix} z(k).
\end{aligned}
\qquad\blacksquare
$$

The robust solution in this case will also be attainable by a controller of the form

$$
\begin{aligned}
z_1\,(k+1) &= (A_d + B_d K_d - G_{d1} C_d)\, z_1\,(k) + G_{d1} e_d\,(k) \\
z_2\,(k+1) &= -G_{d2} C_d z_1\,(k) + \Phi_d z_2\,(k) + G_{d2} e_d\,(k) \\
u_d\,(k) &= K_1 z_1\,(k) + H z_2\,(k),
\end{aligned}
\tag{26}
$$

where all the matrices are calculated as in continuous case.

The previous theorem ensures the regulation goals for a discrete linear system. However, if this discrete system comes from a discretization of a continuous one; that is, if the discrete system (23) is characterized by

$$A_d = e^{A\delta}, \quad B_d = \int_0^{\delta} e^{A\tau} d\tau B, \quad C_d = C, \quad S_d = e^{S\delta}, \quad R_d = R$$

where δ is the sampling time, then a different situation may occur. Indeed, the discrete controller (26) will ensure the regulation goals only at the sampling instants, but not in the intersampling time. To deal with this problem, we also need to ensure that, even in the presence of parameter variations, the discrete controller will stabilize the continuous time system and at the same time guarantee that the continuous output tracking error, and not only the discrete output tracking error, converges asymptotically to zero. This problem will be referred as the Robust Discretized Regulator Problem (RDRP).

The first objective can be reached by using a general property of discretized system; namely, if for the system (23) there exists a matrix K_d such that $A_d + B_d K_d$ has all its eigenvalues inside the unit circle, then the controller $u(k) = K_d x(k)$ will also stabilize the continuous system. Thus, the closed-loop stability can be then assured by properly assigning the discrete poles inside the unit circle.

Concerning the regulation of the output tracking error, the situation is different. In general, the discrete controller cannot ensure convergence of the error to zero, since it is based on the use of zero order holders. To force the system to stay on the zero-error submanifold even in the intersampling periods, we propose the use of an *exponential holder* which is an analog device that produces a continuous signal from a discrete signal, whose structure may be obtained as follows.

Consider the solution of equation (21) given by

$$\xi(t) = e^{\Phi(t-t_0)} \xi(t_0)$$
$$u_{ss}(t) = H\xi(t).$$

Setting $t = k\delta + \theta$, $\theta \in [0, \delta)$, and $t_0 = k\delta$, we get

$$\xi(k\delta + \theta) = e^{\Phi\theta} \xi(k\delta)$$
$$u_{ss}(k\delta + \theta) = H\xi(k\delta + \theta),$$

and then the mapping for $u_{ss}(t) = u_{ss}(k\delta + \theta)$ can be written as

$$u_{ss}(t) = He^{\Phi\theta} \xi(k\delta), \quad t \in [k\delta, (k+1)\delta), \quad \theta \in [0, \delta), \quad k = 1, 2, 3, \ldots \quad (27)$$

which describes *exactly* the steady-state input not only at the sampling time instants but also in the intersampling period. Notice that the values of $\xi(k\delta)$ at the sampling instants are obtained from the solution of the discrete time dynamics

$$\xi(k\delta + \delta) = e^{\Phi\delta} \xi(k\delta).$$

Remark 1. Notice that the signal θ in the exponential holder is a periodic sawtooth signal, depicted in Figure 8. A method of constructing such an exponential holder is described in [10].

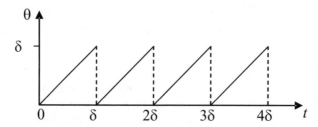

Fig. 8. Periodic sawtooth signal (θ) for the exponential holder

As in the case of continuous linear systems, the exponential holder will then ensure the fulfilment of the regulation conditions for a continuous linear system with a discrete controller. This result is summarized in the following theorem.

Theorem 3. *Assume that for the linear system (1) the following conditions hold:*

1. *The pair (A, B) is controllable.*
2. *The matrix S is neutral stable.*
3. *For all values of the parameters in the neighborhood P there exist solutions $\Pi(\mu)$ and $\Gamma(\mu)$ to the equations*

$$\Pi(\mu) S = A(\mu) \Pi(\mu) + B(\mu) \Gamma(\mu), \tag{28}$$

$$0 = C(\mu) \Pi(\mu) - R(\mu). \tag{29}$$

In this case, the RDRP is solvable by the controller

$$z_1(k+1) = (A_d + B_d K_d - G_{d1} C_d) z_1(k) + G_{d1} e(k)$$
$$z_2(k+1) = -G_{d2} C_d z_1(k) + e^{\Phi\delta} z_2(k) + G_{d2} e(t) \tag{30}$$
$$u_d(k) = K_1 z_1(k) + H e^{\Phi\theta} z_2(k),$$

where $G = \begin{pmatrix} G_1 & G_2 \end{pmatrix}^T$ makes stable the pair

$$\begin{pmatrix} A_d & -B_d H \\ 0 & \Phi \end{pmatrix}, \quad \overline{C} = \begin{pmatrix} C_d & 0 \end{pmatrix}. \quad \blacksquare$$

Remark 2. Condition 1 states the well-known fact that if the pair (A, B) is controllable, then the discretized pairs (A_d, B_d) is generically controllable. Condition 2 will ensure that the discretized exosystem matrix $e^{S\delta}$ will have all its eigenvalues in the unit circle, as required for the discrete regulator solution. Condition 3 guarantees that the immersion (21) exists and then the exponential holder can be calculated.

Corollary 1. *The Robust Discretized Regulator Problem for linear systems is solvable if and only if the robust regulation problem for the continuous linear case is solvable.*

2.4   Robust Regulation Problem for Nonlinear Systems

Let us consider the following nonlinear time-invariant system

$$\dot{x} = f(x, u, \omega, \mu)$$
$$\dot{\omega} = s(\omega) \tag{31}$$
$$e = h(x, \omega, \mu),$$

where $x \in \mathbb{R}^n$, $u \in \mathbb{R}^m$, $\mu \in \mathbb{R}^p$ and $\omega \in \mathbb{R}^r$ are defined as in the linear case. Define for this system:

$$A_0 = \left[\frac{\partial f(x, \omega, u, 0)}{\partial x} \right], \quad B_0 = \left[\frac{\partial f(x, \omega, u, 0)}{\partial u} \right], \quad C_0 = \left[\frac{\partial h(x, \omega, 0)}{\partial x} \right]$$

for $x = 0$, $\omega = 0$, and $u = 0$.

In this setting, the exosystem may be also nonlinear and it is supposed that it does not depend on μ, which is often verified in practical problems since the exosystem models the reference and disturbance signals affecting the plant. Moreover, the third nonlinear function in (31) describes the output tracking error $e \in \mathbb{R}^p$, which, in many cases, is given as a difference between the system output $h(x, \mu)$ and the reference signal, described by $r(\omega)$, namely

$$e(t) = h(x, \mu) - r(\omega).$$

The *State or Error Feedback Regulation Problem* for this system is defined as the problem of tracking the reference signals and/or rejecting the disturbance signals, and maintaining the closed-loop stability property. For the case of robust regulation, we also impose the requirement that these conditions hold when the parameters vary in a neighborhood of the nominal values. As in the case of linear systems, the regulator problem can be formulated as the problem of determining a certain submanifold of the state space (x, ω), where the tracking error is zero, which is rendered attractive and invariant by feedback.

Consider here the nonlinear robust regulation problem (NRRP), which consists in finding, if possible, a dynamic controller of the form

$$\dot{z} = \varphi(z, e)$$
$$u = \vartheta(z)$$

such that, for all admissible values μ in a neighborhood $\wp$ of the nominal values, the following conditions hold

N1 **Stability:** The equilibrium point $(x, z) = (0, 0)$ of the closed-loop system without disturbances

$$\dot{x} = f(x, \vartheta(z), 0, \mu)$$
$$\dot{z} = \varphi(z, 0)$$

is asymptotically stable.

N2 **Regulation:** For each initial condition $(x\,(0)\,,z\,(0)\,,\omega\,(0))$ in a neighborhood of the origin, the solution of the closed-loop system

$$\dot{x} = f\,(x, \vartheta\,(z)\,,\omega,\mu)$$
$$\dot{z} = \varphi\,(z, h\,(x,\omega,\mu))$$
$$\dot{\omega} = s\,(\omega)$$

satisfies the condition $\lim_{t\to\infty} e\,(t) = 0$.

The next theorem states the conditions for the existence of a solution to the NRRP.

Theorem 4. *[24] The Nonlinear Robust Regulation Problem is solvable if and only if there exist mappings*

$$x_{ss} = \pi\,(\omega,\mu)\,, \quad and \quad u_{ss} = \gamma\,(\omega,\mu) = \begin{pmatrix} \gamma_1\,(\omega,\mu) \\ \vdots \\ \gamma_m\,(\omega,\mu) \end{pmatrix},$$

with $\pi\,(0,\mu) = 0$ and $\gamma\,(0,\mu) = 0$, both defined in a neighborhood of the origin, satisfying the equations

$$\frac{\partial\pi\,(\omega,\mu)}{\partial\omega}s\,(\omega) = f\,(\pi\,(\omega,\mu)\,,\gamma\,(\omega,\mu)\,,\omega,\mu)\,,$$
$$0 = h\,(\pi\,(\omega,\mu)\,,\omega,\mu)\,, \tag{32}$$

for all (ω,μ) and such that for each $i = 1,\ldots,m$ the exosystem is immersed into a system

$$\dot{\xi} = \phi\,(\xi)$$
$$\gamma\,(\omega,\mu) = h\,(\xi) \tag{33}$$

defined on a neighborhood Ξ^0 of the origin, in which $\phi\,(0) = 0$ and $h\,(0) = 0$, and the two matrices

$$\Phi = \left[\frac{\partial\phi}{\partial\xi}\right]_{\xi=0}, \quad H = \left[\frac{\partial h}{\partial\xi}\right]_{\xi=0}$$

are such that the pair

$$\begin{pmatrix} A_0 & 0 \\ NC_0 & \Phi \end{pmatrix}, \quad \begin{pmatrix} B_0 \\ 0 \end{pmatrix}$$

is stabilizable for some choice of the matrix N, and the pair

$$(C_0\ 0)\,, \quad \begin{pmatrix} A_0 & B_0H \\ 0 & \Phi \end{pmatrix}$$

is detectable. ∎

Remark 3. Equation (32) is known as the Francis-Isidori-Byrnes equation (FIB) [8] and is the nonlinear version of equation (10) used to find the subset Z on the Cartesian product $\mathbb{R}^n \times \mathbb{R}^m$ called, so far, the zero tracking error submanifold.

The nonlinear immersion (47) is a general case that expresses the fact that the steady state input can be generated, independently of the values of the parameter vector μ, by a nonlinear dynamical system. For the particular case of a linear immersion we present the following corollary.

Corollary 2. *The Nonlinear Robust Regulation Problem is solvable by means of a linear controller if the pair (A_0, B_0) is stabilizable, the pair (C_0, A_0) is detectable, there exist mappings $x_{ss} = \pi(\omega, \mu)$, and $u_{ss} = \gamma(\omega, \mu)$, with $\pi(0, \mu) = 0$ and $\gamma(0, \mu) = 0$, both defined in a neighborhood of the origin, satisfying the conditions (32) and such that, for some set of r_i real numbers $a_{0,i}, a_{1,i}, \ldots, a_{r_i-1,i}$, it holds that*

$$L_s^{r_i}\gamma_i(\omega, \mu) = a_{0,i}\gamma_i(\omega, \mu) + a_{1,i}L_s\gamma_i(\omega, \mu) + \cdots + a_{r_i-1,i}L_s^{r_i-1}\gamma_i(\omega, \mu) \quad (34)$$

(where $L_s^k \gamma_i(\omega, \mu)$ stands for the Lie derivative defined as $L_s^k \gamma_i = \left[\frac{\partial L_s^{k-1}\gamma_i}{\partial \omega}\right]s(\omega)$, $k \geq 1$ with $L_s^0 \gamma_i = \gamma_i$), for all (ω, μ) and – moreover – the matrix

$$\begin{pmatrix} A_0 - \lambda I & B_0 \\ C_0 & 0 \end{pmatrix}$$

is nonsingular for every λ which is a root of the polynomials

$$p_i(\lambda) = a_{0,i} + a_{1,i}\lambda + \cdots + a_{r_i-1,i}\lambda^{r_i-1} - \lambda^{r_i}$$

having non-negative real parts. ∎

Remark 4. Notice that for a linear system the coefficients $a_{0,i}, a_{1,i}, \ldots, a_{r_i-1,i}$ in equation (48) represent the coefficients of the characteristic equation of matrix S. For the nonlinear case, these coefficients do not represent a generalization of the Cayley-Hamilton theorem; hence the assumption is necessary for the existence of the solution of the NRRP.

The mapping $x_{ss} = \pi(\omega, \mu)$ represents the steady state zero output submanifold and $u_{ss} = \gamma(\omega, \mu)$ is the steady state input which makes invariant this steady state zero output submanifold. Condition (48) expresses the fact that this steady state input can be generated, independently of the values of the parameter vector μ, by the linear dynamic system

$$\dot{\xi} = \Phi\xi$$
$$u_{ss} = H\xi \quad (35)$$

where

$$\xi = \begin{pmatrix} \xi_1 \\ \vdots \\ \xi_m \end{pmatrix}, \quad \Phi = diag\,(\Phi_1, \cdots, \Phi_m), \quad H = diag\,(H_1, \cdots, H_m)$$

$$\xi_i = \begin{pmatrix} \xi_{i1} \\ \vdots \\ \xi_{ir_i} \end{pmatrix}, \quad \Phi_i = \begin{pmatrix} 0 & 1 & 0 & \cdots & 0 \\ 0 & 0 & 1 & \cdots & 0 \\ \vdots & \vdots & \vdots & \ddots & \vdots \\ 0 & 0 & 0 & \cdots & 1 \\ a_{0,i} & a_{1,i} & a_{2,i} & \cdots & a_{r_i-1,i} \end{pmatrix}, \quad H_i = \begin{pmatrix} 1 & 0 & \cdots & 0 \end{pmatrix}_{i \times r_i}$$

and $\xi_{ij} = L_s^{j-1}\gamma_i\,(w, \mu)$, $i = 1, 2, 3, \ldots, m$, $j = 1, 2, 3, \ldots, r_i$. This system can be viewed as an immersion of the exosystem (3) into a linear observable system and is similar to system (19). It is also worth noticing that the necessary and sufficient conditions for the solution of the NRRP may be given in terms of a nonlinear immersion as presented in Theorem 4.

The controller (see Figure 9) which solves the NRRP is given by

$$\begin{aligned} \dot{\zeta}_1\,(t) &= (A_0 + B_0 K_1 - G_1 C_0)\,\zeta_1\,(t) + G_1 e\,(t) \\ \dot{\zeta}_2\,(t) &= -G_2 C_0 \zeta_1\,(t) + \Phi \zeta_2\,(t) + G_2 e\,(t) \\ u\,(t) &= K\zeta_1\,(t) + H\zeta_2\,(t), \end{aligned} \tag{36}$$

where K and G_1, G_2 make stable matrices $(A_0 + B_0 K)$ and

$$\begin{pmatrix} A_0 & -B_0 H \\ 0 & \Phi \end{pmatrix} - \begin{pmatrix} G_1 \\ G_2 \end{pmatrix} \begin{pmatrix} C & 0 \end{pmatrix},$$

respectively. Obviously, the stabilizability an detectability of the pairs

$$[A_0, B_0] \quad \text{and} \quad \left[\begin{pmatrix} A_0 & -B_0 H \\ 0 & \Phi \end{pmatrix}, \begin{pmatrix} C & 0 \end{pmatrix} \right]$$

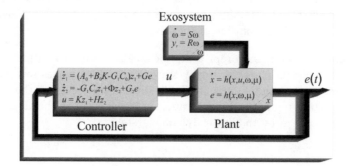

Fig. 9. The robust controller scheme for nonlinear systems

respectively, are the necessary conditions for the NRRP solvability. Notice that the controller is linear but it guarantees robust regulation for the nonlinear system. The main feature guaranteeing the zero output tracking error is the immersion (35) which incorporates the nonlinearity of the steady state input.

The following example illustrate the calculations involved in the construction of an error feedback regulator for nonlinear systems.

Example 3. Isothermal nonlinear CSTR control

To illustrate the features of the nonlinear regulator, let us consider the following consecutive reactions

$$A \xrightarrow{k_1} B \xrightarrow{k_2} C$$

that take place in an isothermal CSTR. It is assumed that $A \to B$ has first-order kinetics whereas $B \to C$ has second-order kinetics. The reactor is well mixed and has constant volume. A dimensionless model for such a reactor is given by [41]

$$\begin{aligned}
\dot{x}_1 &= -x_1 - Da_1 x_1 + u \\
\dot{x}_2 &= Da_1 x_1 - x_2 - Da_2 x_2^2 \\
y &= x_2
\end{aligned} \tag{37}$$

where, x_1 and x_2 are the dimensionless concentrations of A and B, respectively. The control variable u corresponds to the inlet concentration of A and the Damkhöler numbers, Da_1 and Da_2, are used to relate the effect of the reaction terms to the residence time (the reader is referred to [41] for further details on the reacting system model and the definition of variables and parameters).

In this case, we propose an oscillatory profile represented by $y_r = a + b \sin(ct + \theta)$ which can be generated by the exosystem

$$\begin{aligned}
\dot{\omega} &= s(\omega) \\
y_r &= \omega_1
\end{aligned}$$

where

$$s(\omega) = \left(\omega_2, \omega_3, -c^2\omega_2\right)^T.$$

In this case, the error is $e = y - y_r = x_2 - \omega_1$ which is zero when $x_{2ss} = \omega_1 =: \pi_1(\omega)$. Then by using equation (51), both the steady state input u_{ss} and the steady state x_{1ss} can be obtained from the following expressions

$$x_{1ss} = \frac{\omega_2 + \omega_1 + Da_2\omega_1^2}{Da_1} =: \pi_2(\omega)$$

$$u_{ss} = \frac{1}{Da_1}\left[\omega_3 + \omega_2 + 2Da_2\omega_1\omega_2 + (1 + Da_1)\left(\omega_2 + \omega_1 + Da_2\omega_1^2\right)\right] =: \gamma(\omega)$$

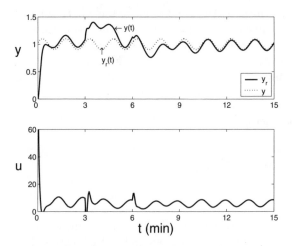

Fig. 10. Nonlinear robust control simulation (initial values:$x_1 = x_2 = 0$)

By using Lie algebra, it is computed $z_k = L_s^{k-1} \gamma(\omega)$, $k = 1, 2, \ldots, 5$, to generate the immersion

$$\dot{z} = \Phi z$$
$$\gamma(\omega) = Hz$$

where

$$\Phi = \begin{pmatrix} 0 & 1 & 0 & 0 & 0 \\ 0 & 0 & 1 & 0 & 0 \\ 0 & 0 & 0 & 1 & 0 \\ 0 & 0 & 0 & 0 & 1 \\ 0 & -4c^4 & 0 & -5c^2 & 0 \end{pmatrix}, \quad H = \begin{pmatrix} 1 & 0 & 0 & 0 & 0 \end{pmatrix}. \tag{38}$$

For simulations purposes, let us choose $y_r = \omega_1 = 1 + \frac{1}{10} \sin\left(\frac{4}{3}\pi t + \frac{\pi}{9}\right)$ as the periodic reference signal. Then, choosing K and G such that the eigenvalues for the matrices $(A_0 + B_0 K)$ and $\left[\begin{pmatrix} A_0 & -B_0 H \\ 0 & \Phi \end{pmatrix} - G \begin{pmatrix} C & 0 \end{pmatrix}\right]$ are $(-10 \pm 2i)$ and $(-10 \pm 2i, -5 \pm i, -2, -1, -4)$, respectively, the following controller can be found

$$\begin{pmatrix} \dot{z}_{11} \\ \dot{z}_{12} \\ \dot{z}_{21} \\ \dot{z}_{22} \\ \dot{z}_{23} \\ \dot{z}_{24} \\ \dot{z}_{25} \end{pmatrix} = \begin{pmatrix} -16.932 & -448.32 & 0 & 0 & 0 & 0 & 0 \\ 1 & -35 & 0 & 0 & 0 & 0 & 0 \\ 0 & 1052 & 0 & 1 & 0 & 0 & 0 \\ 0 & -23653 & 0 & 0 & 1 & 0 & 0 \\ 0 & -94447 & 0 & 0 & 0 & 1 & 0 \\ 0 & 1551200 & 0 & 0 & 0 & 0 & 1 \\ 0 & 7012000 & 0 & -\frac{1024}{81}\pi^4 & 0 & -\frac{80}{9}\pi^2 & 0 \end{pmatrix} \begin{pmatrix} z_{11} \\ z_{12} \\ z_{21} \\ z_{22} \\ z_{23} \\ z_{24} \\ z_{25} \end{pmatrix} + \begin{pmatrix} 396.27 \\ 31.932 \\ -1052 \\ 23653 \\ 94447 \\ -1551200 \\ -7012000 \end{pmatrix} e(t)$$

Table 1. Parameter values for simulation

Value	Da_1	Da_2
Nominal	1	2
$0 \leq t < 3$	0.8	1.8
$3 \leq t < 6$	1.4	1.8
$t \geq 6$	1.4	2.6

In order to test the aforementioned nonlinear controller in the face of parameter uncertainty, we introduced changes on the Damkhöler parameters during the course of the reactions. These changes are reported in Table 1.

From inspection of Figure 10, it can be seen that the nonlinear regulator shows excellent tracking properties under the influence of initial error and parameter variations. As expected, the nonlinear regulator compensates the effect of erroneous initial conditions and particularly, the simultaneous changes on Da_1 and Da_2. Notice that as a result of the parameter variations, x_2 deviates from the desired oscillatory profile and then rapidly approaches the oscillatory reference signal once the nonlinear controller compensates the aforementioned changes (the controller even saturates from below to compensate the effect of the first unexpected change on Da_1).

2.5   The Ripple Free Nonlinear Robust Regulation Problem

When dealing with controllers implemented by digital devices and zero order holders, it is well-known that the sampled version of the continuous time controller (36) may introduce instability in the closed-loop system [32]. For the case of nonlinear systems, as in the linear case, a discretized model based controller using zero order holder guarantee a zero output tracking error only at the sampling instants. In the intersampling time, the output tracking error will generally present ripple, due to the fact that the internal model cannot be reproduced when zero order holders are used (except in the particular case of constant reference signals). To achieve a zero-error submanifold even in intersampling periods, a robust controller based on the discretized linear approximation of system (31), and the internal model dynamics obtained from the continuous version (35) has been developed by using an *exponential holder* [9]. This leads to an hybrid controller that involves a digital compensator, a state and internal model estimator and an analog internal model input.

The *Ripple Free Robust Regulation Problem* (RFRRP) can be formulated as the problem of finding a dynamic controller of the form

$$z_d(k+1) = \varphi_d(z_d, e_d)$$
$$u(t) = \vartheta(z_d, \theta) \qquad \theta \in [0, \delta) \qquad (39)$$

such that, for all admissible parameter values μ, the following conditions are satisfied

(S) *Stability.* The solution of the system

$$x_d (k+1) = F_d (\delta, x_d, \vartheta_d (z_d, \theta), 0, \mu)$$
$$z_d (k+1) = \varphi_d (z_d, h (x_d, \omega, \mu))$$

at the sampling instants goes asymptotically to zero. Here, $F_d (\delta, x_d, \vartheta_d (z_d, \theta), 0, \mu)$ is the discrete version of system (31).

(R) *Regulation.* For each initial condition $(x (0), z (0), \omega (0))$ in a neighborhood of the origin, the solution of the closed-loop system

$$\dot{x} = f (x, \vartheta_d (z_d, \theta), 0, \mu)$$
$$z_d (k+1) = \varphi_d (z_d, h (x, \omega, \mu))$$
$$\dot{\omega} = s (\omega)$$

guarantees that $\lim_{t \to \infty} e (t) = 0$.

It has been shown in [9] that if an immersion (35) for system (31) is found, then the RFRRP is solvable by the following controller

$$\zeta_{d1} (k+1) = (A_{d0} + B_{d0} K_d - G_{d1} C_{d0}) \zeta_{d1} + G_{d1} e_d \tag{40a}$$

$$\zeta_{d2} (k+1) = -G_{d2} \zeta_{d1} + e^{\Phi \delta} \zeta_{d2} + G_{d2} e_d \tag{40b}$$

$$u (k\delta + \theta) = K_d \zeta_{d1} + H e^{\Phi \theta} \zeta_{d2} \tag{40c}$$

where K_d and $G_d = \left(G_{d1} \; G_{d2} \right)^T$ render stable the matrices

$$(A_{d0} + B_{d0} K_d) \quad \text{and} \quad \left[\begin{pmatrix} A_{d0} & -M_{d0} \\ 0 & \Phi_d \end{pmatrix} - \begin{pmatrix} G_{d1} \\ G_{d2} \end{pmatrix} (C_{d0} \; 0) \right]$$

respectively, namely the stabilizability and detectability of the pairs

$$(A_{d0}, B_{d0}) \quad \text{and} \quad \left[\begin{pmatrix} A_{d0} & -M_{d0} \\ 0 & \Phi_d \end{pmatrix}, (C_{d0} \; 0) \right]$$

is a necessary condition. Here,

$$M_{d0} = \int_0^\delta e^{A_0 \tau} H e^{\Phi (\delta - \tau)} d\tau, \qquad \Phi_d = e^{\Phi \delta}$$

whereas A_{d0}, B_{d0} and C_{d0}, are the discretized nominal matrix of the linear approximation of system (31) given by A_0, B_0 and C_0.

$$A_{d0} = e^{A_0 \delta}, \qquad B_{d0} = \int_0^\delta e^{A_0 \tau} d\tau B_0, \qquad C_{d0} = C_0$$

As it can be observed, equation (40a) represents the error-state observer, while (40b) is the internal model observer; both are discrete systems that

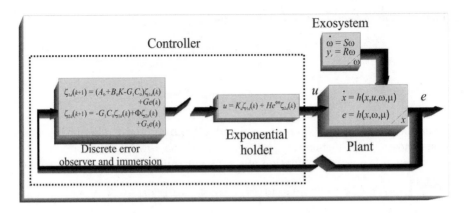

Fig. 11. Free ripple robust controller for nonlinear systems)

depend on the output error measured. On the other hand, the first term of (40c) is a discrete error compensator, while the second term is the internal model exponential holder producing a continuous input. Figure 11 depicts the proposed controller scheme.

Remark 5. In order to construct the controller (22), (30), (36) and (54), it is not necessary to know neither the continuous steady state x_{ss} nor the discrete one x_{dss}, but only the continuous steady state input $u_{ss}(t)$. However, for nonlinear systems, in the particular case of polynomials describing function $\gamma(\omega, \mu)$ (such as triangular systems describing by polynomial terms), Φ can be determined without knowing exactly $\gamma(\omega, \mu)$, but only the maximum degree of the polynomial. Another remarkable feature of controller (54) is that it is based on the discretized linear part of the system description. The complete continuous controller (36) is equivalent to the discrete (54) only in the sense that both provide the exact steady state input needed to maintain the dynamics of the system within the zero output submanifold. Clearly, the transient behavior provided by both controllers are different.

Example 4. Calculation of the exponential holder

Consider again the CSTR where the consecutive reactions $A \xrightarrow{k_1} B \xrightarrow{k_2} C$ take place. The required discrete nonlinear control law is found by obtaining a discrete version of the immersion developed in Example 3 and its associated exponential holder which has the form $He^{\Phi\theta}$, and in this particular example is a 1×5 dimension vector, where Φ and H are presented in matrices (38). The exponential holder can be calculated by using the definition of an exponential matrix,

$$He^{\Phi\theta} = H + H\Phi\theta + H\frac{(\Phi\theta)^2}{2!} + H\frac{(\Phi\theta)^3}{3!} \cdots = H \sum_{k=0}^{\infty} \frac{(\Phi\theta)^k}{k!}. \qquad (41)$$

The solution of the linear differential equation

$$\dot{z}(t) = \Phi z(t) \qquad z(0) = z_0$$
$$y = Hz(t)$$

is given by

$$y = He^{\Phi t}z_0.$$

Since the eigenvalues of Φ are 0, $\pm ci$ and $\pm 2ci$, from linear differential equations theory, each element i, $i = 1, \ldots, 5$, of the row vector $He^{\Phi\theta}$ can be expressed as a linear combination of a constant, sine and cosine functions; i.e.

$$\left[He^{\Phi\theta}\right]_i = a_{1i} + a_{2i}\cos(c\theta) + a_{3i}\sin(c\theta) + a_{4i}\cos(2c\theta) + a_{5i}\sin(2c\theta), i = 1, 2, \ldots, 5.$$

Moreover, by using the series expansion of the sine and cosine functions, it is obtained that

$$\left(He^{\Phi\theta}\right)_i = a_{1i} + a_{2i}\left[1 - \frac{(c\theta)^2}{2!} + \frac{(c\theta)^4}{4!} - \cdots\right] + a_{3i}\left[c\theta - \frac{(c\theta)^3}{3!} + \frac{(c\theta)^5}{5!} - \cdots\right] +$$
$$+ a_{4i}\left[1 - \frac{(2c\theta)^2}{2!} + \frac{(2c\theta)^4}{4!} - \cdots\right] + a_{5i}\left[2c\theta - \frac{(2c\theta)^3}{3!} + \frac{(2c\theta)^5}{5!} - \cdots\right]$$
$$= (a_{1i} + a_{2i} + a_{4i}) + c(a_{3i} + 2a_{5i})\theta - \frac{c^2}{2!}(a_{2i} + 2^2 a_{4i})\theta^2 -$$
$$- \frac{c^3}{3!}(a_{3i} + 2^3 a_{5i})\theta^3 + \frac{c^4}{4!}(a_{2i} + 2^4 a_{4i})\theta^4 + \cdots$$

which after rearrangement, and when compare to each element of the i-th column, $i = 1, \ldots, 5$, of the row vector expansion (41) $([H]_i, [H\Phi]_i, \ldots, \left[H\frac{\Phi^4}{4!}\right]_i)$ yields the following set of 5 linear equations for the constants $a_{1i}, \ldots, a_{5i}$ of each i-th element of $He^{\Phi\theta}$

$$\begin{pmatrix} 1 & 1 & 0 & 1 & 0 \\ 0 & 0 & c & 0 & 2c \\ 0 & -\frac{c^2}{2!} & 0 & -\frac{(2c)^2}{2!} & 0 \\ 0 & 0 & -\frac{c^3}{3!} & 0 & -\frac{(2c)^3}{3!} \\ 0 & -\frac{c^4}{4!} & 0 & -\frac{(2c)^4}{4!} & 0 \end{pmatrix} \begin{pmatrix} a_{1i} \\ a_{2i} \\ a_{3i} \\ a_{4i} \\ a_{5i} \end{pmatrix} = \begin{pmatrix} [H]_i \\ [H\Phi]_i \\ \left[H\frac{\Phi^2}{2!}\right]_i \\ \left[H\frac{\Phi^3}{3!}\right]_i \\ \left[H\frac{\Phi^4}{4!}\right]_i \end{pmatrix}.$$

Thus, after solving these linear equations the exponential holder for the immersion is obtained,

$$He^{\Phi\theta} = \begin{pmatrix} 1 \\ \frac{4}{3c}\sin(ct) - \frac{1}{6c}\sin(2ct) \\ \frac{5}{4c^2} - \frac{4}{3c^2}\cos(ct) + \frac{1}{12c^2}\cos(2ct) \\ \frac{1}{3c^3}\sin(ct) - \frac{1}{6c^3}\sin(2ct) \\ \frac{1}{4c^4} - \frac{1}{3c^4}\cos(ct) + \frac{1}{12c^4}\cos(2ct) \end{pmatrix}^T$$

which can be directly implemented in the design of the discrete-time regulator.

where x_{2r} is the reference conversion, x_{3r} denotes the reference temperature (which is a constant value, x_{3s}, the actual operating temperature) and x_{2s} is the average conversion around which x_2 will oscillate with given amplitudes (A) and frequencies ($c_T = 2\pi/P_T$ where P_T is the period of oscillation). These reference signals can be also expressed as the output of an exosystem defined by

$$\dot{\omega} = s\left(\omega\right) = \begin{pmatrix} c_T \omega_2 \\ c_T\left(\omega_3 - \omega_1\right) \\ 0 \\ 0 \end{pmatrix} \tag{50}$$

Let the initial conditions be $\omega_1\left(0\right) = x_{2s}$, $\omega_2\left(0\right) = A$, $\omega_3\left(0\right) = x_{2s}$, and $\omega_4\left(0\right) = x_{3s}$, such that $\omega_1\left(t\right) = x_{2r}\left(t\right)$ and $\omega_4\left(t\right) = x_{3r}\left(t\right)$. Thus, the tracking errors can be expressed as:

$$e_1\left(t\right) = x_2\left(t\right) - \omega_1\left(t\right) \tag{51a}$$

$$e_2\left(t\right) = x_3\left(t\right) - \omega_4\left(t\right) \tag{51b}$$

To find the steady state mappings, we proceed as follows. The tracking errors are zero when

$$\pi_2\left(\omega\right) = \omega_1\left(t\right), \tag{52}$$

$$\pi_3\left(\omega\right) = \omega_4\left(t\right). \tag{53}$$

These new definitions can be substituted in Equation (48b) to find the initiator concentration (x_{1ss}) required to track the oscillatory conversion trajectory; this substitution yields

$$\pi_1\left(\omega, \mu\right) = \left[\frac{c_T \omega_2 + D\omega_1}{g\left(\omega_1, \omega_4\right)\left(1 - \omega_1\right)}\right]^2 \tag{54}$$

Equations (52)-(54) represent the steady state zero output submanifold. Once the mapping $x_{ss} = \pi\left(\omega, \mu\right)$ is obtained, we need to find the mapping $u_{ss} = \gamma\left(\omega, \mu\right)$ in order to apply the control strategy. These mappings are

$$u_{1ss} = \dot{\pi}_1\left(t\right) + \left[k_d\left(\omega_4\right) + D\right]\pi_1\left(\omega\right) =: \gamma_1\left(\omega, \mu\right) \tag{55}$$

$$u_{2ss} = \frac{\left(\omega_4 - T_f\right)D - a_1\left(c_T\omega_2 + D\omega_1\right)}{a_2\left(T_c - \omega_4\right)} =: \gamma_2\left(\omega, \mu\right), \tag{56}$$

then $\dot{\pi}_1\left(t\right)$ can be obtained by the Lie derivative, $\dot{\pi}_1\left(t\right) = L_s \pi_1\left(t\right)$.

3.3 Continuous Control

It can be shown that u_{2ss} can be generated by a linear immersion of dimension 3 obtained by using Lie algebra. However, u_{1ss} cannot be generated by a linear immersion. In fact, if one applies the Lie algebra, the resulting immersion is nonlinear due to the complex mathematical structure of the polymerization system. Instead, one can find such an immersion by finding a differential equation whose solution, for certain initial conditions, renders the immersion

$$\dot{z}_1(t) = \Phi_1(\omega, z_1) \tag{57a}$$

$$\gamma_1(\omega, \mu) = H_1 z_1(t) \tag{57b}$$

and

$$\dot{z}_2(t) = \Phi_2(z_2) \tag{58a}$$

$$\gamma_2(\omega, \mu) = H_2 z_2(t) \tag{58b}$$

where $z_1 = \left(z_{11}\ z_{12}\ z_{13}\ z_{14}\right)^T$, $z_2 = \left(z_{21}\ z_{22}\ z_{23}\right)^T$,

$$\Phi_1(\omega, z_1) = \begin{pmatrix} R_2(\omega) z_{11} + \frac{c_T R_1(\omega)[R_3(\omega) z_{12} + z_{14}]}{D} + \frac{z_{11} z_{13}}{z_{12}} \\ z_{13} \\ z_{14} + \frac{z_{13}^2}{z_{12}} \\ -z_{13} + \frac{z_{13} z_{14}}{z_{12}} \end{pmatrix}, \quad H_1 = \begin{pmatrix} 1 \\ 0 \\ 0 \\ 0 \end{pmatrix}^T$$

$$\Phi_2(z_2) = \begin{pmatrix} z_{12} \\ z_{13} \\ -z_{12} \end{pmatrix}, \quad H_2 = \begin{pmatrix} 1 \\ 0 \\ 0 \end{pmatrix}^T$$

$$R_1(\omega) = \left[\frac{c_T \omega_2 + D \omega_1}{1 - \omega_1}\right]^2$$

$$R_2(\omega) = 2\left[\frac{c_T(\omega_3 - \omega_1) + D \omega_2}{c_T \omega_2 + D \omega_1} + \frac{\omega_2}{1 - \omega_1}\right]$$

$$R_3(\omega) = 2\left[\frac{D(\omega_3 - \omega_1) - c_T \omega_2}{c_T \omega_2 + D \omega_1} - \left(\frac{c_T(\omega_3 - \omega_1) + D \omega_2}{c_T \omega_2 + D \omega_1}\right)^2 + \frac{\omega_1 + \omega_2^2 - \omega_1^2}{(1 - \omega_1)^2}\right]$$

Summarizing, both $\gamma_1(\omega, \mu)$ and $\gamma_2(\omega, \mu)$ can be expressed as

$$\gamma = Hz$$

where the dynamics of z is given by

$$\dot{z} = \Phi(\omega, z) \tag{59}$$

and

$$\Phi(\omega, z) = \begin{pmatrix} \Phi_1(\omega, z_1) \\ \Phi_2(z_2) \end{pmatrix}, \quad H = \begin{pmatrix} H_1 & 0 \\ 0 & H_2 \end{pmatrix}, \quad z = \begin{pmatrix} z_1 \\ z_2 \end{pmatrix}$$

A detailed description of the procedure to obtain the immersion is given in [19].

Finally, by following the procedure outlined in section 2.4, the controller that solves the problem is

$$\dot{\zeta}_1(t) = (A_0 + B_0 K_1 - G_1 C_0)\,\zeta_1(t) + G_1 e(t), \tag{60a}$$

$$\dot{\zeta}_2(t) = \Phi\,(\omega(t), \zeta_2(t)) - G_2\,[C_0 \zeta_1(t) - e(t)], \tag{60b}$$

$$u(t) = K \zeta_1(t) + H \zeta_2(t), \tag{60c}$$

where A_0, B_0, and C_0, are the nominal matrices of the linear approximation of system (48) presented in Appendix A, while K and G_1, G_2 make stable the matrices $(A_0 + B_0 K)$ and

$$\begin{pmatrix} A_0 & -B_0 H \\ 0 & \Phi_0 \end{pmatrix} - \begin{pmatrix} G_1 \\ G_2 \end{pmatrix}(C\,0),$$

respectively. Here Φ_0 is obtained from the linear representation of immersion (59), i.e., $\Phi_0 = \left[\frac{\partial \Phi(\omega, z)}{\partial z}\right]_{\substack{z=0 \\ \omega=0}}$.

Application of the robust continuous controller

This controller was applied to a methylmethacrylate (MMA) solution homopolymerization conducted in a continuous stirred tank reactor. The solvent and initiator are ethyl acetate and benzoyl peroxide, respectively. The polymerization system parameters for numerical simulations have been taken from [41], and are detailed in Table 3.

Table 3. MMA kinetic and physicochemical parameter values, [41]

Parameter	Value	Parameter	Value
k_{d0}	$5.95 \times 10^{13}\,s$	E_d	$29.6 kcal$
k_{p0}	$7.0 \times 10^6\,s$	E_p	$6.3 kcal$
k_{t0}	$1.76 \times 10^9\,s$	E_t	$2.8 kcal$
b_1	54.5	b_2	-0.13
ρC_p	$400 kcal/l$	$(-\Delta H_p)$	$13.5 kcal/mol$
k_{tc}	$k_t/9.23s$	k_{td}	$8.23 k_t/9.23s$
f	0.6		

Several changes in different kinetic parameter values and inlet conditions were introduced in the system to test the disturbance and parametric changes rejection properties of the robust controller (60). These changes took into account the presence of realistic perturbations that are unavoidable in industry. A formal distinction of perturbations into unmodeled disturbances and modeling errors were made for the purpose of listing the most relevant sources. Among unmodeled disturbances: variations in the temperature of the main feed stream entering the reactor (T_f) which has a direct influence on the reactor temperature; presence of impurities in the feed which reduce the initiator efficiency, and variations on the solvent concentration (S_f) which has a direct influence in both the rate of polymerization and the gel effect. Among modeling errors: errors in physicochemical and thermodynamic properties (such as ρC_p); errors in the parameters of the model describing the gel effect; error in the values of the kinetic constants of chain propagation which play a very important role in the modeling of the polymerization reactor because they are contained both in the mass balance equation for monomer and the energy balance equation for the reactor (the use of inaccurate E_p and k_{p0} may cause a poor prediction of the monomer concentration and the reactor temperature and consequently, the control performance would deteriorate since the controlled variables are more strongly dependent on these two variables than others). Table 4 lists the changes of the aforementioned parameters (with respect to the nominal values) and the time at which these changes were introduced during the progress of the MMA polymerization.

Table 4. Step changes on parameters and inlet conditions for the continuous case

Parameter	Previous value	New value	Time of change (min.)
T_f	263	355	30
ρC_p	400	350	50
b_1	54.5	5.4	75
S_f	5.4	5.1	90
k_{p0}	7.0×10^6	6.6×10^6	100
E_p	6.3	6.5	120
f	0.6	0.5	150
b_2	-0.13	-0.135	180
T_f	355	380	210

Figure 13 shows the simulation results for a periodic conversion profile (around 70%, with an amplitude of 3% and a period of 9000 seconds) and constant temperature (350 K). The matrix feedback gain K was chosen such that the eigenvalues of $(A_0 + B_0 K)$ were $[50 + i\ 50 - i\ 70]$ while the observer gain G was calculated by solving an algebraic Ricatti equation. Notice that the temperature loop is insensitive to all changes in the parameters and disturbances. The conversion loop, although is more sensitive to such changes, is able to overcome saturation and to achieve stabilization even when the gel effect is more significant due to the consecutive changes on b_2 and T_f. The slight decrease of conversion (after the change on b_2) can be explained by the addition of high amounts of initiator that delays the onset of autoacceleration; such an addition, however, gives lower molecular weight polymers as the monomer concentration decreases or drifts with conversion. This is compensated by depletion of the control input ($u_1 = 0$) which tends to increase the chain length. Moreover, notice that in order to achieve the desired periodic

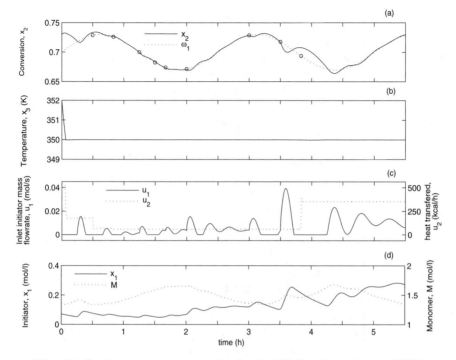

Fig. 13. Continuous robust regulation for the MMA polymerization CSTR

conversion trajectory, u_1 saturates from below but rarely exhibits excursions towards large initiator feedrates.

3.4   Discrete Control

In the implementation of the continuous version of the robust regulator, we assumed that all outputs were continuously available. For the implementation of the discrete version, we are assuming that both temperature and conversion are sampled at fixed sampling times. In order to avoid ripple, we implement the control scheme presented in section 2.5.

We have previously mentioned that the nonlinear nature of the polymerization reactor gives rise to a nonlinear immersion and as a consequence, it is impossible to construct its corresponding exponential holder. To avoid this problem, it is necessary to analyze the mathematical structure of u_{1ss}, to conclude that the nonlinear term is mainly due to the function that describes the effect gel. A suitable solution is given by finding a simpler mathematical function to satisfactorily describe the gel effect phenomena (we should recall that the most common gel effect functions such as (47), are actually given by empirical correlations). We propose the following function for g_t to represent the diffusional limitation of the polymerization reaction

$$g_t = (1 - x_2)^2 \left(c_1 x_2^2 + c_2 x_2 + c_3 \right) \tag{61}$$

which renders a linear immersion.

It is important to remark that in order to obtain the immersion of u_{1ss}, one needs only the general structure of function (61) and not the exact values of the parameters c_1, c_2 and c_3. Hence, this structure yields

$$\pi_1 (\omega, \mu) = \frac{1}{K} \exp \left(\frac{\beta_2}{\omega_4} \right) \left(c_1 x_2^2 + c_2 x_2 + c_3 \right) \left(c_T \omega_2 + D \omega_1 \right)^2 \tag{62}$$

By using Lie algebra one obtains

$$\dot{\nu} = \Phi_1 \nu$$
$$\pi_1 (\omega, \mu) = H_1 \nu$$

where $\pi_1 (\omega, \mu)$ is represented by an linear immersion of length 9, and $\nu_i = L_s^{i-1} \pi_1 (\omega, \mu)$, $i = 1, 2, \ldots, 9$. Here,

$$\Phi_1 = c_T \begin{pmatrix} 0 & 1 & 0 & 0 & 0 & 0 & 0 & 0 & 0 \\ 0 & 0 & 1 & 0 & 0 & 0 & 0 & 0 & 0 \\ 0 & 0 & 0 & 1 & 0 & 0 & 0 & 0 & 0 \\ 0 & 0 & 0 & 0 & 1 & 0 & 0 & 0 & 0 \\ 0 & 0 & 0 & 0 & 0 & 1 & 0 & 0 & 0 \\ 0 & 0 & 0 & 0 & 0 & 0 & 1 & 0 & 0 \\ 0 & 0 & 0 & 0 & 0 & 0 & 0 & 1 & 0 \\ 0 & 0 & 0 & 0 & 0 & 0 & 0 & 0 & 1 \\ 0 & -576 & 0 & -820 & 0 & -273 & 0 & -30 & 0 \end{pmatrix} \quad \text{and} \quad H_1 = \begin{pmatrix} 1 \\ 0 \\ 0 \\ 0 \\ 0 \\ 0 \\ 0 \\ 0 \\ 0 \end{pmatrix}^T$$

Notice that the immersion does not contain uncertain parameters.

Now, looking at Equation (55), $\gamma_1 (\omega, \mu) = \nu_2 + [k_d (\omega_4) + D] \nu_1$, one concludes that $\gamma_1 (\omega, \mu)$ is a linear combination of ν_1 and ν_2 which have the same dynamics. Thus, $\gamma_1 (\omega, \mu)$ can be generated by the immersion

$$\dot{z}_1 = \Phi_1 z_1 \tag{63a}$$
$$\gamma_1 (\omega, \mu) = H_1 z_1 \tag{63b}$$

with $z_1 = \left(z_{11} \ z_{12} \ \cdots \ z_{19} \right)^T$. On the other hand, $\gamma_2 (\omega, \mu)$ may be represented by immersion (59) which is linear; that is

$$\dot{z}_2 = \Phi_2 z_2 \tag{64a}$$
$$\gamma_2 (\omega, \mu) = H_2 z_2 \tag{64b}$$

where

$$\Phi_2 = \begin{pmatrix} 0 & c_T & 0 \\ 0 & 0 & c_T \\ 0 & -c_T & 0 \end{pmatrix}$$

Summarizing, both $\gamma_1(\omega, \mu)$ and $\gamma_2(\omega, \mu)$ can be expressed as

$$\gamma = Hz \tag{65}$$

where the dynamics of z is given by

$$\dot{z} = \Phi z \tag{66}$$

with

$$\Phi = \begin{pmatrix} \Phi_1 & 0 \\ 0 & \Phi_2 \end{pmatrix}, \quad H = \begin{pmatrix} H_1 & 0 \\ 0 & H_2 \end{pmatrix} \quad \text{and} \quad z = \begin{pmatrix} z_1 \\ z_2 \end{pmatrix}$$

Exponential holder

The exponential holder for this system is calculated by using the definition of the exponential matrix, $e^A = I + A + \frac{A^2}{2!} + \frac{A^3}{3!} \cdots = \sum_{k=0}^{\infty} \frac{A^k}{k!}$, to obtain

$$He^{\Phi\theta} = \begin{pmatrix} 1 & 0 \\ \frac{8\sin(c_T\theta)}{5} - \frac{2\sin(2c_T\theta)}{5} + \frac{8\sin(3c_T\theta)}{105} - \frac{\sin(4c_T\theta)}{140} & 0 \\ \frac{205}{144} - \frac{8\cos(c_T\theta)}{5} + \frac{\cos(2c_T\theta)}{5} - \frac{8\cos(3c_T\theta)}{315} + \frac{\cos(4c_T\theta)}{560} & 0 \\ \frac{61\sin(c_T\theta)}{90} - \frac{169\sin(2c_T\theta)}{360} + \frac{\sin(3c_T\theta)}{10} - \frac{7\sin(4c_T\theta)}{720} & 0 \\ \frac{91}{192} - \frac{61\cos(c_T\theta)}{90} + \frac{169\cos(2c_T\theta)}{720} - \frac{\cos(3c_T\theta)}{30} + \frac{7\cos(4c_T\theta)}{2880} & 0 \\ \frac{29\sin(c_T\theta)}{360} - \frac{13\sin(2c_T\theta)}{180} + \frac{\sin(3c_T\theta)}{40} - \frac{\sin(4c_T\theta)}{360} & 0 \\ \frac{5}{96} - \frac{29\cos(c_T\theta)}{360} + \frac{13\cos(2c_T\theta)}{360} - \frac{\cos(3c_T\theta)}{120} + \frac{\cos(4c_T\theta)}{1440} & 0 \\ \frac{\sin(c_T\theta)}{360} - \frac{\sin(2c_T\theta)}{360} + \frac{\sin(3c_T\theta)}{840} - \frac{\sin(4c_T\theta)}{5040} & 0 \\ \frac{1}{576} - \frac{\cos(c_T\theta)}{360} + \frac{\cos(2c_T\theta)}{720} - \frac{\cos(3c_T\theta)}{2520} + \frac{\cos(4c_T\theta)}{20160} & 1 \\ 0 & \sin(c_T\theta) \\ 0 & 1 - \cos(c_T\theta) \end{pmatrix} \tag{67}$$

Application of the robust discrete controller

The tracking performance of the discrete error feedback regulator was tested (as in the continuous case) in the face of parameter variations and load disturbances. These changes are listed in Table 5. In this case, an average conversion of 70% with oscillations of 3% amplitude and 9000 seconds period and a constant temperature ($360K$) were chosen. The matrix feedback gain (K_d) and the observer gains (G_{d1} and G_{d2}) were calculated by solving algebraic Ricatti equations. Figure 14 illustrates the performance of the discrete robust regulator. As can be observed, the changes introduced during the polymerization are rapidly compensated to maintain the reactor temperature at its set point. This figure also shows that the conversion loop is more sensitive to those changes. However, it is able to cope with the difficulties imposed by the load changes and parameter variations. Moreover, notice that u_1 overcome the saturation problems shown by its continuous counterpart, and provided less abrupt responses even at the intersampling periods.

Table 5. Step changes on parameters and inlet conditions for discrete case

Parameter	Previous value	New value	Time of change (min)
T_c	363	310	20
b_1	54.5	54	40
f	0.6	0.5	60
ρC_p	400	360	90
ρC_p	360	440	240
k_{p0}	7.0×10^6	7.7×10^6	240
k_{d0}	5.96×10^6	5.40×10^6	270

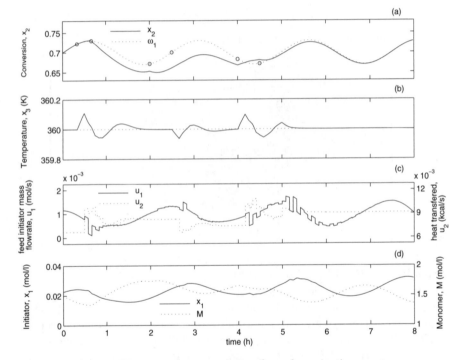

Fig. 14. Discrete robust regulation for polymerization reactor

4   Conclusion

In this chapter, an unifying methodology for solving the output tracking problem in oscillatory chemical reactors has been presented. This methodology consists on the design of an error feedback controller and a steady state input estimator. The performance of both continuous and discrete structures of the robust regulator have been examined through extensive numerical simulations under various uncertainties and external disturbances. The proposed structures have shown to maintain good stability properties even when confronting significant modeling errors and load disturbances.

Acknowledgments. This work was partially supported by CONACYT, grants 36687-A and 163390. Also, the authors would like to thanks the anonymous reviewers for their suggestions to improve this paper.

References

[1] J. Alvarez, R. Suarez, and A. Sanchez. Nonlinear decoupling control of free radical polymerization continuous stirred tank reactors. *Chem. Eng. Sci.*, 45:3341–3354, 1990.

[2] J.E. Bailey. Periodic operation of chemical reactors: A review. *Chem. Eng. Commun.*, 1:111–124, 1973.

[3] S. Bittanti, P. Coaneri, and G.D. Nicolao. The difference periodic Riccati equation for the periodic prediction problem. *IEEE Trans. Automat. Contr.*, 33:706–712, 1988.

[4] S. Bittanti and G. Guardabassi. Optimal cyclastationary control: The LQG approach. In *IEEE Conf. Decision and Control (CDC)*, pages 166–167, Piscataway, NJ, 1981.

[5] D.D. Bruns and J.E. Bailey. Process operation near an unstable steady state using nonlinear feedback control. *Chem. Eng. Sci.*, 30:755–762, 1975.

[6] D.D. Bruns and J.E. Bailey. Nonlinear feedback control for operating a nonisothermal CSTR near an unstable steady state. *Chem. Eng. Sci.*, 32:257–264, 1977.

[7] D. Butler, E. Friedler, and K. Gatt. Characterizing the quantity and quality of domestic wastewater inflows. *Wat. Sci. Technol.*, 31:13–24, 1995.

[8] C.I. Byrnes and A. Isidori. Output regulation for nonlinear systems: An overview. *Int. J. Robust Nonlinear Control*, 10:323–337, 2000.

[9] B. Castillo-Toledo and S. Di Gennaro. On the nonlinear ripple free sampled-data robust regulator. *Eur. J. Control*, 8:44–55, 2002.

[10] B. Castillo-Toledo and G. Obregon-Pulido. Guaranteeing asymptotic zero intersampling tracking error via a discretized regulator and exponential holder for nonlinear systems. *J. App. Reserch and Tech.*, 1:203–214, 2003.

[11] M. Chang and S. Schmitz. Feedback control of unstable state in a laboratory reactor. *Chem. Eng. Sci.*, 30:837–846, 1975.

[12] C.C. Chen, C. Hwang, and R.Y.K. Yang. Performance enhancement and optimization of chemostat cascades. *Chem. Eng. Sci.*, 50:485–494, 1995.

[13] A. Cinar, K. Rigoponlos, S.M. Meerkov, and X. Shu. Vibrational Control of an exothermic CSTR. In *American Control Conference (ACC)*, pages 593–598, Seattle, WA, 1986.

[14] A. Cinar, K. Rigoponlos, S.M. Meerkov, and X. Shu. Nonlinear Vibrational Control of an exothermic CSTR. In *American Control Conference (ACC)*, pages 1017–1022, Mineapolis, MN, 1987.

[15] R.E. Claybugh, J.R. Griffin, and A.T. Watson. Process for broadening the MWD in polymers. *U.S. Patent*, 3:472–829, 1969.

[16] R. Femat. Chaos in a class of reacting systems induced by robust asymptotic feedback. *Physica D*, 136:193–204, 2000.

[17] B.A. Francis. The linear multivariable regulator problem. *SIAM J. Control Optim.*, 15(3):486–505, 1977.

[18] N. Friis and A.E. Hamielec. Gel effect in free radical polymerization. In *Am. Chem. Soc. Symp. Ser.*, volume 24, page 883, 1976.

[19] J.P. García-Sandoval. Nonlinear robust control for a class of polymerization reactors. Master's thesis, Universidad de Guadalajara, Mexico, 2003.

[20] O.M. Grasselli and S. Longhi. Robust tracking and regulation of linear periodic discrete-time systems. *Int. J. Control*, 54:613–633, 1991.

[21] J. Hamer, T. Akramov, and W. Ray. The dynamic behavior of continuous polymerization reactors II, nonisothermal solution homopolymerization and copolymerization in a CSTR. *Chem. Eng. Sci.*, 36:1897–1914, 1981.

[22] O.E. Hansen and S.B. Jorgensen. Control of forced cyclic processes. In *IFAC DYCORD*, volume 95, pages 21–26, 1995.

[23] S. Hara, Y. Yamamoto, T. Omata, and N. Nakano. Repetitive control system. *IEEE Trans. Automat. Contr.*, 33:659–668, 1988.

[24] A. Isidori. *Nonlinear Control Systems.* Springer Verlag, third edition, 1995.

[25] A. Isidori and C.I. Byrnes. Output regulation of nonlinear systems. *IEEE Trans. Automat. Contr.*, 35:131–140, 1990.

[26] R.L. Laurence and G. Vasudevan. Belief function combination and conflict management. *I&EC Process Design and Development*, 7(3):427–433, 1968.

[27] J.G. Lazar and J. Ross. Experiments on the effects of external periodic variations on constrains on the thermodynamics of an oscillatory reactors. *J. Phys. Chem.*, 92:3579–3589, 1990.

[28] J.H. Lee, S. Natarajan, and K.S. Lee. A model-based predictive control approach to repetitive control of continuous processes with periodic operations. *Int. J. Control*, 11:195–207, 2001.

[29] Yu. Sh. Matros. *Unsteady Processes in Catalytic Reactors.* Elsevier, 1985.

[30] Yu. Sh. Matros. *Catalytic Processes Under Unsteady Conditions.* Elsevier, 1986.

[31] G.R. Meira. Forced oscillations in continuous polymerization reactors and molecular weight distribution control. A survey. *J. Macromol. Sci.- Rev. Macromol. Chem.*, 20(2):207–241, 1981.

[32] S. Monaco and D. Normand-Cyrot. Minimum phase nonlinear discrete-time systems and feedback stabilization. In *IEEE Conf. Decision and Control (CDC)*, pages 979–986, Los Angeles, USA, 2002.

[33] K. Otawara and L.T. Fan. Enhaming the performance of spontaneously oscillatory chemical reactions. *J. Phys. Chem. A*, 101:9678–9680, 1997.

[34] K. Otawara and L.T. Fan. Increasing the yield from a chemical reactor with spontaneously oscillatory chemical reactions by a nonlinear feedback mechanism. *Comput. Chem. Eng.*, 25:333–335, 2001.

[35] F. Ozgulsen, R.A. Adomaitis, and A. Cinar. A numerical method for determining optimal parameter values in forced periodic operation. *Chem. Eng. Sci.*, 47:605–613, 1992.

[36] M. Perez and P. Albertos. Self-oscillating and chaotic behaviour of a PI-controlled CSTR with control valve saturation. *J. Process Control*, 14:51–59, 2004.

[37] F. Delli Priscoli. Robust regulation for nonlinear systems subject to measurable disturbances. In *European Control Conference (ECC)*, pages 3056–30061, 1995.

[38] W.F. Ramirez. *Process control and identification*. Academic Press, London, 1994.

[39] W.H. Ray. Periodic operation of polymerization reactors. *I&EC Process Design and Development*, 7(3):422–426, 1968.

[40] W.H. Ray. On the mathematical modeling of polymerization reactors. *J. Macromol. Sci.- Rev. Macromol. Chem.*, 8:1–56, 1972.

[41] W.H. Ray. *Advanced process control*. McGraw-Hill, New York, 1981.

[42] D.M. Ruthven, S. Farooq, and K.S. Knaebel. *Pressure Swing Adsorption*. VCH Publishers, New York, 1994.

[43] F.J. Schork, R.B. Deshpander, and K.W. Leffew. *Control of Polymerization Reactors*. Marcel Dekker, New York, 1993.

[44] J. Spitz, R.L. Laurence, and D.C. Chappalear. An experimental study of a polymerization in periodic operation, in continuous polymerization reactors (Edited by T.C. Bouton and D.C. Chappalear). *AIChE Symposium Series*, 66(160):86–101, 1976.

[45] L.E. Sterman and B.E. Ydstie. Periodic forcing of the CSTR: an application of the generalized π-criterion. *A.I.Ch.E. Journal*, 37:936–996, 1991.

[46] F. Vargas, J. Alvarez, and R. Suarez. Nonlinear study of the periodic operation for free-radical homopolymerization reactors. In *IEEE Int. Conf. Control Applications*, volume 1, pages 84–89, 1989.

[47] N. Watanabe, H. Kurimoto, and M. Matsubara. Periodic control of continuous stirred tank reactor III: Case of multistage reactor. *Chem. Eng. Sci.*, 39:31–36, 1984.

[48] N. Watanabe, H. Kurimoto, M. Matsubara, and K. Onogi. Periodic control of continuous stirred tank reactor II: Case of a nonisothermal single reactor. *Chem. Eng. Sci.*, 37:745–752, 1982.

[49] D.J. Wu, Y. Xie, Z. Ma, and N.H.L. Wang. Design of SMB chromatography for amino acid separations. *Ind. Eng. Chem. Res.*, 37:4023–4035, 1998.

Appendix

Definition of variables in equation (48)

$$x_1 = I, \quad x_2 = \frac{M_f - M}{M_f}, \quad x_3 = T, \quad u_1 = qI_f, \quad u_2 = Q, \quad D = \frac{q}{V}$$

$$a_1 = M_f \frac{(-\Delta H_p)}{\rho C_p}, \quad a_2 = \frac{1}{\rho C_p}$$

$$\beta_1 = \frac{E_d}{R}, \quad \beta_2 = \frac{2E_p + E_d - E_t}{R}, \quad \beta_3 = b_1 \left(\frac{M_f}{M_f + S_f} \right), \quad \beta_4 = b_2 \left(\frac{M_f}{M_f + S_f} \right)$$

$$k_d(x_3) = k_{d0} \exp \left[-\frac{\beta_1}{x_3} \right], \quad k_p(x_3) = k_{p0} \exp \left[-\frac{E_p}{Rx_3} \right], \quad K = k_{p0}^2 \frac{2fk_{d0}}{k_{t0}}$$

$$g(x_2, x_3) = k_p \left(\frac{2fk_d}{k_t} \right)^{1/2} = \sqrt{K} \exp \left[-\frac{1}{2} \left(\frac{\beta_2}{x_3} - (\beta_3 + \beta_4 x_3) x_2 \right) \right]$$

Part II

Control and Diagnosis of Biological Processes

Robust Nonlinear Observers for Bioprocesses: Application to Wastewater Treatment

V. Alcaraz-González and V. González-Álvarez

Department of Chemical Engineering. University of Guadalajara. Blvd. Marcelino Garcia Barragán 1451, 44430 Guadalajara Jalisco, Mexico
{victor,victorga}@ccip.udg.mx

Summary. In this chapter, some state estimation schemes used in bioprocesses engineering are firstly reviewed with particular emphasis on the so-called nonlinear observers. Second, two simple robust nonlinear observers are proposed and applied in a number of bioprocesses. The first estimation scheme is a generalized asymptotic observer which has demonstrated to be robust in the face of a complete lack of knowledge of the process nonlinearities whereas the second one has shown to be robust in the face of both, the system nonlinearities and the input uncertainty and yields guaranteed intervals for the non-measured variables from the available measurements. The design of both observer schemes are detailed for a general dynamical system which may be applied in a number of bioprocesses. The interval observer is further developed for fault detection and diagnosis purposes. In addition, detailed stability analysis are developed for both observers. Finally, these observers are tested in an actual experimental pilot plant used for the anaerobic digestion of wine vinasses. Key results are presented and discussed.

1 Introduction

Because of the increasing complexity and necessity for safety of industrial processes, efficient monitoring and decision support systems are becoming more and more important. Indeed, even in normal operational conditions, several types of disturbances may occur with serious consequences in the performance of the process. Hence, there is a clear need for advanced control in order to keep the system performance as close as possible to optimal.

This is particularly true in the case of bioprocesses where the state of the living part of the system must be closely monitored. Extensive surveys have been published and several international conferences have been held on this topic. Furthermore, the last two decades have seen an increasing interest to improve the operation of bioprocesses by applying advanced control schemes. In particular, biological Wastewater Treatment Processes (WWTP's), more efficient than the traditional physico-chemical methods but at the same time

H.O. Méndez-Acosta et al. (Eds.): Dyn. & Ctrl. of Chem. & Bio. Proc., LNCIS 361, pp. 119–164, 2007.
springerlink.com

more complex, call for a good performance to be guaranteed consistently, which has major consequences for instrumentation, control and automation [50]. Two main factors (which can be interpreted as both, incentives and constraints) have contributed to this new paradigm. The first factor has been the increasing demand for high quality purified wastewater in the effluent of a WWTP which has led to even more increasing strict norms to comply with environmental regulations [39]. The second factor has been the ever-present financial constraints of plant operation [22]. None of these two factors can be dissociated and hence, they must be taken into account in the instrumentation, control and automation of WWTP's. Indeed, incomplete treatment may induce severe process malfunction in the plant, which not only reduces the quality of the treatment, but they may also have a major economic impact. Thus, the need for optimally controlled plants will increase due to tightening permit requirements and the need to reduce cost [22], [50]. In order to fulfill the requirements related to water quality and the new tight ecological norms, as well as to reduce costs, the optimal control of WWTP faces important uncertainties arising from the intrinsic complexity of the plant design.

In this chapter, the attention is mainly focused in the anaerobic digestion process because is one of the most uncertain bioprocesses. Used in wastewater treatment, this process transforms the influent organic matter into carbon dioxide and methane through a multi-step set of self-catalytic heterogeneous reactions [38]. The following six principal factors of uncertainty are true for both wastewater treatment in general, and anaerobic digestion in particular. They clearly show the need for monitoring systems and automatic control in order to optimize the process operation or to detect disturbances.

- *Parametric uncertainty*: A great number of bacterial species carry out the transformations of organic load and nutrients in wastewater treatment processes without direct or easily comprehensible relationships between the microbial populations and viability. The role of each bacterial species is fuzzy [30], and aspects such as cellular physiology and its modeling are not easily understood from external measurements [18], [68]. As a first consequence, the kinetics of these transformations is often poorly or inadequately known [66]. Extensive efforts to model the kinetics have been undertaken, but these have not been successful to elucidate how yield coefficients, kinetic parameters and the bacterial population distribution change as a function of both, the influent composition and the operating conditions.

- *Nonlinearity*: In addition, it is well known that the process kinetics shows a highly nonlinear behavior. This a serious drawback in instrumentation and automatic control because, in contrast to linear systems where the observability can be established independently of the process inputs, the nonlinear systems must accomplish with the detectability condition depending on the available on-line measurements, including process inputs in the case of non autonomous systems [23].

- *Stability when dealing with time varying systems*: Finding the transition matrix needed for establishing the stability conditions is not an easy task. Indeed, even in the theory of linear systems, there is usually no trivial choice for solving the required conditions and the solution rather comes from numerical integration [57]. This hinders to guarantee the stability beforehand.
- *Lack of on-line sensors*: A third factor of uncertainty arises from the lack of adequate sensors to measure on-line all the important variables in the process. Even when some of these sensors are available commercially nowadays, they are still expensive, time consuming and require additional expenses for the installation and use on-site [39]. For instance, in order to prevent the analyzers becoming clogged, the wastewater samples must be pre-treated so that they are free of suspended solids. Even when ultrafiltration technologies are efficiently used for this purpose, the cleaning of membranes becomes a tedious and time consuming task. Furthermore, the range of operation between the input stream and the output stream is quite different. In consequence, if one has the possibility of using an online analyzer, a common practice is to place it in the output stream rather than in the input stream. Even when this installation offers a smaller range of variation given the smaller concentration in suspended solids, the price to pay for such an advantage is an always present uncertainty about the influent composition.
- *Influent uncertainty*: Closely related to the preceding point is the fact that the composition of the influent is highly influenced by constraints, which may vary in a random manner depending on human industrial or environmental activities [72]. Again, without suitable on-line sensors to measure these variations, only estimates based on statistical confidence intervals may be used in some cases.
- *Initial conditions uncertainty*: The uncertainty issue is further compounded by the simple task of determining the necessary process initial conditions to study the dynamics of the overall process.

To deal with these problems, several solutions have been proposed in the past such as the well known classical extended Kalman filters (EKF) and Luenberger observers (ELO) which allow the estimation of both the parameters and the state of the system. One of the reasons for the popularity of these estimators is that they are easy to implement since the algorithm can be derived directly from the state space model. However, since these estimators are based on a linearized model of the process, the stability and convergence properties are essentially local and valid only around an operating point and it is rather difficult to guarantee its stability over wide ranges of operation. One reason for the problem of convergence of EKF/ELO is that, in order to guarantee the exponential convergence of the observer, the process must be locally observable which, as it turns out, is restrictive in many practical situations and may account for the failure of these state estimators to find widespread application. For instance, the implementation of EKF/ELO in bioreactors is based on the *a priori* knowledge of the observability of the process. Because of the nonlinear aspects of their dynamics, the observability analysis is rather complex in biochemical process applications; and the usually large uncertainty in the kinetics of the biochemical reactions and analytical expressions used to describe them makes the approach even more difficult. As a matter of fact, very few works deal with the observability of nonlinear biochemical processes (e.g., [17], [24]) and they are usually concerned with particular process applications. Another problem is that the theory for the EKF/ELO is developed using a perfect knowledge of the system model and parameters, in particular of the process kinetics, and as a consequence, it is difficult to develop error bounds to take into account the large uncertainty of these parameters.

Several linearization methods have been also proposed [7], [11], [49]. Nevertheless, only local behavior can be guaranteed as they miss practical results on performance and stability. Other approaches are adaptive observers [6], [29] and sliding observers based on the theory of variable structure systems [70] but their design involve conditions that must be assumed *a priori* or that are usually hard to verify [48]. All these approaches solve —in one sense or another— some of the problems described above but in the most of the cases, the complexity of the resulting estimating algorithms is a limitation for real time computation. Indeed, monitoring algorithms will prove to be efficient if they are able to incorporate the important well-known information on the process while being able to deal with the missing information (lack of on-line measurements, uncertainty on the process dynamics, etc.) in a robust way.

In this chapter two new robust nonlinear observers for a class of lumped time varying non-linear models useful in chemical and biochemical engineering processes are described. Such robust observers are capable of coping simultaneously with the aforementioned problems while remaining easy to implement with a minimum number of straightforward conditions to verify. The first one is the so-called *asymptotic observer*, which although requires the knowledge of the process inputs, it has the main advantage to permit the exact cancellation of the nonlinear terms of the systems that facilitates its design, stability analysis and implementation. Then, based upon the structural basis of the asymptotic observer, a second one, the *interval observer* is introduced. Under some structural and operational conditions, and assuming that only guaranteed lower and upper limits on the inputs are available, this observer allows us the reconstruction of a guaranteed interval on the unmeasured states instead of reconstructing their precise numerical values. Finally, an original diagnosis system design is presented. Based on the set-valued information provided precisely by interval observers, this approach is capable of detecting and isolating a number of sensor faults, always in the presence of the maximum uncertainty scenarios described above.

This chapter is organized as follows: In section 2, the theoretical basis of observers are briefly reviewed. Section 2 also serves to present the state of the art of successful observer applications in bioprocesses. In this section the principles of both, the asymptotic observer and the interval observer, are illustrated with their design and implementation upon a typical fermentation process. In section 3, a general nonlinear dynamical model is presented and then, upon the basis of this model, generalized versions of the asymptotic observer and the interval observer are detailed in sections 4 and 5, respectively. In section 6, the design of both observers is developed and implemented in an actual pilot 1 m^3 fixed bed anaerobic digester for the treatment of wine vinasses. Finally, the current tendencies on robust nonlinear observers and their application on bioprocesses are discussed before the chapter conclusions are drawn.

2   Fundamentals

The efforts to meet the specific needs for the management and the processes optimization can be classified according to the type of approach employed. On the one hand, it is possible to distinguish the approaches based on data sets, those founded on expert knowledge (in the broad sense of the term) and those founded on an analytical description of the system. Particularly, this work is interested in the estimation problems and the diagnosis problems in the context of the third quoted family, *i.e.*, inside an analytical context. On the other hand, one can distinguish the static and dynamical approaches according to whether the relations between the measured variables and those to estimate are static or dynamic. In this chapter the attention is focused on the dynamical case, namely on the efficient development and implementation of state estimation schemes (estimators, "software sensors", state observers or simply observers) that can be used to design control and optimization strategies in chemical and biochemical processes.

Three important reasons justify the resorting to an estimator:

- *Knowledge acquisition and modelling.* For example, when one must determine a variable, which cannot be measured directly, or to estimate key process parameters.
- *Process control.* For example,when one wishes to control a variable not directly measured, or which is not readily available or that unmeasured variables intervene in the design of a control law.
- *Supervision and diagnosis.* For example, when it is necessary to supervise a microorganism concentration in a biofilm to avoid unstable operating conditions.

From a systematic point of view, the inputs of an estimator are the process inputs (*i.e.*, control variables and measured disturbances) as well as measurements available on-line in the process, while the outputs of an estimator are the estimates of the unmeasured variables. A diagram illustrating this concept is shown in Figure 1.

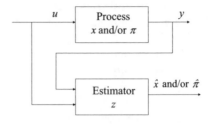

Fig. 1. A general estimation scheme (x: state variables, π: parameters, z: auxiliary variables)

The internal structure of an observer is based on the model of the considered system. Of course, the model can be extremely simple or reduced to a simple algebraic relationship binding available measurements. However, when the model is of the dynamical type, the value of a variable is no longer influenced uniquely by the inputs at the considered moment but also by the former values of the inputs as well as by other system variables. These phenomena are then described by differential equations. Since these models carry information on the interactions between the inputs and the state variables, they are used to estimate unmeasured variables from the readily available measurements.

In terms of methods, synthesis approaches can be classified according to the model type on which they are based. Thus, one distinguishes — among those applied to the bioprocesses — two great classes: linear and non-linear approaches. In the linear approaches, one distinguishes the Luenberger observer (LO) and the Kalman filter (KF) according to the deterministic or stochastic context, respectively. In the non-linear approaches, one can distinguish the LO and KF in their extended form (*i.e.*, ELO/EKF), the observers with unknown kinetics (called also asymptotic observer), the High Gain observers (HGO) and the interval observer (IO). These estimation schemes will be described later in this chapter.

A review of state-of-the-art of the state estimation in bioprocesses is presented in Table 1 that includes the advantages and disadvantages of the most common estimation schemes (*i.e.*, EKF and ELO) and the more specific estimation approaches that have been applied in such dynamical systems.

2.1   Observers for Bioprocesses

A Brief Description of the Observers

In order to describe the guiding principles of the observers considered in this chapter, let us use the following general model without noise

$$\begin{aligned} \frac{dx}{dt}(t) &= f\left(x\left(t\right), u\left(t\right)\right), \quad x\left(t_0\right) = x_0 \\ y\left(t\right) &= h\left(x\left(t\right)\right) \end{aligned} \tag{1}$$

Table 1. State estimation and parameter estimation in bioprocesses

Principles and methods	Necessary knowledge	Advantages	Disadvantages	References
Application of linear methods				
Kalman Filter	Process model, (including process kinetics, but it is possible to estimate some kinetic parameters on-line), process inputs, statistical noise properties.	Well known approach. It takes into account the measuring noise as well as process inputs noise.	Model linearization; Inputs knowledge; Stability and convergence are only locally valid.	[6][1]
Extended Luenberger Observer	Process model, (including process kinetics, but it is possible to estimate some kinetic parameters on-line), process inputs.	Well known approach. It allows tuning the convergence rate by pole placement.	Model linearization; Inputs knowledge; Stability and convergence are only locally valid.	[6]
Development of specific estimation methods				
Asymptotic Observer	Process model, (process kinetics as well as yield coefficients can be estimated on-line), process inputs.	It takes specifically into account the nonlinear structure of the system;Simplicity of the method; Stability and convergence are guaranteed if the inputs are persistent and bounded.	Partial model knowledge; Inputs knowledge; Non-adjustable convergence rate.	[6]
High Gain Observer	Complete process model, process inputs.	It takes specifically into account the nonlinear structure of the system; Adjustable convergence rate.	Good model knowledge; Inputs knowledge; Complex method; High sensitivity to noise.	[23]

[1] For an extensive review of the application of the KF and the LO in bioprocesses, the reader is refered to [6].

Table 1. (*continued*)

Generalized Asymptotic Observer	Process model, (process kinetics as well as yield coefficients can be estimated on-line), process inputs.	It generalizes the class of considered biological systems (including interconnected systems); Simplicity of the method; Stability and convergence are guaranteed if the inputs are persistent[2] and bounded[3].	Partial model knowledge; Inputs knowledge; Non-adjustable convergence rate.	[13]
Robust Generalized Asymptotic Observer with partially adjustable convergence rate	Process model, (process kinetics as well as yield coefficients can be estimated on-line), process inputs.	It generalizes the class of considered biological systems (including interconnected systems); Simplicity of the method; Stability and convergence are guaranteed if the inputs are persistent and bounded; Adjustable convergence rate.	Partial model knowledge; Partially adjustable convergence rate.	[31]
	Complete process model, process inputs.		Model knowledge; Inputs knowledge.	[32]
Interval Observer	Process model, (process kinetics as well as yield coefficients can be estimated on-line), process inputs.	It generalizes the class of considered biological systems (including interconnected systems); Simplicity of the method; Stability and convergence are guaranteed if the inputs are persistent and bounded; Robust in the face of unknown inputs; **Partially adjustable convergence rate.	*Partial model knowledge; Non adjustable convergence rate. **[46]	*[2],[3],[33]

[2] The dilution rate cannot be equal to zero for a long time period since the estimator convergence rate depends on this variable. This is the actual meaning of persistent.

[3] The dilution rate is bounded by the maximum capacity of the pump used in the feed.

where $x(t) \in \Omega \subset \Re^n$, $u \in \Re^m$, $y \in \Re^p$ represent the state vector, the input vector and the output vector, respectively. x_0 denotes the initial conditions at the instant t_0. $f : \Re^n \times \Re^m \rightarrow \Re^n$ and $h : \Re^n \rightarrow \Re^p$ are functions that comply with the existence and the unicity properties of the solutions.

Some Classical Observers

The usefulness of the classical observers lies on the dynamic characteristics of the estimation error. If the user is able to regulate the error decrease rate, the direct consequence is that the estimated value will converge as fast as desired towards the actual value of the considered variable.

A general description of a classical observer is given by

$$\tfrac{d\hat{x}}{dt}(t) = f(\hat{x}(t), u(t)) + k(h(\hat{x}(t)) - y(t)) \tag{2}$$

where f describes exactly and without any modeling error the actual system. Then, the model predictions are corrected by the difference between the true output (measurements) and the theoretical output given by the observer. Afterwards, one may tune the amplitude of the correction by means of k, which is called the *observer gain*. The EKF, ELO and HGO belong to this class of observers but they differ essentially on the assumptions and on the procedures that make possible the calculation of their respective gain. For the EKF, the gain is calculated such that the statistical properties of the variables to be estimated are guaranteed in spite of the uncertainties accounted for, which can affect the measurements or the state of the system. It is worth mentioning from a stochastic point of view that the EKF is a filter rather than an observer. In contrast, the ELO lies within a deterministic scope where the uncertainties and measurement noise are no longer characterized by statistical properties. The gain of this observer may be then calculated such that the properties related to its convergence rate and to its stability are guaranteed. In both cases, nevertheless, the calculation of their gain requires a linearization of the system model, which is only valid around a given operating condition and requires a good characterization of non-linearities which in most bioprocesses imply the complete knowledge of process kinetics. On the other hand, the HGO's do not require the linearization of the system, which makes possible to obtain a convergence rate higher than the EKF and ELO. However, this observer is very difficult to synthesize in practice since it remains very sensitive with respect to both, the noise and the quality of the used model.

Asymptotic Observers

With regard to the observers known as asymptotic, introduced by [6], their design is done differently insofar as they do not have a gain, making possible to tune its convergence rate. Moreover, they have the advantage of being completely independent of the nonlinearities of the system (*i.e.*, the kinetic terms) provided that they satisfy certain assumptions during their syntheses, which will be detailed later on. In order to illustrate the basis of these observers, let us consider a simple bioprocess in which a biomass X grows on a substrate S in a continuous stirred tank reactor (CSTR):

$$\begin{aligned} \dot{X} &= (\mu(t) - D(t))\, X \\ \dot{S} &= -k\mu(t)X + D(t)\,(S_{in}(t) - S) \end{aligned} \tag{3}$$

where $\mu(t)$: biomass specific growth rate
        k : yield coefficient for substrate conversion into biomass
        $S_{in}(t)$: input substrate concentration
        $D(t)$: dilution rate ($D(t) = Q(t)/V \geq 0$, where $Q(t)$ is the flux rate and V is the reactor volume)

Now, assume that it is desired to estimate the biomass concentration X by using the measurements of S —presumably continuous—, without the knowledge of the growth rate. With this aim, let us introduce an auxiliary variable ζ defined by:

$$\zeta \equiv kX + S \tag{4}$$

so that the dynamics of ζ may be written in the following form:

$$\dot{\zeta} = -D(t)\,(\zeta - S_{in}(t)) \tag{5}$$

Then, an estimate of the microorganisms concentration can be easily obtained by solving (4) for $\hat{X}$; that is

$$\hat{X} = \tfrac{1}{k}\left(\hat{\zeta} - S\right) \tag{6}$$

where $\hat{\zeta}$ is the solution of:

$$\dot{\hat{\zeta}} = -D(t)\left(\hat{\zeta} - S_{in}(t)\right) \tag{7}$$

In this example, it is easy to see that, the dynamics of ζ is quite independent of the kinetics μ, which gives a specific solution to the well known problem of the biomass growth rate ignorance in the bioprocess [66]. It is also verified that the convergence rate cannot be tuned because it depends exclusively on the value of the dilution rate. Indeed, Bastin and Dochain [6] have shown that a condition to guarantee the convergence of the observer is that the dilution rate D must be a persisting input, i.e., that there are two positive constants c_1 and c_2 such as, at every time t:

$$0 < c_1 \leq \int_t^{t+c_2} D(\tau)d\tau \tag{8}$$

In practice, it has been shown that c_2 must be small compared to the process time-constants and that $\frac{c_1}{c_2}$ is sufficiently large to guarantee the convergence rate of the observer [9].

With regard to the interval observers, they will be described in detail further on. However, their use for the robust observation of unmeasured variables in the context of a biological WWTP is justified in the following section.

Interval Observers

In the approaches presented above, the system inputs are necessary to synthesize the observers. Moreover, even when the asymptotic observers allow robustifying the HGO's — in the sense that they limit the knowledge, necessary *a priori* to synthesize them —, they do not allow the tuning of the convergence rate, which is completely determined by the process conditions (*i.e.*, the dilution rate). This property can involve important convergence times.

The greatest difficulty limiting the applicability of the methods presented previously in the biological WWTP's lies obviously in the difficulty (and sometimes in the impossibility) of measuring the input process concentrations. Even when in certain industries (pharmaceutical, food, etc.,), the inputs of the system are often control variables, it is not the case in WWTP's, where variations on the composition of the effluent to be treated is one of the principal sources of disturbance and uncertainty. A first solution to this constraint relays in the use of approaches known as "robust" respect to unknown disturbances. Among those that make possible to rebuild univocally the state of the dynamical system, two approaches are distinguished: i) the sliding mode observers [20] and, ii) the unknown inputs observers [15]. However, these approaches require the knowledge of the kinetics — or at least partially —, which is a difficult constraint in the wastewater treatment field. Several reasons have limited the capacity to measure the process inputs:

- First of all, the installation of advanced sensors generally requires the implementation of a filtration system. However, the presence of suspended matter in a very large quantity in the process input limits the possibility of resorting to such a system, which are, in addition, extremely expensive in both investment and maintenance;
- In many practical applications, the measurement devices cover a specific range of operation, which may be used in the input or in the output of the bioprocess but not in both places. Input concentrations are usually higher than the output concentrations and thus they require two different sensors with different sensitivities which may cause an additional cost of measuring devices. This remark prevents the use of a single sensor that must be placed at the influent stream to measure a certain variable and the placed it at the output to record some ather variable or magnitud;
- Finally, if these sensors are necessary under a process control point of view, and by taking into account the fact that in order to install two sensors, one in the input and the other at the output, is not financially possible, the limited knowledge of the bioprocess and their dubious character encourage control engineers to consider feedback control rather than feed-forward control. A sensor, if it is available, will preferably be installed at the output of the process rather than at the input.

Thus, it clearly seems that the approaches described above cannot be rigorously applied to the biological WWTP because they are not detectable for such unknown inputs. Indeed, in the general context of the observer's

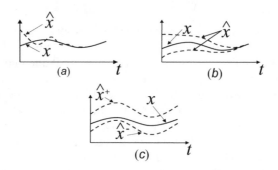

Fig. 2. Observability and detectability for nonlinear systems (see body text)

theory, and from a non-linear point of view, three situations may intervene (see Figure 2):

a) Observable systems where one can tune the convergence rate;
b) Non observable but detectable systems — the non observable modes of the system are stable — ;
c) Undetectable systems, where it is possible to rebuild only an interval in which it is possible to guarantee that the true values of the unmeasured variables really are.

The first two cases correspond to univocally valued type approaches while the third one corresponds to the set approach in which one can classify the interval observer.

Consequently, several ways are offered to us:

• First of all, certain industries (for example, those in charge of the urban wastewater processing) have an influent composition varying in a periodic way. It is thus possible to identify a function making possible to simulate these variations. However, the risk of falling in a fault situation is higher if the composition of the effluent deviates significantly from the one simulated;
• The second possibility consists by assuming these variations to behave as a succession of steps whose amplitude is updated by measurements carried out off-line. If this method is indeed very simple, it proves practically impossible to implement in the most of the installed facilities because the variations are sometimes relatively sudden and unforeseen and the number of off-line essays to be realized may become too large to handle;
• The third possibility refers to a set of research topics on the development of methods making possible to estimate on-line — for control or diagnosis purposes — the input process concentrations. Starting from the knowledge

of the model and the available measurements, such approaches consist of synthesizing observers for unknown inputs[1] (*Cf.*, for example [4] or [36]).

- Another possibility consists in the set-membership approaches aiming at the study of methods which directly allow the handling of $\Re^n$ sub-sets. These sets are generally characterized by their bounds and then are usually enclosed in simple geometric shapes such as ellipsoids, parallelotopes, polytopes or more complex fields. These tools were firstly developed for mathematicians and data processing engineers within the framework of arithmetic of intervals where interesting solutions were brought for the resolution of non-linear equations systems in a bounded-error set context (see for example [41]). Nevertheless, these approaches, in their algorithmic philosophy, are beyond the scope of this work, and hence is up to the reader to refer to [40], [42] or [44], for further information.

- Last but not least, one can also resort to other robust approaches by trying to limit the available knowledge on the process inputs. Consider for instance that only certain bounds, between which the input concentrations are likely to vary, are known. By using only this knowledge, it is no longer feasible to reconstruct the exact value of the variables to be estimated, but it is sometimes possible, to rebuild guaranteed intervals for the variables to estimate (*i.e.*, intervals between which it is possible to guarantee that the true values of the considered variables really are).

This last approach is illustrated in the following bioprocess (later it will be extended to cover more complex systems). In fact, the observers by intervals, such as the one developed here, are based on the asymptotic observers.

Let us take the example developed previously for the asymptotic observers (*Cf.*, equations (3) to (7)). Notice that, in order to estimate $\hat{X}$ by using equations (6) and (7), a readily measurement of S_{in}, is needed. Let us suppose that this measurement is not available but only bounds, inside which this variable is ensured to evolve, are known such as $S_{in}^-(t) \leq S_{in}(t) \leq S_{in}^+(t)$. Then, it is possible to synthesize two observers:

$$
\begin{aligned}
\dot{\hat{\zeta}}^+ &= -D(t)\left(\hat{\zeta}^+ - S_{in}^+(t)\right) \\
\dot{\hat{\zeta}}^- &= -D(t)\left(\hat{\zeta}^- - S_{in}^-(t)\right)
\end{aligned}
\tag{9}
$$

which — provided that they are correctly initialized — will allow us to rebuild the bounds $\hat{X}^+$ and $\hat{X}^-$ as:

$$
\begin{aligned}
\hat{X}^+ &= \tfrac{1}{k}\left(\hat{\zeta}^+ - S\right) \\
\hat{X}^- &= \tfrac{1}{k}\left(\hat{\zeta}^- - S\right)
\end{aligned}
\tag{10}
$$

---

[1] Do not be confused with the unknown inputs observers theory in which the goal is to estimate state variables of a system subjected to unmeasured inputs. Here, the goal is precisely to estimate these unknown inputs and not only certain state variables.

within which it is guaranteed that the true value of X will be, *i.e.*, $\hat{X}^-(t) \leq \hat{X}(t) \leq \hat{X}^+(t)$. To show this, let us introduce the estimation error $e = \hat{X}^+ - X$. Hence, its dynamics can be written as:

$$\dot{e} = -D(t)e + D(t)\left(S_{in}^+(t) - S_{in}(t)\right) \tag{11}$$

By hypothesis, $D(t)\left(S_{in}^+(t) - S_{in}(t)\right) \geq 0$. Consequently, and provided that $e(0) \geq 0$, then $e(t) \geq 0, \forall t$. This shows that whatever $0 < S_{in}(t) \leq S_{in}^+(t)$, it is guaranteed that $X(t) \leq \hat{X}^+(t)$ if the observer is adequately initialized, namely $0 < X(0) \leq \hat{X}^+(0)$. A similar reasoning would enable us to show that by choosing $\hat{X}^-(0) \leq X(0)$, it is possible to guarantee that $\hat{X}^-(t) \leq \hat{X}(t)$. Finally, it has been thus established that if $\hat{X}^-(0) \leq \hat{X}(0) \leq \hat{X}^+(0)$, and being given $S_{in}^-(t) \leq S_{in}(t) \leq S_{in}^+(t)$, the use of the interval observer given by the equations (9-10) enable us to rebuild lower and higher limits of X so that $\hat{X}^-(t) \leq \hat{X}(t) \leq \hat{X}^+(t)$. This approach, was introduced in 1998 by Rapaport and coworkers [53], [55], and tested through numerical simulation for two bioprocess: a nitrification and an activated sludge process.

2.2 Fault Detection and Diagnosis for Biological Systems

A fault can be regarded as a not allowed deviation of at least one property or one characteristic parameter of the system compared to the normal operating conditions. The occurrence of a fault may have serious consequences in the process and may cause the facilities standstill, even to damage them. Several examples illustrating the gravity of the occurrence of sensor or actuators faults are reported in [25], [26], [62]. The fault detection refers to the determination of the presence of faults as well as their occurrence moment while the diagnosis is the determination of its amplitude and its behavior.

Related studies to the diagnosis of bioprocess have been limited and have used mainly heuristic approaches. Moreover, they have been concerned with the detection of a disfunction of the bioprocess (detection of a desestabilization, state of the biomass, etc.) rather than the detection and the location of sensor and/or actuators faults. The interested reader will be able to refer to the following references: [4], [5], [12], [14], [27], [28], [43], [45], [47], [52], [59], [60], [61], [63], [64], [71].

Interval Observer Based Diagnosis

System diagnosis frequently lies on a model that represents the normal behavior of a particular process to be supervised. The fundamental problem comes then from the inaccuracies associated with the model, either related to the ignorance of the kinetics or its parameters, or related to the ignorance of its inputs. Within the framework of this chapter, the interest is focused on the detection and location of sensor faults in the presence of unknown inputs. Among the existing solutions based on observers, one can distinguish the approaches based on non-linear unknown inputs observers (see for example, [21],

[56]) and those based on adaptive non-linear observers ([16]). More recently, a new strategy of diagnosis, —the bounding approach—, has been proposed. Such a strategy has made possible to discriminate a fault of a modeling error. This strategy is based on the interval philosophy and takes into account uncertainties, which may vary as a function of time and may affect any parameters of the model that are represented by bounded variables [1], [51].

In this section an innovating approach based on the use of a battery of interval observers functioning in parallel is presented. Such an approach allows us to detect a violation of the assumptions related to the unknown inputs (substrates concentrations at the input of the process). In order to illustrate the principle of this approach, consider again a mono-biomass, mono-substrate bioprocess within a CSTR, described by equations (3) to which the dynamics of one of the products, represented by P, has been added:

$$\begin{aligned}
\dot{X} &= (\mu(t) - D(t))\, X \\
\dot{S} &= D(t)\,(S_{in}(t) - S) - k_1\mu(t)X \\
\dot{P} &= -PD(t) + k_2\mu(t)X
\end{aligned} \qquad (12)$$

Let us suppose that X, S and P are measured on-line. Since these measurements are given, the goal is to synthesize a system of diagnosis taking explicitly into account the non-linear character of the system (12), which is able to detect and locate a sensor fault (on the measures of X, S or P) without any knowledge on the biomass specific growth rate expression and knowing only lower and higher limits, $S_{in}^-(t)$ and $S_{in}^+(t)$, respectively, within S_{in} actually evolves. For this purpose, an interval observer following the procedure described in Section II.4.5. is first synthesized. With this aim, let us introduce the linear transformations $Z_1 = X - \frac{1}{k_2}P$ and $Z_2 = X - \frac{k_1}{k_2}P$. Then, for this bioprocess, the interval observer takes the following form:

$$\begin{aligned}
\dot{\hat{Z}}_1 &= -D(t)\,\hat{Z}_1 & \hat{X} &= \hat{Z}_1 + \frac{P}{k_2} \\
\dot{\hat{Z}}_2^- &= -D(t)\,\hat{Z}_2^- + D(t)\,S_{in}^-(t) & \hat{S}^- &= \hat{Z}_2^- - \frac{k_1}{k_2}P \\
\dot{\hat{Z}}_2^+ &= -D(t)\,\hat{Z}_2^+ + D(t)\,S_{in}^+(t) & \hat{S}^+ &= \hat{Z}_2^+ - \frac{k_1}{k_2}P
\end{aligned} \qquad (13)$$

The reader will note that this observer is quite independent of the kinetics. Now, let us consider the following residuals:

$$\begin{aligned}
r_1(t) &= X(t) - \hat{X}(t) \\
r_2(t) &= S(t) - \hat{S}^-(t) \\
r_3(t) &= \hat{S}^+(t) - S(t)
\end{aligned} \qquad (14)$$

Residuals r_2 and r_3, when they are negative, highlight essentially the fact that the measurement of the substrate concentration S in the reactor exits out of its lower and upper estimated bounds, $\hat{S}^-$ and $\hat{S}^+$, respectively. Thus, in order to determine the presence of a fault, it is simply enough to check the sign of r_2 and r_3 whereas r_1 is compared with a threshold fixed by the user. Next, if it is kept in mind that a fault on a variable would activate only a certain number of residuals, then the combination of active and inactive residuals would be, in principle, unique for each kind of fault. Therefore, for the fault location, the determination is made by using the following table, in which, each column represent a signature that clarifies the dependence of r_i, for $i = 1, 2, 3$, on each variable to be detected:

Table 2. Sensor fault signature

	S	X	P
r_1	0	1	1
r_2 or r_3	1	0	1

Remember that one of the synthesis hypotheses of this diagnosis approach is that S_{in} must be within bounds (i.e., $S_{in}^-(t) \leq S_{in}(t) \leq S_{in}^+(t)$). Now, let us suppose that this assumption is violated. Given that the bounds $S_{in}^-(t)$ and $S_{in}^+(t)$ intervene directly in equations (14), it is clear that the incidence of this fault will be detected by r_2 or by r_3 (according to whether S_{in} passes under its lower bound or above its upper bound). Since the fault signatures associated with the residuals r_2 and r_3 are identical, it will not be possible to distinguish a violation of the hypothesis on the unknown input S_{in} assumptions, of a fault of the sensor that measures S. However, by taking into account the fact that both r_2 and r_3 cannot be negative at the same time, it will be possible to distinguish between the two following cases:

- if r_2 is negative, it is possible to detect either, a sensor fault on the measuring of S or a violation of the assumptions on the unknown input such as $S_{in}(t) < S_{in}^-(t)$.
- if r_3 is negative, it is possible to detect either, a sensor fault on the measuring of S or a violation of the assumptions on the unknown input such as $S_{in}^+(t) < S_{in}(t)$.

Numerical Implementation

Let us consider a bioprocess whose dynamics is governed by the equation system (12) with Monod type kinetics and the parameters $\mu_{max} = 1.25 \text{ h}^{-1}$, $K_s = 7.65 \text{ Kg/m}^3$, $k_1 = 12.1 \text{ Kg/Kg}$, $k_2 = 7.65 \text{ mol/m}^3$.

Following the observer synthesis procedure described above — with the known input $S_{in}(t)$ —, an observer defined by the following system is obtained:

$$
\begin{aligned}
\dot{Z}_1 &= -DZ_1 & \hat{X} &= Z_1 + \frac{P}{k_2} \\
\dot{Z}_2 &= -D(t)Z_2 + DS_{in} & \hat{S} &= Z_2 - \frac{k_1}{k_2}P
\end{aligned}
\tag{15}
$$

Then, in accordance with the assumptions defined in the preceding paragraph, it is supposed that $S_{in}(t)$ is not measured on line but that lower and upper bounds (within which, $S_{in}(t)$ evolves) are available, i.e., $S_{in}^-(t) \leq S_{in}(t) \leq$

$S_{in}^+(t)$) where $S_{in}^-(t)$ and $S_{in}^+(t)$ are given (see Figure 4) and that the dilution rate is known (see Figure 3). Then, an interval observer identical to that defined by (13) may be synthesized. The residuals defined by (14) may be generated on-line (see Figures 11 to 13). Simulations are carried out over 100 hours of fermentation. Several successive faults upon the measures of X, S and P are applied to the nonlinear system (12). More specifically, the protocol described in Table 3 is used.

Table 3. Sensor fault protocole

N° Fault	Time (h)	Affected variable	Perturbation type	Perturbation
1	10 to 15	P	additive	$P_m = P + 25$
2	20 to 25	S	additive	$S_m = S + 5$
3	26 to 31.5	X	constant	$X_m = 0.4$
4	35 to 45	X	ramp	$X_m = 0.07X - 1.85$
5	45 to 55	S	false curve	$S_m = 1/(0.0183S - 0.7583)$
6	65 to 70	P	constant	$P_m = 350$
7	80 to 95	S	ramp	$S_m = 0.1773S - 5.1867$

By inspecting Figure 8, it can be seen that four faults were detected along the simulations. It should be noticed that the first detection was, in fact, due to a bad initialization of the observers (see Figures 5 and 11) but disappeared once the necessary time to convergence was reached. This figure also shows that faults 3 and 4 of Table 3 (corresponding to 2nd and 3rd faults in Figure 8) were detected and located in a completely satisfactory way, whereas the last fault in the figure corresponds to a false location of fault 6. Indeed, this short signal located on the measurement of X corresponded to a indecision period during which the measurement of S (see Figure 6) did not go beyond its estimated bounds. Once one of the bounds was crossed, the location was perfectly tracked as it is shown in Figures 8 and 10. The difference at $t = 10$ h between the measurement of X and its estimate (which results in another difference also on S in Figure 6) corresponds in fact, to the influence of the fault 1 (upon the measurement of P) on the estimate of X. As it is shown in Figures 8 and 10, this fault was detected and perfectly located. Concerning the faults affecting the measurement of S, only 2 (faults 2 and 7) out of 3 faults (faults 2, 5 and 7) were actually detected and located (see Figure 9). This is explained by the fact that the measurement of S was not used to rebuild other variables and that the amplitude of the fault 5 was not enough to force the S measurement to go beyond its estimated bounds. Consequently, this defect passed completely unnoticed. Indeed, the convergence rate of observers used to generate the residues played a paramount role in the good performance of this algorithm. Then, a residual that requires a long time to be active can

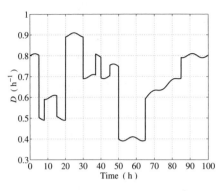

Fig. 3. Dilution rate

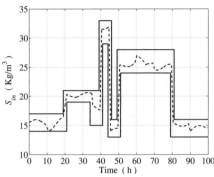

Fig. 4. Input substrate concentration (- -) between the upper and the lower bounds (—)

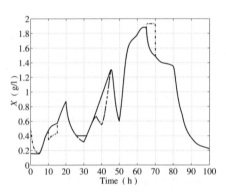

Fig. 5. Biomass concentration in the reactor (- -), its estimation (- . -) and its measurement (—)

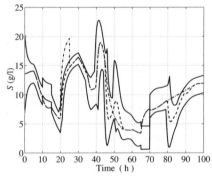

Fig. 6. Substrate concentration in the reactor (- -), its measurement (- . -) and the estimated upper bound and lower bound (—)

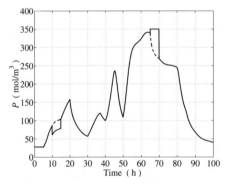

Fig. 7. Product concentration in the reactor (- -) and its measurement (—)

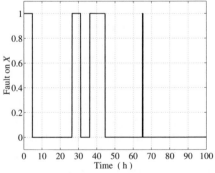

Fig. 8. Fault signals on the measurement of X

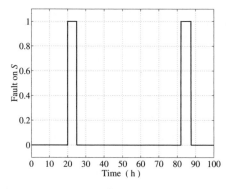

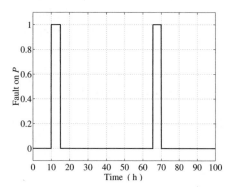

Fig. 9. Fault signals on the measurement of S

Fig. 10. Fault signals on the measurement of P

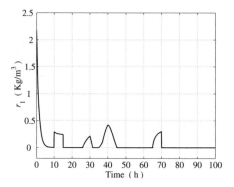

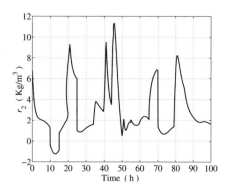

Fig. 11. Residual r_1

Fig. 12. Residual r_2

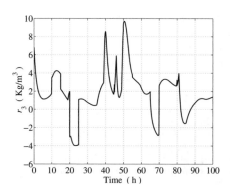

Fig. 13. Residual r_3

involve the no detection of the fault, and a residual which requires a long time
to disappear once the corresponding fault ceases, can lead to a false alarm for
another fault.

3   A General Model

Here, a class of biochemical processes whose mass balances can be described
by the following nonlinear ordinary differential equations are considered.

$$\dot{x}(t) = Cf(x(t),t) + A(t)x(t) + b(t) \tag{16}$$

where $x(t) \in \Re^n$ is the state vector, $C \in \Re^{n \times r}$ represents a matrix of coefficients (e.g. stoichiometric or yield coefficients) and $f(x(t),t) \in \Re^r$ denotes
the vector of nonlinearities (including reaction rates). The time varying matrix
$A(t) \in \Re^{n \times n}$ is the state matrix while $b(t) \in \Re^n$ belongs to a vector gathering
the inputs (i.e., the mass feeding rate vector) and/or other functions possibly
time-varying (e.g., the gaseous outflow rate vector if any).

Notice that the structure of the model (16) is not restricted to bioprocesses
and it can be used to describe a very large number of chemical processes
as well. Examples of this class of processes are continuous reactors, recycle
reactors and interconnected reactors where the matrix $A(t)$ is normally a
function of the plant operating conditions (e.g., the dilution rate(s)). Now, the
framework of uncertainties and the minimum knowledge on the system that
are necessary to design the observers is formally described. For this purpose
the following hypothesis is introduced.

Hypothesis H1:

a) $f(x(t),t)$ is fully unknown.
b) $A(t)$ is known for each $t \geq 0$.
c) m state variables are measured on-line.
d) C is constant and known.
e) $A(t)$ is bounded, that is, there exist two constant matrices A^- and A^+ such
    as $A^- \leq A(t) \leq A^+$.
f) Initial conditions of the state vector are unknown but guaranteed bounds
    are given as $x^-(0) \leq x(0) \leq x^+(0)$.
g) The input vector $b(t)$ is unknown but guaranteed bounds, possibly time
    varying, are given as $b^-(t) \leq b(t) \leq b^+(t)$.

Note 1. The operator $\leq$ applied between vectors and between matrices should
be understood as a collection of inequalities between components.

The hypotheses *H1a-d* will be used for the design of the asymptotic observer
while the hypothesis *H1e* will be used for stability analysis. Hypotheses *H1f*
and *H1g* will be used later in the design of the interval observer.

By using the hypothesis *H1c*, it is assumed that the state space can be split in such a way that (16) can be rewritten as:

$$\begin{aligned}
\dot{x}_1(t) &= C_1 f\left(x(t), t\right) + A_{11}(t) x_1(t) + A_{12}(t) x_2(t) + b_1(t) \\
\dot{x}_2(t) &= C_2 f\left(x(t), t\right) + A_{21}(t) x_1(t) + A_{22}(t) x_2(t) + b_2(t)
\end{aligned} \tag{17}$$

where the m measured state variables have been grouped in the $x_2(t)$ vector $(dim\ x_2(t) = m)$ and the variables that have to be estimated are represented by $x_1(t)$ $(dim\ x_1(t) = s = n - m)$. Matrices $A_{11}(t) \in \Re^{s \times s}$, $A_{12}(t) \in \Re^{s \times m}$, $A_{21}(t) \in \Re^{m \times s}$, $A_{22}(t) \in \Re^{m \times m}$, $C_1 \in \Re^{s \times r}$, $C_2 \in \Re^{m \times r}$, $b_1(t) \in \Re^s$ and $b_2(t) \in \Re^m$ are the corresponding partitions of $A(t)$, C and $b(t)$, respectively.

In the following sections, the asymptotic observer as well as the interval observer developed previously will be extended to this general dynamical system.

4 The Asymptotic Observer

Asymptotic observers were firstly synthesized in [6] for the case where the dilution rate is constant and the A matrix is diagonal with the dilution rate present on each element of its main diagonal. In this case, condition (8) is sufficient to guarantee the convergence of the observer. These observers were then extended by [13] to cover a more general case of equation (16) where the A matrix may be non-diagonal. In such a case, the regular persistence of the dilution rate, when it is time varying, is not sufficient to guarantee the convergence and stability of the observer. Then, sufficient conditions to guarantee the stability of the observer in the case of a constant A matrix were proposed in [13]. This condition establishes that A must be of the Metzler type[2]. Further on, in order to guarantee the stability of the asymptotic observers if a non-diagonal A matrix is a function of time, some conditions at least sufficient, will be shown in this section.

4.1 Design of the Asymptotic Observer

Given that the nonlinearities $f(x(t), t)$ are unknown, the asymptotic observer is designed in such a way that it enables us the reconstruction of the unmeasured states from the measured ones, whatever the unknown nonlinearities are. This can be done by finding a suitable linear combination of the state variables $w(t) = Nx(t)$ with $N \in \Re^{s \times n}$, such that:

$$NC = 0 \tag{18}$$

The following addidtional hypothesis is recalled:

Hypothesis H2:

$$rank\ C_2 = c$$

---

[2] A matrix is Meztler if all its off-diagonal elements are positive or nulls.

which implies the following three important properties. The first one is related to the partition (17) while the next two allows us to find at least one nontrivial solution of (18).

1. In order to fulfill this hypothesis, m and s are fixed as $m \geq c$, and thus, $s \leq n-c$. In other words, the kind and the number of states that should be measured and estimated are fixed by hypothesis $H2$. Thus, this hypothesis also establishes how the partition (17) features.
2. The second one is that C_1 can be written as a linear combination of C_2 (*i.e.*, $C_1 = KC_2$ with $K \in \Re^{s \times m}$)).
3. The third property is that the non-trivial solution of (18) admits at least s columns of N to be arbitrary chosen.

Once m and s are fixed by the property (1) and, considering the other two properties (2) and (3), a non-trivial solution of (18) can be established as follows:

Let N be split as $N = [N_1 \vdots N_2]$ with $N_1 \in \Re^{s \times s}$, $N_2 \in \Re^{s \times m}$ and (18) be consequently rewritten as $N_1 C_1 + N_2 C_2 = 0$. Then, due to the properties (2) and (3), the simplest solution of (18) is that N_1 can be arbitrarily chosen to compute N_2 as $N_2 = -N_1 C_1 C_2^{\S}$ where $C_2^{\S}$ is a generalized pseudo-inverse of C_2 satisfying $C_2 C_2^{\S} C_2 = C_2$. Notice, however, that in general $C_2^{\S}$ is not necessarily unique (this fact will be used later for stability analysis). Thus, under hypotheses $H1a - d$ and $H2$, the following system [13]:

$$
\begin{aligned}
\dot{\hat{w}}(t) &= W(t)\,\hat{w}(t) + X(t)\,x_2(t) + Nb(t) \\
\hat{w}(0) &= N\hat{x}(0) \\
\hat{x}_1(t) &= N_1^{-1}\left(\hat{w}(t) - N_2 x_2(t)\right)
\end{aligned}
\tag{19}
$$

with

$$
\begin{aligned}
W(t) &= \left(N_1 A_{11}(t) + N_2 A_{21}(t)\right) N_1^{-1} \\
X(t) &= N_1 A_{12}(t) + N_2 A_{22}(t) - W(t) N_2
\end{aligned}
$$

is an asymptotic nonlinear observer for the nonlinear time-varying model (16). Notice that system (19) requires the full knowledge of $A(t)$ and $b(t)$, which is ensured by hypotheses $H1a$-b. Notice also that, due to hypotheses $H1c$-d and $H2$, the system (19) effectively permits the exact cancellation of the non-linear terms contained in $f(x(t), t)$.

Remark 1

- It should be noticed that a choice of s columns of N other than N_1 is also possible and that, whatever the choice, the existence of N_1^{-1} is necessary for rebuilding the unmeasured states. However, without loss of generality, it will be assumed that $N_1 = kI_s$ where k is an arbitrary, real and positive constant parameter. Other reason for this choice will be detailed below.

- In order to eliminate the dependence of $f(x(t), t)$, Chen [13], distinguishes between two cases: $c \leq r$, and b) $c < r$. For each case, different solutions of (18) are proposed. In this chapter, by using $C_2^\S$ in the solution of (18), both cases are covered.
- A particular case of the asymptotic observer (19) occurs when $A(t)$ is proportional to the identity matrix. Bastin and Dochain [6] have shown that in this particular case, it is possible to make another(s) partition(s) $x(t) = [x_a^T(t) \vdots x_b^T(t)]^T$ (where the unmeasured and the measured states are arbitrarily distributed between $x_a(t)$ and $x_b(t)$); and, thus, to propose another linear combination of the state $w(t) = N_a x_a(t) + N_b x_b(t) =$ $N_1 x_1(t) + N_2 x_2(t)$ such that $N'C = 0$, $N' = [N_a \vdots N_b]^T$. This allows $\dot{w}(t)$ to be fully independent of the original state (which is possible because of the exact cancellation of the matrix $X(t)$ in (19)). Then, N' may be used to cancel the non-linear terms in the $w(t)$ dynamics and N may be used to rebuilding the unmeasured state exactly as in (4). Furthermore, in such a case, a matrix N_1 only left invertible is sufficient. Of course, this allows a more flexible design of the asymptotic observer. Nevertheless, in a more general case, when $A(t)$ is not proportional to the identity matrix, any partition of the state other than $x(t) = \left[x_1^T(t) \vdots x_2^T(t) \right]^T$ would imply the presence of the unmeasured states in the dynamics of $w(t)$, because of the no cancellation of $X(t)$ in (19), which is, of course, unacceptable.

4.2   Stability of the Asymptotic Observer

As it was mentioned earlier, Bastin and Dochain [6] have provided sufficient conditions for the stability of (19) when $A(t)$ is proportional to the identity matrix. In such a case, the stability of (19) depends exclusively upon the properties of $A(t)$. It was also previously established that when A is not proportional to the identity matrix and it has a defined structure but it is constant, Chen, [13], provided sufficient conditions for the stability of (19). Then, in this section an alternative issue for the stability of (19) is proposed when $A(t)$ is both not proportional to the identity matrix and time-depending.

On the one hand, a property called *cooperativity* will be used. This property must hold upon the dynamics of the observation error associated to (19). The cooperative system theory enables to compare several solutions of a differential equation. More particularly, if a considered system $\dot{\xi} = f(\xi, t)$ is cooperative, then it is possible to show that given two different initial conditions defined term by term as $\xi_1(0) \leq \xi_2(0)$; then, solutions to this system will be obtained in such a way that $\xi_1(t) \leq \xi_2(t)$, where ξ_1 and ξ_2 are the solutions of the differential equations system with the initial conditions $\xi_1(0)$ and $\xi_2(0)$, respectively. This is exactly the same result established previously in the case of simple mono-biomass/mono-substrate systems. With regard to this property the following lemma is recalled.

Lemma 1. *[58]: A system* $\dot{\zeta} = f(\zeta, t)$ *is said to be cooperative if* $\frac{\partial f_i(\zeta, t)}{\partial \zeta_j} \geq 0, \forall i \neq j$, *which implies that if* $\zeta(0) \geq 0$ *then* $\zeta(t) \geq 0, \forall t \geq 0$.

Notice that this condition means simply that the Jacobian matrix must be a Metzler matrix. Notice also that this condition fulfil the one proposed in [13] that has been evoked before. The reader will note that any system of order one is cooperative; as well as the asymptotic observer developed in [6] insofar as the matrix A is diagonal.

On the other hand, notice that in general, as has been stated above, $C_2^\S$ may not be unique. The only case when $C_2^\S$ is unique is when C_2 is a full rank square matrix and then $C_2^\S$ becomes C_2^{-1}. In such a case, the stability properties of (19) are strictly governed by the process system operating conditions. Moreover, if $C_2^\S$ is not unique, the stability properties of (19) are a function of both, the operational limits and a suitable choice of $C_2^\S$.

Now, let us consider the matrix

$$W_e(t) = N_1^{-1} W(t) N_1 = A_{11}(t) - C_1 C_2^\S A_{21}(t) \tag{20}$$

which is associated to the dynamics of the observation error of (19). Notice that under the hypothesis *H1e* it is possible to compute two constant matrices, W_e^- and W_e^+, such that $W_e^- \leq W_e(t) \leq W_e^+ \forall t \geq 0$. Then, the following hypotheses are introduced.

Hypotheses H3:

a) $W_{e,ij}^- \geq 0, \forall i \neq j$.
b) W_e^- and W_e^+ are Hurwitz stable.

Proposition 1. *Whatever* $C_2^\S$ *is (i.e.,* $C_2^\S$ *is a generalized pseudo-inverse (no unique solution) or* $C_2 = C_2^{-1}$ *(unique solution)), the observer (19) is asymptotically stable if the hypotheses H3 hold.*

Proof. Firstly, with reference to the previous lemma, the first condition of this proposition simply states that W_e^- and thus, $W_e(t)$ and W_e^+ are cooperative. Secondly, let $e(t) = \hat{x}_1(t) - x_1(t)$ if $\hat{x}(0) - x(0) \geq 0$ or $e(t) = x_1(t) - \hat{x}_1(t)$ if $\hat{x}(0) - x(0) \leq 0$ be the observation error associated to (19). It is easy to verify that e follows the dynamics: $\dot{e}(t) = W_e(t)e(t)$. Then, due to the cooperativity of $W_e(t)$, $W_e(t)^-$ and $W_e(t)^+$, $e(t)$ is bounded by $e_1(t)$, solution of $\dot{e}_1(t) = W_e^- e_1(t)$, $e_1(0) = e(0)$, and $e_2(t)$, solution of $\dot{e}_2(t) = W_e^+ e_2(t)$, $e_2(0) = e(0)$ (*i.e.,* $e_1(t) \leq e(t) \leq e_2(t)$, $\forall t \geq 0$). In addition, if W_e^- and W_e^+ are Hurwitz stable, then $e(t)$ necessarily converges asymptotically to zero as $t \to \infty$ for any $e(0)$. Therefore, $\hat{x}_1(t)$ also converges asymptotically towards $x_1(t)$ as $t \to \infty$ for any initial conditions.

Remark 2. Besides the role that $C_2^\S$ plays in the stability of the asymptotic observer, notice that the structure of the matrix associated with the error estimation, $W_e(t)$, suggests that $C_2^\S$ could also play an important role on the tuning of the asymptotic observer. Indeed, if $C_2^\S$ is not unique and if some parameters other than the dilution rate(s) are present in the sub-matrix A_{21}, then $C_2^\S$ could be used for placing the desired poles in W_e^- and W_e^+ independently of the process operating conditions. However, this problem is beyond the framework of this chapter.

5   The Interval Observer

By generalizing the reasoning that has been previously presented in the case mono-biomass/mono-substrate, it is relatively easy to show that one can surround the unmeasured states of (16), *i.e.*, to rebuild bounds on the unmeasured states provided that adequate initial conditions are chosen. The robustification of this type of asymptotic observers (those originally presented in [6]), in the case of diagonal A, was in fact developed and relatively easy established by Rapaport and Harmand [53], [55]. However, when the matrix A of the model (16) is not diagonal, the problem is much less obvious because the associated asymptotic observer is written this time in a different form (as it has been underlined in the preceding section). It is then necessary to recall again the cooperative systems theory to synthesize an interval observer in a more general class of systems [3], [33], [34], [46].

5.1   Design of the Interval Observer

The theory of interval observers first introduced by Rapaport *et al.*, [35], [55], establishes that, a necessary condition for designing such interval observers is that a known-inputs observer exists (*i.e.*, any observer that can be derived if $b(t)$ is known). If such an observer exists and if $b(t)$ is unknown (*i.e.*, only lower and upper bounds are known), the structure of this observer may be used to build an interval observer. In this section, this first requirement is cover by choosing an asymptotic observer as a basis for the interval structure. Indeed, in addition to be a known-inputs observer, the asymptotic observer has the property to be robust in the face of uncertainties on nonlinearities (*i.e.*, it permits the exact cancellation of the non-linear terms).

On the other hand, the main characteristic of interval observers is the use of the aforementioned cooperativity property, which must hold on the estimation error dynamics. Now, let hypotheses *H1f-g* be verified. That means that, on the one hand, some bounds are now available on the initial conditions and, on the other hand, $b(t)$ is considered in the following as unmeasured, but some lower and upper bounds — possibly time varying — are known. In such a situation, notice that model (16) may not be observable. Consequently, it may not possible to design an asymptotic observer such as (19). Nevertheless, its basic exponentially stable structure and its property of being independent of the nonlinearities may be used. The main idea developed in the following is to design a set-valued observer in order to build guaranteed intervals for the unmeasured variables instead of estimating them precisely. Then, let us consider the following proposition.

Proposition 2. *Consider the following dynamical system:*

For the upper bound:
$$\dot{w}^+(t) = W(t)\,w^+(t) + X(t)\,x_2(t) + Mv^+(t)$$
$$w^+(0) = Nx^+(0)$$
$$\widehat{x}_1^+(t) = N_1^{-1}\left(w^+(t) - N_2x_2(t)\right)$$

(21)

For the lower bound:
$$\dot{w}^-(t) = W(t)\,w^-(t) + X(t)\,x_2(t) + Mv^-(t)$$
$$w^-(0) = Nx^-(0)$$
$$\widehat{x}_1^-(t) = N_1^{-1}\left(w^-(t) - N_2x_2(t)\right)$$

with $v^\pm(t) = \left[b^\pm(t) \quad \frac{1}{2}\left(b_2^+(t) + b_2^-(t)\right) \quad \pm\frac{1}{2}\left(b_2^+(t) - b_2^-(t)\right)\right]^T$, $\tilde{N}_2 = [\lvert N_{2,ij} \rvert]$ *and* $M = \left[N_1 \vdots N_2 \vdots \tilde{N}_2\right]$.

If $W_e(t) = N_1^{-1}W(t)N_1$ *is cooperative then, under hypotheses H1-H3, the dynamical system (21) is a robust stable interval observer for the system (16) and it guarantees that* $x_1^-(t) \leq x_1(t) \leq x_1^+(t)$ *given* $.\, x^-(0) \leq x(0) \leq x^+(0)$.

Proof. Let $e^+ = x_1^+ - x_1$ and $e^- = x_1 - x_1^-$ be the observation errors associated to (21) which are related to the unmeasured state variables for the upper and lower bounds, respectively. For simplicity, the notation e^* is sused here to represent any of the errors e^+ or e^- since their dynamics have the same mathematical structure. Then, it is straightforward to verify that:

$$\dot{e}^* = W(t)\,e^* + V^*$$

(22)

with $V^* = N_1^{-1}\left(Mv^+ - Nb\right)$ in the upper bound case and $V^* = N_1^{-1}\left(Nb - Mv^-\right)$ in the lower bound case and where $W_e(t)$ is stable due to hypothesis *H3*. It is obvious (when considering the definitions of e^+ and e^-) that if hypothesis *H1f* holds, then $e^*(0) \geq 0$. Thereafter, the positivity of V^* is ensured by hypothesis *H1g* and the choice of $N_1 = kI_s$ as established in the *Remark 1*. Thus, from hypotheses *H3* and the application of *Lemma 1*, the system (22) is cooperative and stable. Therefore, it is guaranteed that $e^* \geq 0, \forall t \geq 0$ and thus, $x_1^-(t) \leq x_1(t) \leq x_1^+(t), \forall t \geq 0$.

However, it should be noticed that normally the interval observer (21) cannot be tuned. In fact, since the interval observer is structurally based on the asymptotic version, it has the same drawback, *i.e.*, when the sub-matrix A_{21} is identically null, the convergence rate will depend exclusively on the process operating conditions. It should be also taken into account that, at the steady state, the interval amplitude obtained for each estimated state is a function of the corresponding interval amplitude of the process input. Large intervals considered on the process inputs will produce large intervals of the estimated states. Thus, in practical situations concerning the hypothesis *H1g*, it is then desirable to consider smaller intervals of the process inputs as much as possible. Recent studies for the synthesis of interval exponential observers are being undertaken in this direction (see for example [3], [46], [54]).

6   A Study Case: A Wastewater Treatment Process

Anaerobic Digestion (AD) is a series of multi-substrate multi-organism biological processes that take place in the absence of oxygen and by which organic matter (expressed as COD, the Chemical Oxygen Demand) is decomposed and converted into biogas, a mixture of mainly carbon dioxide and methane, microbial biomass and residual organic matter [19]. Several advantages are recognized to AD processes when used in WWTP's: high capacity to treat slowly degradable substrates at high concentrations, very low sludge production, potential for production of valuable intermediate metabolites, low energy requirements and the possibility for energy recovery through methane combustion. AD is indeed one of the most promising options for delivery of alternative renewable energy carriers, such as hydrogen, through conversion of methane, direct production of hydrogen, or conversion of by-product streams. However, despite these large interests and few thousands commercial installations refereed world-wide [65], many industries are still reluctant to use AD processes, probably because of the counterpart of their efficiency: they can become unstable under some circumstances. Hence, actual research aims not only at extending the potential of anaerobic digestion [69], but also to optimize AD processes and increase their robustness towards disturbances [67]. The design of efficient state estimators clearly goes in these two last directions since instrumentation is usually scarce at industrial scale.

6.1   The Anaerobic Digester Model

In the following, a model of an anaerobic digestion process carried out in a continuous fixed bed reactor for the treatment of industrial wine distillery vinasses is considered [10]:

$$
\begin{aligned}
\dot{X}_1 &= (\mu_1 - \alpha D)X_1 \\
\dot{X}_2 &= (\mu_2 - \alpha D)X_2 \\
\dot{Z} &= D(Z^i - Z) \\
\dot{S}_1 &= D(S_1^i - S_1) - k_1\mu_1 X_1 \\
\dot{S}_2 &= D(S_2^i - S_2) + k_2\mu_1 X_1 - k_3\mu_2 X_2 \\
\dot{C}_{TI} &= D(C_{TI}^i - C_{TI}) + k_7\left(k_8 P_{CO_2} + Z - C_{TI} - S_2\right) + k_4\mu_1 X_1 + k_5\mu_2 X_2
\end{aligned}
\tag{23}
$$

where X_1, X_2, S_1, S_2 and C_{TI} are respectively, the concentrations of acidogenic bacteria, methanogenic bacteria, COD, Volatile Fatty Acids (VFA) and total inorganic carbon. The variable Z is linked to the total alkalinity and represents the sum of strong ions in the medium. At chemical equilibrium at pH≈7 it can be reasonably assumed that $Z = S_2 + HCO_3^-$ where HCO_3^- is the bicarbonate ion concentration. It is supposed that no other ion significantly influences Z. The variable $D = D(t)$ is the dilution rate. In all cases, the upper index i indicates "influent concentration".For a fixed bed reactor, the biomass is attached on a support. It is therefore not affected by the dilution rate as in a continuous reactor. Nevertheless some bacteria are detached by the liquid flow. This effect may be incorporated in the hydrodynamic modeling of the biomass if it is considered that only a fraction α of the biomass is affected by the dilution. The parameter $\alpha(0 \leq \alpha \leq 1)$ reflects this process heterogeneity: $\alpha = 1$ corresponds to an ideal fixed bed reactor, whereas $\alpha = 0$ corresponds to an ideal CSTR.

As in any other mass balance model of bioprocesses, a strongly nonlinear kinetic behavior is present due to the reaction rates. These rates are given by:

$$\mu_1 = \frac{\mu_{1\,max}S_1}{S_1 + K_{S_1}} \quad \text{and} \quad \mu_2 = \frac{\mu_0 S_2}{S_2 + K_{S_2} + \frac{S_2^2}{K_{I_2}}} \tag{24}$$

The CO_2 partial pressure P_{CO_2} is expressed as function of the states as:

$$P_{CO_2} = \frac{\Phi - \sqrt{\Phi^2 - 4k_8 P_T\,[CO_2]}}{2k_8} \tag{25}$$

with $\Phi = [CO_2] + k_8 P_T + \frac{k_6 \mu_2 X_2}{k_7}$ and $[CO_2] = C_{TI} + S_2 - Z$.

Definition of parameters and their values are listed in Table 4. For simplicity in the notation, the time dependence of state variables as well as the dilution

Table 4. Model parameters from [8]

Parameter	Meaning	Value
k_1	Yield coefficient for COD degradation	12.1 Kg COD/Kg X_1
k_2	Yield coefficient for fatty acid production	181.2 mol VFA/Kg X_1
k_3	Yield coefficient for fatty acid consumption	1640 mol VFA/Kg X_2
k_4	Yield coefficient for CO_2 production due to X_1	169 mol CO_2/Kg X_1
k_5	Yield coefficient for CO_2 production due to X_2	273 mol CO_2/Kg X_2
k_6	Yield coefficient for CH_4 production	1804 mol CH_4/Kg X_2
k_7	Liquid/gas transfer rate	200 d^{-1}
k_8	Henry's constant	0.2201 mol CO_2/m^{3}KPa
α	Proportion of dilution rate for bacteria	0.5 (dimensionless)
μ_{max_1}	Maximum acidogenic biomass growth rate	1.25 d^{-1}
μ_0	Maximum methanogenic biomass growth rate	0.85 d^{-1}
K_{S_1}	Saturation parameter associated with S_1	7.65 Kg COD/m^3
K_{S_2}	Saturation parameter associated with S_2	18 (mol VFA/m^3)$^{1/2}$
K_{I_2}	Inhibition constant associated with S_2	25 mol VFA/m^3
P_T	Total pressure in the reactor	105.72 KPa

rate D and the CO_2 partial pressure P_{CO_2} has been omitted in equations (23) to (25).

The nonlinear observers developed in the previous sections may be then applied to the dynamical process model (23) defining the state vector in the following way:

$$\xi_1 = X_1 \qquad \xi_3 = C_{TI} \qquad \xi_5 = S_1$$
$$\xi_2 = X_2 \qquad \xi_4 = Z \qquad \xi_6 = S_2 \tag{26}$$

Thus, the model (23) takes the following matrix form:

$$
\begin{bmatrix} \dot{\xi_1} \\ \dot{\xi_2} \\ \dot{\xi_3} \\ \dot{\xi_4} \\ \dot{\xi_5} \\ \dot{\xi_6} \end{bmatrix}
=
\begin{bmatrix} 1 & 0 \\ 0 & 1 \\ k_4 & k_5 \\ 0 & 0 \\ -k_1 & 0 \\ k_2 & -k_3 \end{bmatrix}
\begin{bmatrix} \mu_1\xi_1 \\ \mu_2\xi_2 \end{bmatrix}
+
\begin{bmatrix}
-\alpha D & 0 & 0 & 0 & 0 & 0 \\
0 & -\alpha D & 0 & 0 & 0 & 0 \\
0 & 0 & -(D+k_7) & k_7 & 0 & -k_7 \\
0 & 0 & 0 & -D & 0 & 0 \\
0 & 0 & 0 & 0 & -D & 0 \\
0 & 0 & 0 & 0 & 0 & -D
\end{bmatrix}
\begin{bmatrix} \xi_1 \\ \xi_2 \\ \xi_3 \\ \xi_4 \\ \xi_5 \\ \xi_6 \end{bmatrix}
$$
$$
+
\begin{bmatrix}
D\xi_1^i \\
D\xi_2^i \\
D\xi_3^i + k_7 k_8 P_{CO_2} \\
D\xi_4^i \\
D\xi_5^i \\
D\xi_6^i
\end{bmatrix}
\tag{27}
$$

or simply $\dot{\xi} = Cf(\xi(t),t) + A(t)\xi(t) + b(t)$ which is identical to model (16) with $x(t) = \xi(t)$.

6.2 Design of the Observers

It is then straightforward to apply the observers (19) and (21) described the in previous sections. From (27), it is clear that $rankC = 2$. Hence, from hypothesis H2, only a minimum of two measurements is required to reconstruct the state of the AD process. So, in the following, the two substrate concentrations S_1 and S_2, will be used to estimate X_1, X_2, C_{TI} and Z.

Thus, A is split in the following form:

$$
A(t) =
\begin{bmatrix} A_{11} & \vdots & A_{12} \\ \cdots & \cdot & \cdots \\ A_{21} & \vdots & A_{22} \end{bmatrix}
=
\begin{bmatrix}
-\alpha D & 0 & 0 & 0 & \vdots & 0 & 0 \\
0 & -\alpha D & 0 & 0 & \vdots & 0 & 0 \\
0 & 0 & -(D+k_7) & k_7 & \vdots & 0 & -k_7 \\
0 & 0 & 0 & -D & \vdots & 0 & 0 \\
\cdots & \cdots & \cdots & \cdots & \cdot & \cdots & \cdots \\
0 & 0 & 0 & 0 & \vdots & -D & 0 \\
0 & 0 & 0 & 0 & \vdots & 0 & -D
\end{bmatrix}
\tag{28}
$$

Notice that in (28) the only time depending variable is the dilution rate D. In the four examples developed in this section, D fluctuates between $0.05 \leq D(t) \leq 1.15$. (see Figures 14, 27, 39 and 53). Then, in agreement with hypothesis $H1e$, $A^- = A \mid_{D=1.15}$ and $A^+ = A \mid_{D=0.05}$.

Without loss of generality and in agreement with the *Remark R1*, N_1 can be arbitrarily chosen as $N_1 = I_4$. Thus, N takes the form:

$$N = \left[N_1 \vdots N_2 \right] = \frac{1}{k_1 k_3} \begin{bmatrix} k_1 k_3 & 0 & 0 & 0 & \vdots & k_3 & 0 \\ 0 & k_1 k_3 & 0 & 0 & \vdots & k_2 & k_1 \\ 0 & 0 & k_1 k_3 & 0 & \vdots & (k_3 k_4 + k_2 k_5) & k_1 k_5 \\ 0 & 0 & 0 & k_1 k_3 & \vdots & 0 & 0 \end{bmatrix} \tag{29}$$

It is easy to verify that $W(t)$ and $W_e(t)$ are described by:

$$W(t) = W_e(t) = \begin{bmatrix} -\alpha D & 0 & 0 & 0 \\ 0 & -\alpha D & 0 & 0 \\ 0 & 0 & -(D + k_7) & k_7 \\ 0 & 0 & 0 & -D \end{bmatrix} \tag{30}$$

Notice that hypothesis $H3a$ is automatically fulfilled. Furthermore, due to hypothesis $H1e$ it is possible to find two matrices W_e^- and W_e^+ such that $W_e^- = W_e \mid_{D=1.15}$ and $W_e^+ = W_e \mid_{D=0.05}$, respectively. It is obvious that hypothesis $H3b$ is also fulfilled. Then, clearly *Lemma 1* and propositions $P1$ and $P2$ hold. Finally, it is easy to verify that $X(t)$ and M have the following structures:

$$X(t) = \frac{1}{k_1 k_3} \begin{bmatrix} (\alpha - 1) k_3 D(t) & 0 \\ (\alpha - 1) k_2 D(t) & (\alpha - 1) k_1 D(t) \\ (k_3 k_4 + k_2 k_5) k_7 & k_1 k_7 (k_5 - k_3) \\ 0 & 0 \end{bmatrix} \tag{31}$$

$$M = \frac{1}{k_1 k_3} \begin{bmatrix} k_1 k_3 & 0 & 0 & 0 & k_3 & 0 & k_3 & 0 \\ 0 & k_1 k_3 & 0 & 0 & k_2 & k_1 & k_2 & k_1 \\ 0 & 0 & k_1 k_3 & 0 & (k_3 k_4 + k_2 k_5) & k_1 k_5 & (k_3 k_4 + k_2 k_5) & k_1 k_5 \\ 0 & 0 & 0 & k_1 k_3 & 0 & 0 & 0 & 0 \end{bmatrix} \tag{32}$$

Therefore, using (29) to (32), the observers are exactly set like in equations (19) and (21) respectively.

Example 1: Asymptotic observer implementation

This simulation example was attempted to cover a large range of situations and operating conditions that are effectively met in practice. Simulations were carried out using the parameter values reported in Table 4 for model (23).

They were carried out over a 50 day period at different dilution rates and at different input substrate concentrations. In order to do these simulations more realistic, small fluctuations as well as drastic step perturbations were introduced alternately in the input concentrations (see Figures 15 to 18) and in the dilution rate (see Figure 14). The input variables X_1^i and X_2^i were assumed negligible. The measurements of S_1, S_2 and P_{CO_2} was calculated directly from the model (see Figures 19, 20 and 21).

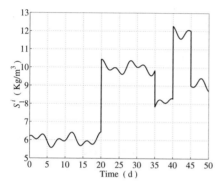

Fig. 14. Influent S_1 concentration

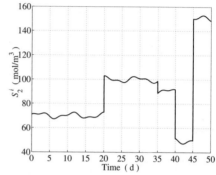

Fig. 15. Influent S_2 concentration

Figures 22, 23, 24 and 25 depict the estimation results (dashed line - -)of unmeasured states; X_1, X_2, Z and C_{TI}, respectively in which two different sets of initial conditions were used (see Table 5). In these figures, it is compared the response of the observer with the "measurements" provided by the model simulations. Notice that the observer exhibited good convergence properties even at the non-steady state despite the fact that the convergence rate depended largely on the dilution rate. One can also see in these figures the excellent stability properties for all the initial conditions tested.

Example 2: Application of the asymptotic observer in an actual anaerobic digestion plant

The asymptotic observer was tested in an experimental 1 m^3 upflow anaerobic fixed bed reactor pilot plant used for the treatment of industrial wine distillery vinasses obtained from local distilleries in the Narbonne area (France) (see Figure 26). These experimental runs were carried out over a 35 day period. Measurements of the dilution rate as well as the S_1 and S_2 concentrations and the partial CO$_2$ pressure were performed on-line (see Figures 27 to 30). The measurements of S_1^i, S_2^i, Z^i and C_{TI}^i were obtained from off-line data and they were assumed constant between assays (see Figures 31 to 34). As in the simulation application, the influent biomass concentrations, X_1^i and X_2^i were assumed negligible.

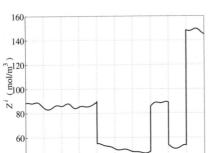

Fig. 16. Influent Z concentration

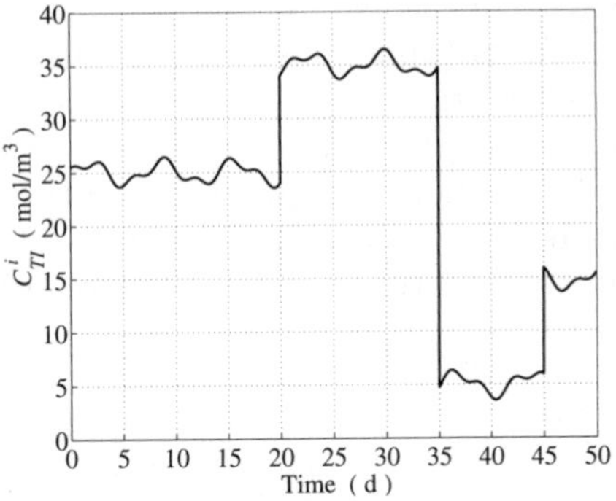

Fig. 17. Influent C_{TI} concentration

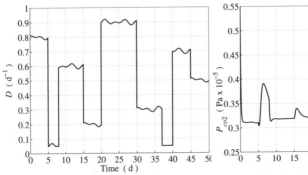

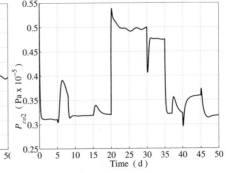

Fig. 18. Dilution rate

Fig. 19. Measurements of partial CO_2 pressure

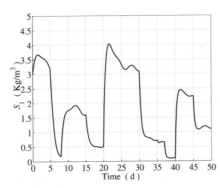

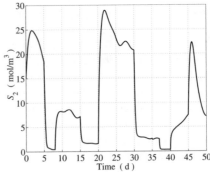

Fig. 20. Effluent measurements of S_1 concentration

Fig. 21. Effluent measurements of S_2 concentration

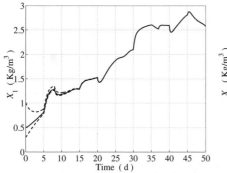

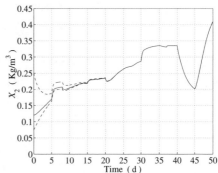

Fig. 22. Estimate of the acidogenic bacteria concentration

Fig. 23. Estimate of the methanogenic bacteria concentration

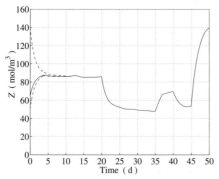

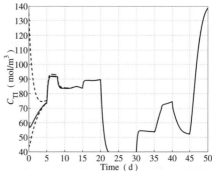

Fig. 24. Estimate of the concentration of Z

Fig. 25. Estimate of the total inorganic carbon concentration

Estimation results for the unmeasured states are depicted in Figures 35 to 38. Notice that these experimental runs were perturbed with real experimental noise in the input concentrations, the dilution rate and the measured states. In despite of these perturbations the asymptotic observer showed excellent convergence and stability properties as it was able to estimate, with a

Table 5. Initial conditions for simulation runs

	$X_1(0)$ (Kg/m^3)	$X_2(0)$ (Kg/m^3)	$Z(0)$ (mol/m^3)	$C_{TI}(0)$ (mol/m^3)	$S_1(0)$ (Kg/m^3)	$S_2(0)$ (mol/m^3)
$\{\xi(0)\}_{model}$	0.5	0.12	65	60	3	15
$\{\xi(0)\}_1$	1	0.25	140	135	3	15
$\{\xi(0)\}_2$	0.3	0.075	50	45	3	15

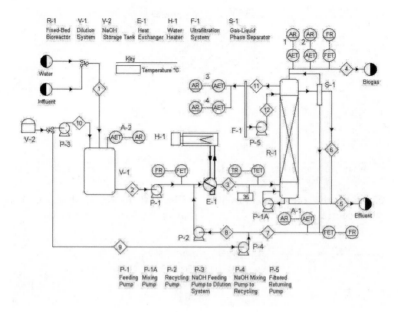

Fig. 26. Schematic representation of the anaerobic fixed bed bioreactor

reasonably good accuracy, the unmeasured states exhibiting a non-divergent behavior.

Example 3: Interval observer implementation

The interval observer implementation was also implemented through numerical simulations which were carried out over a 100 days period at different dilution rates and at different input substrate concentrations but, in this case, it was considered that input concentrations were unknown and only guaranteed intervals of these inputs were known (see Figures 39 to 42).

The "real input concentrations" shown in these figures were only used to validate hypothesis *H1g* and to simulate the model (6) from which the measurements S_1, S_2 and P_{CO_2} were taken directly. These "measurements" are reported in Figures 43 to 45. The variables C_{TI}^i, X_1^i and X_2^i were supposed to be negligible. As in the previous numerical example, small fluctuations as well as drastic step perturbations were alternatively introduced in the dilution rate and in the input concentrations to resemble actual industrial operating conditions.

Inspecting Figures 46 to 49 one can see that the interval observer exhibited fast convergence properties even at the non-steady state. In all cases, the interval observer estimated with precision the bounds within which the real state evolves.

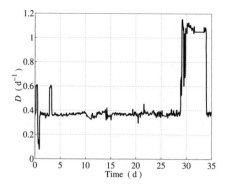

Fig. 27. Experimental on-line measurements of D

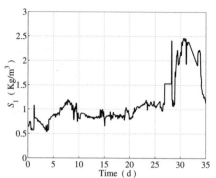

Fig. 28. Experimental on-line effluent S_1 concentration

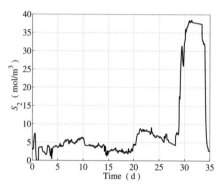

Fig. 29. Experimental on-line effluent S_2 concentration

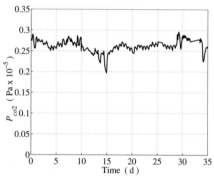

Fig. 30. Experimental on-line P_{CO_2} measurements

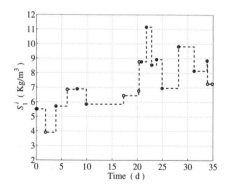

Fig. 31. Experimental influent S_1 concentration

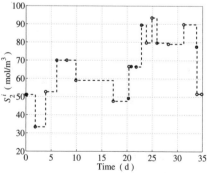

Fig. 32. Experimental influent S_2 concentration

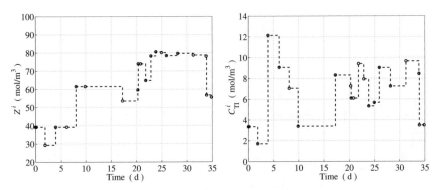

Fig. 33. Experimental influent Z concentration

Fig. 34. Experimental influent C_{TI} concentration

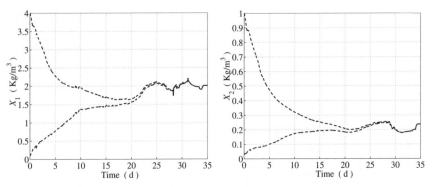

Fig. 35. Estimate of the acidogenic bacteria concentration

Fig. 36. Estimate of the methanogenic bacteria concentration

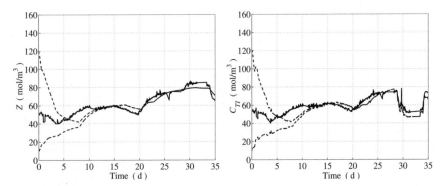

Fig. 37. Estimate of the concentration of Z: (- -) estimated state, (—) experimental on-line measurements

Fig. 38. Estimate of the total inorganic carbon concentration: (- -) estimated state, (—) experimental on-line measurements

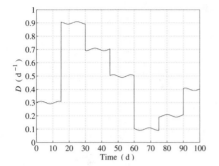

Fig. 39. Dilution rate

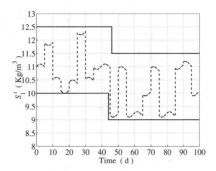

Fig. 40. Influent S_1 concentration: (—) upper and lower bounds, (- -) input concentration

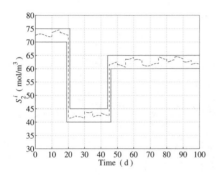

Fig. 41. Influent S_2 concentration: (—) upper and lower bounds, (- -) input concentration

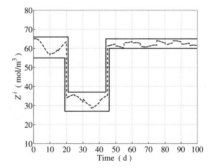

Fig. 42. Influent Z concentration: (—) upper and lower bounds, (- -) input concentration

Example 4: Application of the interval observer in an actual anaerobic digestion plant

The interval observer was finally tested on-line in a series of experimental runs conducted over a 35 days period in the anaerobic fixed bed reactor pilot plant. Parameters used are listed in Table 4. For the experimental implementation of the observer the influent concentrations were considered unknown and only a known boundary region was supplied to the interval observer. Figures 50 to 52 depict the corresponding bounded intervals for S_1^i, S_2^i and Z^i, respectively.

For comparison purposes and to show that hypothesis H1g holds, the off-line readings (taken at irregular time periods) of these inputs have been added. It is worth mentioning that this data set was not used in the on-line implementation of the interval observer. C_{TI}^i, X_1^i and X_2^i were again assumed negligible. Actual experimental data for the dilution rate, $D(t)$, as well as the partial CO_2 pressure, which were measured on-line in the pilot plant, have been also introduced in the interval observer calculations . These on-line readings are

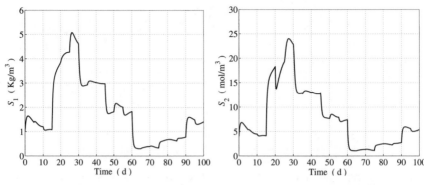

Fig. 43. Model simulation of the S_1 concentration in the reactor (used as on-line measurement by the observer)

Fig. 44. Model simulation of the S_2 concentration in the reactor (used as on-line measurement by the observer)

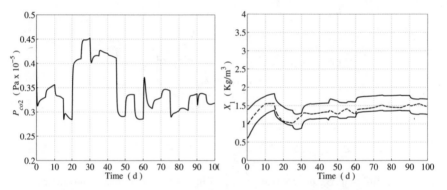

Fig. 45. Model simulation of the partial CO_2 pressure (used as on-line measurement by the observer)

Fig. 46. Estimation of the acidogenic bacteria concentration: (—) upper and lower estimated states, (- -) predictions of the model

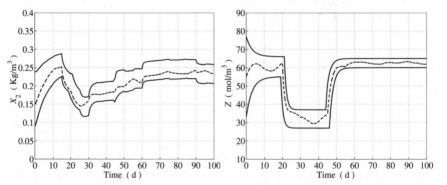

Fig. 47. Estimation of the methanogenic bacteria concentration: (—) upper and lower estimated states, (- -) predictions of the model

Fig. 48. Estimation of the concentration of Z: (—) upper and lower estimated states, (- -) predictions of the model

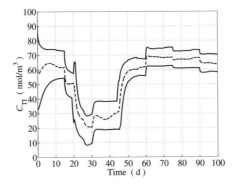

Fig. 49. Estimation of the total inorganic carbon concentration: (—) upper and lower estimated states, (- -) predictions of the model

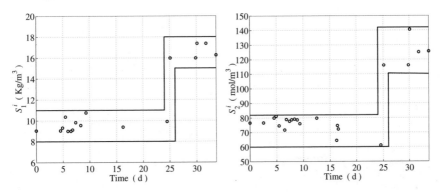

Fig. 50. S_1^i input concentration: (—) bounds, (o) off-line readings

Fig. 51. S_2^i input concentration: (—) bounds, (o) off-line readings

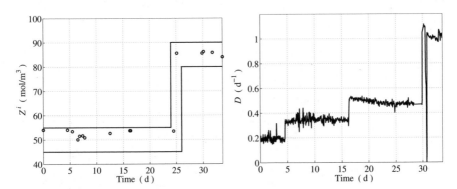

Fig. 52. Z^i input concentration: (—) bounds, (o) off-line readings

Fig. 53. On-line dilution rate measurements

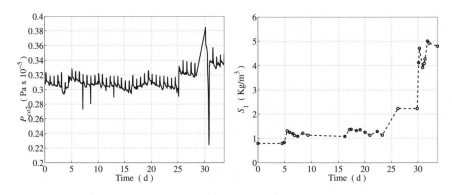

Fig. 54. On-line CO_2 partial pressure measurements

Fig. 55. On-line effluent COD concentration

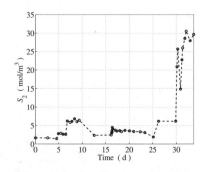

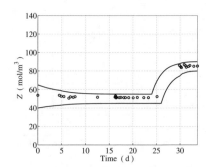

Fig. 56. On-line effluent VFA concentration

Fig. 57. Alkalinity concentration: (—) estimated upper and lower bounds and (o) off-line readings

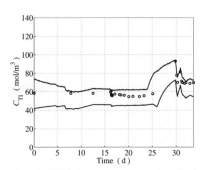

Fig. 58. Total inorganic carbon concentration: (—) estimated upper and lower bounds and (o) off-line readings

depicted in Figures 53 and 54, respectively. Substrate concentration measurements, S_1 and S_2 (shown in Figures 55 and 56) were available from infrequent on-line assays. For the interval observer implementation, these measurements were assumed to remain constant between the assays.

Figures 57 and 58 shows the estimation results for the intervals of the unmeasured states C_{TI} and Z. Notice how the interval bounds estimated by the interval observer envelop correctly these unmeasured states. For all the other unmeasured states, notice that although the interval observer design did not allow us to tune the convergence rate, the interval observer showed excellent robustness and stability properties and provided satisfactory estimation results in the event of highly corrupted measurements and operational failures. Notice in particular, the robustness of the interval observer around day 25 when the inlet concentrations drastically increased and when a major disturbance occurred at day 31, due to an operational failure, resulting in a rapid fall of both, the dilution rate (which actually fell to zero) and the substrate concentration readings. Off-line readings of C_{TI} and Z (not used in the state estimation calculations) were also added to validate the proposed interval observer design (see Figures 57 and 58). It should be noticed that the compromise between the convergence rate and robustness was not fully achieved until the estimation error dynamics reached the steady state.

7 Current Tendencies

Knowledge based approaches such as fuzzy logic, neural networks or multi-agents model currently constitute an important axis of research and application in bioprocesses. They have shown their usefulness particularly when one does not have an analytical model but that a certain expertise is available. Harmand and Steyer [37] have addressed that when this expertise comprises a sufficiently important know-how, approaches such as fuzzy logic will be preferred. If, on the other hand, one has only a limited experience but lays out of a rather important data base, the statistical approaches such as neural networks can be used.

There is no doubt that one of the factors limiting the implementation of the advanced control approaches is the lack of actuators and on-line sensors which can combine performance and weak costs. However, this situation can change quickly if the legislative constraints are confirmed. The sensors used for obtaining the experimental results showed in this chapter testify to these recent developments, even if their use on an industrial scale still requires to be confirmed. Let us notice however that they constitute a way of obligated passage to solve certain industrial problems, which require approaching to the biological reality [62]. However, observers clearly offer an interesting alternative when one has a sufficiently precise model of the system considered.

8   Conclusion

In this chapter two powerful nonlinear observers for bioprocesses, which are robust in the face of a complete lack of knowledge of the nonlinearities of the system as well as in the face of major uncertainties on the process inputs, have been detailed. Besides of the design features of these observers, sufficient conditions for existence, stability and convergence have been established. Particularly, the use of interval observers as a powerful tool for fault detection and diagnosis was also introduced. All the usefulness of these robust nonlinear observers was made manifest in a complete actual studio case: an anaerobic digester for the wastewater treatment of agricultural effluents. In addition, the more successful approaches for estimating parameters and state variables in bioprocesses have been briefly reviewed and compared to these two approaches. Finally, some considerations related to the immediate future of these approaches when applied to the biological WWTP in general, and to anaerobic digestion in particular, were presented and commented on.

References

[1] O. Adrot. *Diagnostic appliqué aux systèmes dont le modèle est à paramètres incertains: L'approche bornante (in French)*. PhD thesis, Institut National Polytechnique de Lorraine, France, 2000.

[2] V. Alcaraz-González, J. Harmand, A. Rapaport, J.P. Steyer, V. González-Álvarez, and C. Pelayo-Ortíz. Software sensors for uncertain wastewater treatment processes: A new approach based on interval observers. *Wat. Res.*, 36(10):2515–2524, 2002.

[3] V. Alcaraz-González, R. Salazar-Pea, V. González-Álvarez, J.L. Gouzé, and J.P. Steyer. A tunable multivariable non-linear robust observer for biological systems. *Comptes Rendus de l'Academie des Sciences Biologies (in Press)*, 2004.

[4] C. Aubrun, J. Harmand, O. Garnier, and J.P. Steyer. Fault detection filter design for an anaerobic digestion process. *Bioprocess Engineering*, 22(5):413–420, 2000.

[5] M.W. Barnett and J.F. Andrews. Expert system for anaerobic digestion process operation. *Journal of Environmental Engineering*, 118(6):949–963, 1992.

[6] G. Bastin and D. Dochain. *On-Line Estimation and Adaptive Control of Bioreactors*. Elsevier, Amsterdam, 1990.

[7] W.T. Baumann and W.J. Rugh. Feedback control of nonlinear systems byextended linearization. *IEEE Trans. Automat. Contr.*, 31(1):–, 1986.

[8] O. Bernard, D. Dochain, A. Genovesi, A. Puñal, D. Perez-Alvarino, J.P. Steyer, and J. Lema. Software sensor design for an anaerobic wastewater treatment plant. In *IFAC-International Workshop on Decision and Control in Waste Bio-Processing*, Narbonne, France, 1998.

[9] O. Bernard and J.L. Gouzé. *Automatique des bioprocédés*, chapter Estimation d'état, pages 87–120. Hermes, France, 2001.

[10] O. Bernard, Z. Hadj-Sadok, D. Dochain, A. Genovesi, and J.P. Steyer. Dynamical model development and parameter identification for anaerobic wastewater treatment process. *Biotechnol. Bioeng.*, 75(4):424–438, 2001.

[11] D. Bestle and M. Zeitz. Canonical form observer design for nonlinear timevariable systems. *Int. J. Control*, 38(2):419–431, 1983.

[12] J. Chen and R.J. Patton. *Robust model-based fault diagnosis for dynamic systems.* Kluwer Academic Publishers, 1999.

[13] L. Chen. *Modelling, identifiability and control of complex biotechnological systems.* PhD thesis, Université Catholique de Louvain, Louvain la Neuve, Belgique, 1992.

[14] D.P. Chynoweth, S.A. Svoronos, G. Lyberatos J.L. Harman, P. Pullammanappallil, and J.M. Owens. Real-time expert system control of anaerobic digestion. *Wat. Sci. Technol.*, 30(12):21–29, 1994.

[15] M. Darouach, M. Zasadzinski, and S.J. Liu. Full-order observers for linear systems with unknown inputs. *IEEE Trans. Automat. Contr.*, 39(3):606–609, 1994.

[16] X. Ding and P.M. Frank. On-line fault detection in uncertain systems using adaptive observers. *European J. of Diagnosis and Safety in Automation*, 3:9–21, 1993.

[17] D. Dochain and L. Chen. Local observability and controllability of stirred tank reactors. *J. Process Control*, 2(3):139–, 1992.

[18] D. Dochain and M. Perrier. *Advanced Instrumentation, Data Interpretation, and Control of Biotechnological Processes*, chapter Monitoring and Adaptive Control of Bioprocesses, pages 347–400. Kluwer Academic Publishers, 1998.

[19] D. Dochain and P. Vanrolleghem. *Dynamical Modelling and Estimation in Wastewater Treatment Processes.* IWA Publishing, 2001.

[20] C. Edwards and S.K. Spurgeon. On the development of discontinuous observers. *Int. J. Control*, 59(5):1211–1229, 1994.

[21] P.M. Frank and X. Ding. Survey of robust residual generation and evaluation methods in observer-based fault detection systems. *J. Process Control*, 7(6):403–424, 1997.

[22] M.T. Garrett. Instrumentation, control, and automation progress in the united states in the last 24 years. *Wat. Sci. Technol.*, 37(12):21–25, 1998.

[23] J.P. Gauthier, H. Hammouri, and S. Othman. A simple observer for nonlinear systems. *IEEE Trans. Automat. Contr.*, 37(6):875–880, 1992.

[24] J.P. Gauthier and I. Kupka. Observability and observers for nonlinear systems. *SIAM J. Control Optim.*, 34(4):975–994, 1994.

[25] A. Genovesi, J. Harmand, and J.P. Steyer. A fuzzy logic based diagnosis system for the on-line supervision of an anaerobic digestor pilot plant. *Biochem. Eng. J.*, 3:171–183, 1999.

[26] A. Genovesi, J. Harmand, and J.P. Steyer. Integrated fault detection and isolation: Application to a winery's wastewater treatment plant. *Applied Intelligence Journal (APIN)*, 13:207–224, 2000.

[27] A.M.D. Genovesi. *Détection de défauts et diagnostic en ligne de procédés biologiques d dépollution (in French).* PhD thesis, Université de Perpignan, France, 1999.

[28] J.J. Gertler. *Fault Detection and Diagnosis in Engineering Systems.* Marcel Dekker Inc., 1998.

[29] M. Gevers and G. Bastin. *Analysis and Optimization of Systems*, chapter A Stable Adaptive Observer for a Class of Nonlinear Second-order Systems, pages 143–155. Springer-Verlag, 1986.

[30] J.J. Godon, E. Zumstein, P. Dabert, F. Habouzit, and R. Moletta. Microbial 16S rDNA diversity in an anaerobic digester. *Wat. Sci. Technol.*, 36(6-7):49–55, 1997.

[31] J.L. Gouzé, O. Bernard, and Z. Hadj-Zadok. Observers with modelling uncertainties for the wastewater treatment process. In *Journées thématiques "Automatique et Environnement"*, Nancy, France, March 2000.

[32] J.L. Gouzé and V. Lemesle. A bounded error observer with adjustable rate for a class of bioreactor models. In *European Control Conference (ECC)*, Porto, Portugal, 2001.

[33] J.L. Gouzé, A. Rapaport, and Z. Hadj-Zadok. Interval observers for uncertain biological systems. *Ecological Modelling*, 133:45–56, 2000.

[34] Z. Hadj-Sadok. *Modélisation et estimation dans les bioréacteurs; prise en compte des incertitudes: application au traitement de l'eau (in French)*. PhD thesis, Université de Nice, France, 1999.

[35] Z. Hadj-Zadok, J.L. Gouzé, and A. Rapaport. State observers for uncertain models of activated sludge processes. In *IFAC-International Workshop on Decision and Control in Waste Bio-Processing*, Narbonne, France, 1998.

[36] J. Harmand. *Identification et Commande des Procédés Biologiques de Dépollution (in French)*. PhD thesis, Université de Perpignan, France, 1997.

[37] J. Harmand and J.P. Steyer. Comparison of several advanced control approaches for anaerobic digestion processes: Towards a new paradigm. In *IWA Conf. Instr., Contr., and Automat. (ICA)*, pages 647–654, Malmö, Sweden, June 2001.

[38] E. Heinzle, I.J. Dunn, and G.B. Ryhiner. Modelling and control for anaerobic wastewater treatment. In R. Aarts, M. Aynsley, J.E. Bailey, P.M. Doran, and I.J. Dunn, editors, *Bioprocess Design and Control*, volume 48 of *Advances in Biochemical Engineering/Biotechnology*, pages 79–114. Springer-Verlag, Berlín, 1993.

[39] R. Huntington. Twenty years development of ICA in a water utility. *Wat. Sci. Technol.*, 37(12):27–34, 1998.

[40] L. Jaulin, M. Kieffer, O. Didrit, and E. Walter. *Applied interval analysis with examples in parameter and state estimation robust control and robotics*. Springer-Verlag, Londres, 1991.

[41] L. Jaulin and E. Walter. Set inversion via interval analysis for nonlinear bounded-error estimation. *Automatica*, 29(4):1053–1064, 1993.

[42] M. Kieffer, L. Jaulin, and E. Walter. Guaranteed recursive nonlinear state estimation using interval analysis. In *IEEE Conf. Decision and Control (CDC)*, pages 16–18, Florida, USA, December 1998.

[43] A. Kobi, S. Nowakowski, and J. Ragot. Fault detection-isolation and control reconfiguration. *Math. Comput. Simulation*, 37:111–117, 1994.

[44] A. Kurzhanski and I. Valyi. *Ellipsoidal calculus for estimation and control*. Birkhäuser, Boston, 1997.

[45] M. Luong, J.M. Paris, D. Maquin, and J. Ragot. Observability, reliability and sensor positioning. In *AICHE Spring National Meeting*, pages 1–10, Houston, USA, 1995.

[46] A. Maloum. *Commandes et observateurs robustes pour les systèmes non-linéaires incertains: Application à la dépollution biologique (in French)*. PhD thesis, Université de Nice, France, 2001.

[47] S. Marsili-Libelli and A. Muller. Adaptive fuzzy pattern recognition in the anaerobic digestion process. *Pattern Recognition Letters*, 17(6):651–659, 1996.

[48] E.A. Misawa and J.K. Hedrick. Nonlinear observers. A state-of-the-art survey. *Trans. ASME J. Appl. Mech.*, 111:344–352, 1989.

[49] S. Nicosia, P. Tomei, and Tornanbé. Feedback control of elastic robots by pseudolinearization techniques. In *IEEE Conf. Decision and Control (CDC)*, Athens, Greece, December 1986.

[50] G. Olsson and B. Newell. *Wastewater Treatment Systems, Modelling, Diagnosis and Control*. IWA Publishing, London, 1999.

[51] S. Ploix, O. Adrot, and J. Ragot. Bounding approach to the diagnosis of a class of uncertain static systems. In *IFAC-SAFEPROCESS*, Budapest, Hungary, June 2000.

[52] P.C. Pullammanappallil, S.A. Svoronos, D.P. Chynoweth, and G. Lyberatos. Expert system for control of anaerobic digesters. *Biotechnol. Bioeng.*, 58(1):13–22, 1998.

[53] A. Rapaport. *Theory in Practice of Control and Systems*, chapter Information state and guaranteed value for a class of min-max nonlinear optimal control problems, pages 391–396. World Scientific, Singapore, 1998.

[54] A. Rapaport and J.L. Gouzé. Practical and polytopic observers for nonlinear uncertain systems. Technical Report RR-4079, INRIA Sophia Anthipolis, France, November 2000.

[55] A. Rapaport and J. Harmand. Robust regulation of a bioreactor in a highly uncertain environment. In *IFAC-International Workshop on Decision and Control in Waste Bio-Processing*, Narbonne, France, 1998.

[56] R. Seliger and P.M. Frank. Fault diagnosis by disturbance decoupled nonlinear observers. In *IEEE Conf. Decision and Control (CDC)*, volume 3, pages 2248–2253, Brighton, Englant, 1991.

[57] R.E. Skelton. *Dynamic Systems Control. Linear Systems Analysis and Synthesis*. John Wiley & Sons, New York, 1988.

[58] H.L. Smith. Monotone dynamical systems. An introduction to the theory of competitive and cooperative systems. *AMS Mathematical Surveys and Monographs*, 41:31–53, 1995.

[59] J.P. Steyer, J. Aguilar-Martin, K. Bousson, A. Charles, J.M. Evrard, L. Leyval, B. Lucas, J. Montmain, N. Rakoto-Ravalontsalama, M. Tomasena, and L. Travé-Massuyès. *Le raisonnement qualitatif*, chapter Supervision et diagnostic des procédés industriels (in French), pages 279–323. Hermes, France, 1997.

[60] J.P. Steyer, N. Bernet, P.N.L. Lens, and R. Moletta. *Environmental Technologies to Treat Sulfur Pollution: Principles and Engineering*, chapter Anaerobic treatement of sulfate rich wastewaters: Process modeling and control, pages 207–235. IWA publishing, London, 2000.

[61] J.P. Steyer, J.C. Bouvier, T. Conte, P. Gras, J. Harmand, and J.P. Delgenes. On-line measurements of COD, TOC, VFA, total and partial alkalinity in anaerobic digestion process using infra-red spectrometry. *Wat. Sci. Technol.*, 45(10):133–138, 2002.

[62] J.P. Steyer, A. Génovési, and J. Harmand. Advanced monitoring and control of anaerobic wastewater treatment plants: Fault detection and isolation. *Wat. Sci. Technol.*, 43(7):183–190, 2001.

[63] J.P. Steyer, A. Génovési, and J. Harmand. *Automatique des bioprocédés*, chapter Outils d'aide au diagnostic et détection de pannes (in French), pages 215–244. Hermes, France, 2001.

[64] J.P. Steyer, D. Roland, J.C. Bouvier, and R. Moletta. Hybrid fuzzy neural network for diagnosis - Application to the anaerobic treatment of wine distillery wastewater in a fluidized bed reactor. *Wat. Sci. Technol.*, 36(6-7):209–217, 1997.

[65] D.E. Totzke. Anaerobic treatment technology overview. Technical report, Applied Technologies Inc., USA, 1999.

[66] J.F. van Impe and G. Bastin. *Advanced Instrumentation, Data Interpretation, and Control of Biotechnological Processes*, chapter Optimal Adaptive Control of Fed-Batch Fermentation Process, pages 401–435. Kluwer Academic Publishers, 1998.

[67] J. Van Lier, A. Tilche, B.K. Ahring, H. Macarie, R. Moletta, M. Dohanyos, L.W. Hulshoff Pol, P. Lens, and W. Verstraete. New perspectives in anaerobic digestion. *Wat. Sci. Technol.*, 43(1):1–18, 2001.

[68] H.M. van Veldhuizen, M.C.M. van Loosdrecht, and J.J. Heijnen. Modelling biological phosphorus and nitrogen removal in a full scale activated sludge process. *Wat. Res.*, 33(16):3459–3468, 2001.

[69] W. Verstraete and P. Vandevivere. New and broader applications of anaerobic digestion. *Critical Reviews in Environmental Science and Technology*, 29(2):151–165, 1999.

[70] B. Walcot and S.H. Zak. Observation of dynamical systems in the presence of bounded nonlinearities/uncertainties. In *IEEE Conf. Decision and Control (CDC)*, pages 961–966, Athens, Greece, 1986.

[71] C.J. Walter, P. Lincoln, and N. Suri. Formally verified on-line diagnosis. *IEEE Trans. Software Eng.*, 23(11):684–721, 1997.

[72] J. Xing, C. Criddle, and R. Hickey. Effects of a long-term periodic substrate perturbation on an anaerobic community. *Wat. Res.*, 31(9):2195–2204, 1997.

Robust Nonlinear Control of a Pilot-Scale Anaerobic Digester

H.O. Méndez-Acosta[1], J.P. Steyer[2], R. Femat[3], and V. González-Alvarez[1]

[1] Depto. de Ingeniería Química, CUCEI-UdG
   hugo.mendez@cucei.udg.mx
[2] Ingénierie des Procédés, INRA-LBE
   steyer@ensam.inra.fr
[3] Depto. de Matemáticas Aplicadas y Sistemas Computacionales, IPICyT
   rfemat@ipicyt.edu.mx

Summary. Recently, anaerobic digestion (AD) has shown potential applicability to reduce organic matter from wastewater. However, AD processes are not widely applied at industrial scale. One of the main factors that explain this phenomenon is related to the lack of control schemes that guarantee a suitable operation of the process. As well, AD control is a non trivial task, because the control schemes must operate under an uncertain environment because of the difficult to monitoring AD processes (on-line measurements are limited and expensive). Particularly, this chapter is devoted to the development and the experimental validation of a robust nonlinear approach to regulate (i) the total amount of decomposable organic pollutant agents measured as total organic carbon (TOC) and, (ii) the concentration of volatile fatty acids (VFA); by using in both cases the input flow rate as manipulated variable. The proposed approach is a model-based controller obtained from geometric control tools and the definition of an uncertain but observable function that lumps the uncertain terms associated to the dynamics of the controlled variables TOC and VFA (e.g., feeding composition, kinetic growth functions and parameters). The robust scheme is composed of a feedback linearizing control law and a high-gain Luenberger observer which adapts the linearizing control law from estimates of the uncertain function. The robust approach is experimentally validated in a fixed-bed AD process used in the wastewater treatment of industrial distillery vinasses. Several experiments are performed to evaluate the controller performance and robustness under different set-points, feeding concentrations and sampling times (2 and 30min). Results show that both, the regulation of COT and VFA can be addressed in spite of the full ignorance of the kinetic growth functions, noisy measurements and unknown feeding composition.

Keywords: Nonlinear Control, Total Organic Carbon, Volatile Fatty Acids, Anaerobic Digestion.

H.O. Méndez-Acosta et al. (Eds.): Dyn. & Ctrl. of Chem. & Bio. Proc., LNCIS 361, pp. 165–199, 2007.
springerlink.com

1   Introduction

In the last few years, there has been an increasing interest to change or modify the traditional wastewater treatment processes. For years, such processes have stronger relied on tight pH and temperature control alternatives. Nowadays, this trend has changed, since the environmental laws have become stricter in order to guarantee a sustainable limit in the quantity of pollutant agents contained in industrial and municipal effluents. Among the available technologies, AD process presents very interesting advantages [12]:

- *Energy production.* AD produces valuable energy (methane, CH_4) from the degradation of the organic matter present in wastewater. For example, $1m^3$ of methane produces 8570kcal of energy which is equivalent to that produced by $0.94m^3$ of natural gas, 1.15l of gasoline, 1.3Kg of charcoal and 9.7kWh of electricity.
- *Good performance under severe operation conditions.* AD presents a great capacity to degrade concentrated and complex organic matter such as carbohydrates, proteins, aromatic compounds and fats under short retention times.
- *Low sludge production.* AD produces 5 to 10 times least sludge than classical aerobic treatment which reduces considerably its maintenance costs.
- *Reduction of pathogenic agents.* AD can be operated under mesophilic (T≈35°C) or thermophilic (T≈55°C) conditions helping to reduce the concentration of weed seeds and disease-producing (pathogenic) organisms. This explains the use of AD processes in the stabilization of primary and secondary sludge.

Figure 1 shows the evolution of AD commercial facilities referenced world-wide in 1999 [36]. As it can be observed, the number of AD processes has

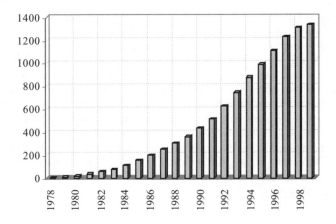

Fig. 1. Evolution of commercial AD processes referenced world-wide (from [36])

grown exponentially from 1978 until 2003 where 2,000 processes were referenced world-wide. However, these statistics evidence that AD processes are not widely applied at the industrial scale. This phenomenon can be explained by the following points [31]:

- The low sludge production is closely linked to the slow growth of microorganisms (e.g., the regeneration time of methanogenic bacteria can vary from 3 days at 35°C to 50 days at 10°C). As a consequence, the star-up phase is frequently time consuming (e.g., 2-4 months or longer in UASB reactors).
- The micro-organisms involved in the AD process are highly sensitive to overloads and disturbances of several causes. For example, the methanogenic bacteria are inhibited by high concentrations of it own substrate (i.e., volatile fatty acids). This explains why AD processes can be easily destabilized.
- AD is a complex biological process carried on by a great diversity of microorganisms which behavior is partially known. In addition, despite recent studies, the spatial distribution of individual organisms in flocks, granules and biofilms is not completely understood even though it has a strong influence on the overall process performance.

1.1 On the Control of AD Processes

Actually, the scientific efforts are focused not only to extend the number of AD applications, but also in increase the process robustness against disturbances [37]. Then, the importance of design efficient control schemes for AD processes is of no doubt. Nevertheless, the AD control is a non trivial task, because the control schemes operates under an uncertain environment due to the lack or the high cost of the devices that allow its on-line monitoring. To address this drawback, the scientific community has had significant advances in the development and validation of relatively inexpensive systems for the monitoring of different variables. For example, an on-line sensor composed by a modified Fourier Transform Infrared (FT-IR) spectrometer working in the mid-infrared range capable to provide precise and reliable measurements of COD, VFA, total organic carbon (TOC) concentrations as well as total (TA) and partial alkalinity (PA) was developed in [32]. More recently, an automatic titrimetric analyzer that measure on-line the concentration of bicarbonates, TA, PA and VFA was developed within the European research project called Telemonitoring and Advanced Telecontrol of High Yield Wastewater Treatment Plants, TELEMAC [24].

The increase in either the availability of reliable on-line measurements or the knowledge of the biological phenomena has allowed the development and validation of different control approaches. One of the first and very interesting works focused in the comparison of different control schemes when applied to AD processes was done by Heinzle et al., 1993 [10]. However, at that time, only 15 control applications were referenced, out of which 6 were done without experimental validation (i.e., numerical simulations). In addition, except two studies, all the other ones were done by using either basic on-off or PI/PID controllers. Recently, a comparison between classical and advanced control schemes has been presented by Steyer et al., 2006 [31]. In this work, the authors summarize the advantages and drawbacks of nine control schemes when applied to different pilot-scale AD processes. One of the most important facts depicted by the authors is that there does not exist a control scheme capable to manage all the disturbances occurring in an AD process. Then, the control approach must be selected from the data, knowledge and model availability. Since the comparison of control schemes applied to AD process is out of the scope of the present chapter, readers interested in this topic are referenced to the before mentioned references.

This chapter is devoted to the development and experimental validation of a robust nonlinear approach in a pilot-scale AD process. Particularly, the regulation of (i) the total amount of decomposable organic pollutant agents measured as total organic carbon (TOC) and, (ii) the concentration of volatile fatty acids (VFA); by using in both cases the input flow rate as manipulated variable are studied. The proposed scheme is a model-based controller obtained from geometric control tools and the definition of an uncertain but observable function that lumps the uncertain terms associated to the dynamics of the controlled variables TOC and VFA (e.g., feeding composition, kinetic growth functions and parameters). The robust scheme is composed of a feedback linearizing control law and a high-gain Luenberger observer which adapts the linearizing control law from estimates of the uncertain function. The chapter is organized as follows. The AD process and the AD dynamical model used in this chapter are briefly described in Section 2. The development and experimental validation of the robust scheme for the TOC regulation is presented in Section 3, whereas in Section 4 it is done for the VFA. In both cases, different operating conditions were used to illustrate the performance and robustness of the proposed approach. Finally, some concluding remarks are pointed out in Section 5.

2   The AD Model

2.1   An Overview of the AD Process

The AD process is composed by a series of complex biological and enzymatic reactions that are carried out by a great and varied group of microorganisms that do not require oxygen to subsist. In this process, the complex organic material is converted to simpler compounds and then degraded by the bacterial culture within the digester to produce a gas mainly composed by methane (CH_4) and carbon dioxide (CO_2) which is known as *biogas*. The biogas production varies with the amount of organic matter feed to the digester and the decomposition rate which is directly influenced by temperature.

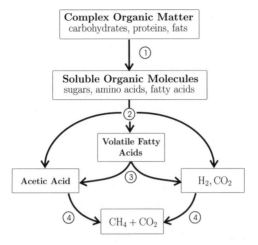

Fig. 2. A simplified scheme of the AD process: 1. Hydrolysis, 2. Acidogenesis, 3. Acetogenesis and 4. Methanogenesis

Figure 2 shows a simplified scheme of the AD process. Note that the AD process can be roughly summarized in four steps (see for more details [19] and references therein).

1. *Hydrolysis.* In this stage, the complex organic matter is decomposed in simple and soluble organic molecules using water to break the chemical branches.
2. *Acidogenesis.* The soluble organic monomers produced in the hydrolysis step are decomposed by means of acidogenic bacteria to produce acetate, VFA, CO_2 and H_2.
3. *Acetogenesis.* Here, the fermentation products (mainly VFA) are converted into acetate, CO_2 and H_2 by the so called acetogenic bacteria.
4. *Methanogenesis.* Finally, the products of either acidogenesis or acetogenesis are transformed into methane and carbon dioxide by the methanogenic bacteria. It is well known that this step is the limiting one in case of non-particulate substrate or non-excessively complex organic matter.

2.2   An Overview on the Modelling of AD Processes

State of the Art
As was pointed out, the most important drawback in the operation of AD processes is related to the instability of the process. However, this drawback can be overcome by associating monitoring procedures with decision support systems that allow the on-line stable operation of the process via a feedback control loop [27, 33]. Nowadays, most of the monitoring and control techniques available in the literature belong to those called "model-based". Such

techniques require a mathematical model that reproduces, in a suitable way, the dynamics of the key variables of the process in order to be applied. As a consequence, the dynamical modeling of AD processes has been an active research area during the last three decades. In 1968, Andrews introduced the Haldane function to characterize growth inhibition in order to emphasize the process instability due to the accumulation of VFA [2]. Then, a model with a single bacterial population was proposed by Graef and Andrews in 1974 [9]. After, the modeling of AD processes was improved by considering three stages: solubilization of organics, acidogenesis and methanogenesis [13]. A four-population model with two acidogenesis reactions and two methanization reactions emphasizing the role of hydrogen was proposed in [25]. Therefore, it is possible to find in the literature several AD models (e.g., [6, 17, 23]). However, most of them describe in detail particular characteristics of the AD process resulting in complex models which are difficult to use for control purposes [5]. For example, recently the AD specialist group of the International Water Association (IWA) proposed a general model for the AD process called ADM1. Such a model includes several bacterial populations and distinct substrates resulting in up to 26 dynamic state variables and 19 biochemical kinetic processes which is a powerful modeling tool but difficult to use for control purposes.

On the other hand, the model validation and identification are rarely done in a wide-range of operation conditions. In addition, most of the proposed models assume that the system behaves as a continuous stirred tank reactor (CSTR). Nevertheless, in practice one of the main objectives is to increase the contact surface between the organic matter and the microorganisms within the reactor in order to improve the process efficiency. This fact has brought a considerable growth in the use of fixed or fluidized bed reactors in the last decade. In this way, the assumption about CSTR-behavior becomes invalid due to the three-phase medium (solid-liquid-gas) and the different flow regimes that can be induced. Therefore, the proposal of dynamic models with monitoring and control aims for different AD systems becomes of great interest.

Recently, a model based in mass-balances was proposed for the monitoring and control of a fixed-bed AD process [7]. This model was born from substantial modifications to the model developed by Graef and Andrews in 1974 [9]. For example, (a) the structural properties of the model were improved in the seek of simplicity; (b) a second bacterial population was introduced to reproduce better the phase of destabilization due to the excessive accumulation of VFA; (c) gaseous flows with respect to the chemical and biological species into the digester were added; and (d) a simplified kinetic model was proposed. This model was successfully validated and identified using a wide range of operating conditions in a semi-industrial fixed-bed AD process located in the Laboratory of Biotechnology and Environment (LBE-INRA) in Narbonne, France. However, such a model can be adapted to describe a wide range of AD systems (e.g., CSTR, UASB, and FBR to mention some).

Description of the Considered Model

In this work, the AD model proposed and validated by Bernard et al., 2001 is used in the development of the robust nonlinear scheme. The use of this model is justified by two facts: i) this model has demonstrated to be useful in the monitoring and control of AD processes and, ii) the semi-industrial fixed-bed anaerobic digester located in the LBE-INRA used in the validation and identification of the model will be also used in the experimental implementation of the robust nonlinear approach here proposed. Therefore, in what follows the model is briefly described.

The model was developed under the following assumptions. First, a parameter α was introduced to consider the process heterogeneity. After, it is considered that the alkalinity is mainly due to the concentration of bicarbonate and VFA. Also, it is assumed that the AD process operates under isothermal conditions (i.e., constant temperature). Finally, a simplified version of the AD process is proposed, where only two steps are included:

- *Acidogenesis.* The population of acidogenic bacteria (X_1) consumes the organic substrate (S_1) and produces CO_2 and volatile fatty acids (S_2)

$$S_1 \rightarrow X_1 + S_2 + CO_2$$

- *Methanogenesis.* The methanogenic population (X_2) uses as substrate the volatile fatty acids (S_2) to produce CO_2 and CH_4

$$S_2 \rightarrow X_2 + CH_4 + CO_2$$

Therefore, the model can be written as follows

$$
\begin{aligned}
\dot{X}_1 &= (\mu_1(S_1) - \alpha D)X_1 \\
\dot{X}_2 &= (\mu_2(S_2) - \alpha D)X_2 \\
\dot{S}_1 &= (S_{1,in} - S_1)D - k_1\mu_1(S_1)X_1 \\
\dot{S}_2 &= (S_{2,in} - S_2)D + k_2\mu_1(S_1)X_1 - k_3\mu_2(S_2)X_2 \\
\dot{Z} &= (Z_{in} - Z)D \\
\dot{C}_{TI} &= (C_{TI,in} - C_{TI})D - q_{CO_2} + k_4\mu_1(S_1)X_1 + k_5\mu_2(S_2)X_2
\end{aligned}
\tag{1}
$$

where the state variables X_1, X_2, S_1 and S_2 are respectively the concentrations of acidogenic bacteria (g/l), methanogenic bacteria (g/l), organic substrate (COD, g/l) and VFA (meq/l). The total inorganic carbon (TIC), C_{TI} is defined in the considered range of pH ($6 \leq pH \leq 8$) as the sum of the dissolved carbon dioxide, C_d and bicarbonate, B concentrations (i.e., $C_{TI} = C_d + B$). The variable Z is linked to the total alkalinity and represents the sum of strong ions in the medium. In the considered pH range, when the chemical equilibrium is reached, it can be reasonably to assume that $Z = S_2 + B$. This means that no other ions significantly affect Z. The subscript *in* represents the concentration of each component in the feeding flow. The dilution rate

$(D,\ \mathrm{h}^{-1})$ is defined as the ratio between the feeding flow $(Q,\ \mathrm{l/h})$ and the digester volume $(V,\ \mathrm{l})$. This model assumes that a constant fraction (α) of the biomass is free in the liquid-phase and that this fraction is not affected by the dilution effect. Thus, the parameter α (of experimental determination) reflects the process heterogeneity and it is defined in the interval $0 \leq \alpha \leq 1$, where $\alpha = 0$ corresponds to an ideal fixed-bed reactor, while $\alpha = 1$ represents a continuous stirred tank reactor (CSTR). q_{CO_2} is the carbon dioxide molar flow rate (mmol/l-h) and it is approached by using the Henrry's law, i.e., $q_{CO_2} = k_L a(CO_2 - K_H P_C)$, where $k_L a$ is the liquid-gas transfer coefficient (hr^{-1}), K_H is the Henrry's constant (mmol/l-atm) and P_C is the partial pressure of CO_2 (atm). Assuming that the gas pressure is in thermodynamic equilibrium with the liquid, $P_c = (\varphi - (\varphi_2 - 4K_H P_T CO_2)^{1/2})/2K_H$ where $CO_2 = C_{TI} + S_2 - Z$, $\varphi = CO_2 + K_H P_T + q_{CH_4}/k_L a$, P_T is the total pressure within the digester and q_{CH_4} is the molar flow rate of methane. The bacterial growth rates for acetogenic (X_1) and methanogenic (X_2) bacteria are approached by the Monod and Haldane kinetics

$$\mu_1(S_1) = \mu_{1,max}\frac{S_1}{S_1 + K_{S1}}, \qquad \mu_2(S_2) = \mu_{2,max}\frac{S_2}{S_2 + K_{S2} + (S_2/K_{I2})^2}$$

where $\mu_{1,max}$ (1/hr), K_{S1} (g/l), $\mu_{2,max}$ (1/hr), K_{S2} (mmol/l) and K_{I2} (mmol/l) are kinetic constants.

In general, when dealing with AD processes two operating conditions can be identified (Méndez-Acosta et al., 2005):

- *Washout Conditions.* Here, the bacterial culture into the digester remains inactive or dead which implies that X_1, $X_2 = 0$ for all $t \geq 0$. Then, it can be proved that the steady state reached by the AD system (1) under this condition is given by the influent concentration (i.e., $S_1 = S_{1,in}$, $S_2 = S_{2,in}$, $Z = Z_{in}$ and $C_{TI} = C_{TI,in}$). This means that the polluting agents within the wastewater are not removed by the digester. As a consequence, this operating condition is undesired and must be avoided to any cost. In practice, this condition can be induced by an overload of organic matter or the presence of toxic agents, which suddenly increase the VFA concentration, decreasing the pH of the system and, consequently, the sludge stability [30].

- *Normal Operating Conditions (NOC).* Under this condition the biomass remains active and the sludge stability is preserved, i.e., X_1, $X_2 > 0$ for all $t \geq 0$. This physically means that part of the pollutant agents entering to the digester are consumed by the bacterial culture (i.e., $(S_1 - S_{1,in}) \neq 0$ and $(S_2 - S_{2,in}) \neq 0$).

From the above, it is immediately obvious that the main control objective when dealing with AD processes is stability, normally as measured by biogas production rate, effluent soluble COD or VFA concentrations. Therefore, the proposal of control schemes that guarantee the process stability by regulating the effluent soluble COD and VFA is of no doubt. In the next sections, the design and experimental validation of a robust nonlinear approach to regulate (i) a biological equivalent of the COD and, (ii) the VFA concentration are presented. It is important to remark that the controller design is carried out by assuming that the AD process is operating under NOC when the control-loop is closed.

3 Regulation of the Pollutant Agents in an AD Process

In general, the main objective of a biological wastewater treatment is to re-duce the concentration of pollutant agents in the wastewater below a specific value, in order to satisfy environmental regulations. Such an objective is ac-complished by manipulating the dilution rate [5]. The amount of decomposable pollutant agents in a wastewater can be measured in terms of its COD, which is defined as the quantity of oxygen required to oxidize the decomposable organic matter into CO_2 and H_2O.

In this section, a robust nonlinear approach is proposed to regulate a bi-ological equivalent of the total amount of organic substrate in the digester, which is defined as:

$$S_T = S_1 + \frac{k_1}{k_2}S_2$$

where S_1 and S_2 are stated in model (1).

Classical pollution measurements using COD are based on yields of chemi-cal reactions involved in organic pollution degradation. Indeed, the theoretical computation of the COD associated to an organic pollutant can be computed form the knowledge of the stoichiometric coefficients involved in the complete oxidation of the considered molecule. Here, the same reasoning is followed but the approach is based on the stoichiometry of the biochemical reactions in-volved in bacterial organic pollution degradation [27]. Thus, regulating S_T is equivalent to regulate the COD in the classical pollution regulation problem. Then, by using model (1), the dynamic behavior of S_T is given by

$$\dot{X}_T = (\mu(.) - \alpha D)X_T$$
$$\dot{S}_T = (S_{T,in} - S_T)D - k'\mu(.)X_T \tag{2}$$

where X_T is the bacterial population involved in the AD process, $S_{T,in}$ is the total concentration of organic substrate at the influent, $\mu(.)$ represents the specific growth rate of the bacterial population and k' is a kinetic constant related to the degradation rate of the total organic substrate. Then, the con-trol objective in this section can be defined as the proposal of a control law that regulates S_T under real operating conditions (i.e., to solve the pollution regulation problem). In the next section, a nonlinear approach is proposed to this end.

3.1 Controller Design

An Overview on the Geometric Control

In the last years, the control based on differential geometry has emerged as a powerful tool to deal with a great variety of dynamic nonlinear systems [11, 15, 26]. This control approach allows the transformation of a nonlinear system into a partially or totally linear one, by means of a nonlinear state transformation, which is obtained from directional derivatives of the output. It is important to remark that geometric control differs totally from the linear approximation of dynamics by calculation of the Jacobian.

On the other hand, the main drawback in the use of this control technique is that it depends on the exact cancellation of the nonlinear dynamics in order to obtain an input-output linear dynamic behavior. As a consequence, the perfect knowledge of the system is required. This means that the presence of modeling errors, unmeasured disturbances and parametric uncertainties are not considered in the controller design. This explains maybe why this control tools have not been applied and validated experimentally in the control of AD processes [3]. Next, some basic concepts of the geometric control theory are reviewed. After, the geometric properties of the AD model (2) are determined. Starting from such properties a classic input-output linearizing control is designed. Then, a number of shortcomings and limitations associated to this controller are highlighted. Finally, a robust approximation is proposed to overcome the drawbacks of the classic scheme.

Basic Concepts of the Geometric Control

Before focusing in the controller design, it is important to review some basic concepts of the geometric control theory. The control tools based in differential geometry are proposed for those nonlinear dynamical systems called affine systems. So, let's star by its definition.

Definition 1. *Let U an open set of $\mathbb{R}^n$. Then, it is said that a dynamic system is affine if it can be written in the following form*

$$\dot{x} = f(x) + \sum_{i=1}^{m} g_i(x)u_i \tag{3}$$

$$y_i = h_i(x); \quad 1 \le i \le p$$

where $x = (x_1, x_2, \ldots, x_n)$ is the state vector, $f, g_1, \ldots, g_m$ are smooth[1] mappings assigning to each point x of U a vector in $\mathbb{R}^n$ (i.e., real-valued smooth vector fields defined on U). Finally, $h_i(x)$ and u_i are smooth real-valued functions defined on U that respectively represent the p-outputs and m-inputs of the system. ∎

---

[1] A map or a function is smooth when they partial derivatives of any order with respect to the state variables $x_1, x_2, \ldots, x_n$ are continuous and exist.

Now, three types of differential operations involving real-valued functions, vector and co-vectors fields are described. These operations are frequently used in the analysis and design of nonlinear control systems (for more details see [15, 26]).

Definition 2. *Let $f(x)$ and $g(x)$ smooth vector fields, $\omega(x)$ a smooth co-vector field and $\lambda(x)$ a smooth real-valued function for all $x \in U$, where U is an open subset of $\mathbb{R}^n$.*

1. ***Directional or Lie Derivative.*** *This operation is defined as the inner product between the partial derivative of a smooth real-valued function and a smooth vector field. As a result, a new smooth real-valued function is obtained. For example, the directional derivative of $\lambda(x)$ along the vector field $f(x)$ is given by*

$$L_f\lambda(x) = \langle d\lambda(x), f(x) \rangle = \sum_{i=1}^{n} \frac{\partial\lambda(x)}{x_i} f_i(x)$$

   If $\lambda(x)$ is being differentiated k-times along $f(x)$, the notation $L_f^k\lambda(x)$ is used. In other words, the function $L_f^k\lambda(x)$ satisfies the recursion

$$L_f^k\lambda(x) = \frac{\partial(L_f^{k-1}\lambda(x))}{\partial x}f(x), \quad where \quad L_f^0\lambda(x) = \lambda(x)$$

2. ***Lie Product or Braked.*** *The second operation involves two smooth vector fields; e.g., $f(x)$ and $g(x)$ both defined on an open set U of $\mathbb{R}^n$. From this operation, a new smooth vector field is constructed by the following inner product*

$$[f(x), g(x)] = \frac{\partial g(x)}{\partial x}f(x) - \frac{\partial f(x)}{\partial x}g(x)$$

   where the terms given by $\partial g(x)/\partial x$ and $\partial f(x)/\partial x$ denote the Jacobian matrix of the vector fields $g(x)$ and $f(x)$, respectively.

3. ***Construction of a Co-vector Field.*** *The third operation is related to the construction of a new smooth co-vector field starting from a smooth vector field and a smooth co-vector field, both defined on an open set U of $\mathbb{R}^n$*

$$L_f\omega(x) = f^T(x)\left[\frac{\partial\omega^T}{\partial x}\right]^T + \omega(x)\frac{\partial f(x)}{\partial x}$$

   where the superscript T denotes a transposition. ∎

Geometric control is based in a coordinate transformation defined in the state space. This coordinate change allows the transformation of the affine system (3) into a called normal form, which can be partially or totally linearizable. However, how to know the degree of linearizability of the affine system? In other words, how to know if the affine system is partially or totally linearizable? Next, some notions are defined in order to answer this question.

Definition 3. *An affine system has a well-defined relative degree r in a point x^0 of U when the following conditions are fulfilled.*

a) $L_g L_f^k h(x) = 0$ *for all x in a neighborhood U^0 of x^0 and all $k < r - 1$*
b) $L_g L_f^{k-1} h(x) \neq 0$ *where*

$$L_g L_f^k h(x) = \sum_{i=1}^{n} \frac{\partial}{\partial x_i} \left(\frac{\partial \left(L_f^{k-1} h(x) \right)}{\partial x_i} f(x) \right) g(x)$$

■

In order to give a simple interpretation to the notion of relative degree, consider that x^0 is the value of the state x at time t^0 (i.e., $x(t^0) = x^0$). Now, assume that it is desired to calculate the value of the output $y(t)$ and its derivatives whit respect to time $y^{(k)}(t)$, for $k = 1, 2, \ldots$ at $t = t^0$. Then, it is obtained that

$$y(t^0) = h(x(t^0)) = h(x^0)$$
$$y^{(1)}(t) = \frac{\partial h}{\partial x} \frac{dx}{dt} = \frac{\partial h}{\partial x} [f(x(t)) + g(x(t))u(t)]$$
$$= L_f h(x(t)) + L_g h(x(t))u(t)$$

If the relative degree r is greater than one, for all t such that $x(t)$ is near of x^0; i.e., for all t near to t^0, it is obtained that $L_g h(x(t)) = 0$ and therefore $y^{(1)} = L_f h(x(t))$. If the same procedure is continued, it is possible to find the following general representation

$$y^{(k)}(t) = L_f^k h(x(t)) \quad \text{for all } k < r \text{ and all } t \text{ near to } t^0$$
$$y^{(r)}(t^0) = L_f^k h(x^0) + L_g L_f^{r-1} h(x^0)u(t^0)$$

Then, the relative degree r is equal to the number of times that the output $y(t)$ must be differentiated at time $t = t^0$ in order to the input $u(t^0)$ explicitly appears.

On the other hand, the Lie bracket of the functions $h(x), L_f h(x), L_f^{r-1} h(x)$ is linearly independent. Therefore, it can be proved that necessarily $r \leq n$ and that the linearly independent functions qualify as a set of new coordinate functions around the point x^0.

Proposition 1. *Suppose that an affine system of order n has relative degree r at x^0. Then, if $r = n$ exists a set of functions*

$$\Phi(x) = \begin{bmatrix} \phi_1(x) = h(x) \\ \phi_2(x) = L_f h(x) \\ \vdots \\ \phi_r(x) = L_f^{r-1} h(x) \end{bmatrix}$$

that allow a local coordinate transformation of the affine system (3) in a neighborhood U^0 of x^0. On the other hand, if r is strictly less than n, it is always *possible to find $n - r$ additional functions (also called complementary functions) $\phi_{r+1}(x), \ldots, \phi_n(x)$ such that the mapping $\Phi(x)$ has a non singular jacobian matrix at x^0 and; as a consequence, qualifies as a local coordinate transformation in a neighborhood U^0 of x^0. In fact, the $n - r$ additional functions can be fixed arbitrarily in such a way that $L_g\phi_i(x) = 0$ for all $r + 1 \leq j \leq n$ and all x around x^0.*

Proof. See [15], page 141.   ∎

Once the mapping $\Phi(x)$ is available, it is possible to find a dynamically-equivalent system by means of the coordinate transformation $z = \Phi(x)$, where z is the independent variable in the new system.

For the $r - 1$ functions it is obtained that

$$\frac{dz_1}{dt} = \frac{\partial \phi_1}{\partial x}\frac{dx}{dt} = \frac{\partial h}{\partial x}\frac{dx}{dt} = L_f h(x(t)) = \phi_2(x(t)) = z_2(t)$$

$$\vdots$$

$$\frac{dz_{r-1}}{dt} = \frac{\partial \phi_{r-1}}{\partial x}\frac{dx}{dt} = \frac{\partial \left(L_f^{r-2}h(x)\right)}{\partial x}\frac{dx}{dt} = L_f^{r-1}h(x(t)) = \phi_r(x(t)) = z_r(t)$$

whereas the r-derivative is given by

$$\frac{dz_r}{dt} = L_f^r h(x) + L_g L_f^{r-1} h(x)u(t).$$

Then, the affine system (3) can be represented in the new coordinates by the following set of equations

$$\dot{z}_i = z_{i+1}, \qquad\qquad 1 \leq i < r \qquad\qquad\qquad (4a)$$
$$\dot{z}_r = L_f^r h(\Phi^{-1}(z)) + L_g L_f^{r-1} h(\Phi^{-1}(z))u(t) \qquad\qquad (4b)$$
$$\quad = b(z) + a(z)u(t)$$
$$\dot{z}_j = L_f \phi_j(\Phi^{-1}(z)), \quad r + 1 \leq j \leq n \qquad\qquad (4c)$$
$$\quad = q_j(z)$$

This representation is also called *normal form* and it is graphically depicted in Figure 3. It can be seen that the normal form is composed of three parts respectively given by the subsystems (4a), (4b) and (4c). The first part presents a linear structure and it is given by a chain of $r - 1$ integrators, whereas the second part has a nonlinear structure, where the input-output relationship explicitly appears. Finally, the last part is conformed by the dynamics of the $n - r$ complementary functions. This part is called *internal dynamics* because it cannot be seen from the input-output relationship (see Figure 3) and whose structure can be linear or nonlinear.

Now, the main results discussed in this section are briefly summarized and used to highlight some interesting properties from the control point of view (see for more details [15, 26]).

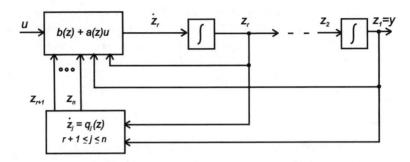

Fig. 3. Block diagram representation of the normal form [15]

- *Relative degree.* An affine system of order n is completely linearizable if $r = n$ and partially linearizable if $r < n$. In other words, the relative degree is directly related to the degree of linearizability of the system.
- *Normal form.* A local coordinate transformation $\Phi(x)$ is defined from the Lie derivatives of the system output $y(t)$. Particularly, if the affine system is partially linearizable, $n - r$ complementary functions must be proposed in order to complete the mapping $\Phi(x)$. These functions are arbitrarily proposed so that the Lie derivative $L_g\phi_j(x) = 0$ holds for $r + 1 \leq j \leq n$. As a consequence, it can be proved that only the r-states of the normal form (i.e., (4a)-(4b)) are observable and controllable when $u(t)$ is used as the control input. Then, in order to guarantee the stability of the closed-loop system, it is required that the internal dynamics of the system be stable, at least, in the BIBO (bounded input - bounded output) sense (see Figure 3). A nonlinear affine system with stable internal dynamics is called a *minimum-phase* system. On one hand, since the normal form results in a dipheomorphic transformation. Therefore, a control law capable to achieve the control objective over the system in the normal form (4), will be also capable to do it over the original system (3). On the other hand, if an affine system is of minimum-phase and its relative degree is well defined for all $x \in U \subset \mathbb{R}^n$ around x^0. As a consequence, it is straightforward to see from Figure 3 that a desired closed-loop linear dynamical behavior $\omega(z)$ can be induced by a control input of the form $u(t) = 1/a(z)[-b(z) + \omega(z)]$.

Model Geometric Properties

In this section, the geometric properties of the AD model (2) are analyzed assuming NOC. Then, by using the geometric control tools described above, a classic geometric controller is designed and used as an intermediary step in the design of the here proposed robust nonlinear approach.

First, let us rewrite the AD model (2) in the affine form

$$\dot{x} = f(x) + g(x)u, \qquad y = h(x)$$

where $x = [X_T \ S_T]^T \in \mathbb{R}^2_+$ is the state vector, $y \in \mathbb{R}_+$ is the system output given by the total concentration of organic substrate (i.e., $y = S_T$) and the dilution rate is the control input (i.e., $u = D$), whereas the vector fields $f(x)$ and $g(x)$ are given by

$$f(x) = \begin{bmatrix} \mu(.)X_T \\ -k'\mu(.)X_T \end{bmatrix}, \qquad g(x) = \begin{bmatrix} -\alpha X_T \\ (S_{T,in} - S_T) \end{bmatrix}$$

where it is non restrictive to assume that the vector fields $f(x), g(x)$, the control input $u = D$ and the system output $y = S_T$ are smooth.

Lemma 1. *Let $y = S_T$ the system output. Then, the AD model (2) has a well-defined relative degree $r = 1$ under NOC, when the dilution rate D is used as the control input.*

Proof. By computing the Lie derivative of the output along the vector fields $f(x)$ and $g(x)$, it is obtained that $L_g L_f^0 h(x) = (S_{T,in} - S_T)$. Remember that under NOC the inequality $(S_{T,in} - S_T) > 0$ is fulfilled. Then, the AD model (2) has a well-defined relative degree $r = 1$ under NOC. ∎

From Lemma 1, it can be concluded that the AD model (2) is partially linearizable, since the relative degree is less than the order of the system. Then, a complementary function ϕ_j for $r + 1 \leq j \leq n$ must be proposed in order to complete the mapping $\Phi(x)$.

Proposition 2. *Let $\phi_j = X_T/(S_{T,in} - S_T)^\alpha$ a complementary function of the AD model (2). Then, the mapping*

$$\Phi(x) = \begin{bmatrix} S_T \\ X_T/(S_{T,in} - S_T)^\alpha \end{bmatrix}$$

defines a local coordinate transformation for each pair $X_T, S_T \in U \subset \mathbb{R}^2$, where U is the space where the trajectories of the AD model (2) live under NOC.

Proof. First, it can be probed that $\phi_j = X_T/(S_{T,in} - S_T)^\alpha$ is a complementary function of the AD model (2), because it is a solution of the Lie derivative

$$L_g \phi_j(x) = (-\alpha X_T)\frac{\partial \phi_j}{X_T} + (S_{T,in} - S_T)\frac{\partial \phi_j}{S_T} = 0, \quad \text{for } r + 1 \leq j \leq n$$

Now, let

$$J_{\Phi(x)} = \begin{bmatrix} 0 & 1 \\ 1/(S_{T,in} - S_T)^\alpha & \alpha X_T/(S_{T,in} - S_T)^{\alpha+1} \end{bmatrix}$$

the Jacobian matrix of the mapping $\Phi(x)$. Then, by computing its determinant, it is obtained that

$$\det\left(J_{\Phi(x)}\right) = -\frac{1}{(S_{T,in} - S_T)^\alpha}$$

Since $(S_{T,in} - S_T) > 0$ under NOC, the existence of an inverse

$$\Phi^{-1}(z) = \begin{bmatrix} z_2(S_{T,in} - z_1)^\alpha \\ z_1 \end{bmatrix}$$

such that $\Phi^{-1}(\Phi(x)) = x$ for each pair $X_T, S_T \in U \subset \mathbb{R}^2$ is guaranteed, which completes the proof. ∎

Once the mapping $\Phi(x)$ is complete, the AD model (2) can be rewritten in the affine form (4) as follows

$$\dot{z}_1 = b(z) + a(z)u \tag{5}$$
$$\dot{z}_2 = q(z)$$

where $b(z) = L_f h(\Phi^{-1}(z)) = -k'\mu(.)X_T$, $a(z) = L_g L_f^0 h(\Phi^{-1}(z)) = (S_{T,in} - S_T)$ and

$$q(z) = L_f \phi_j(\Phi^{-1}(z)) = \mu(.)\phi_j \left[1 - \frac{k'\alpha X_T}{(S_{T,in} - S_T)}\right]$$

Proposition 3. *The AD model (2) is a minimum-phase system under NOC.*

Proof. By analyzing the steady-state of the AD model (2) under NOC, the following relationships are found

$$\mu^{eq}(.) = \alpha D^{eq} \quad \text{and} \quad D^{eq} = \frac{k'\mu^{eq}(.)X_T^{eq}}{(S_{T,in} - S_T)}$$

Let $\Psi(x) = 1/2\phi_j^2$ a candidate Lyapunov function (CLF) of the internal dynamics of the AD model (2). From Proposition 2, it is observed that the CLF is positive defined for all the pairs X_T, S_T contained into the set of NOC (i.e., $\Psi > 0 \ \forall \ X_T, S_T \in U \subset \mathbb{R}^2$). Now, by differentiating the CLF with respect to time, it is obtained that

$$\dot{\Psi}(x) = \frac{k'\mu^{eq}(.)X_T^{eq}}{(S_{T,in} - S_T)}\left[1 - \frac{k'\alpha X_T}{(S_{T,in} - S_T)}\right] = \phi_j^{eq}\left[\mu^{eq}(.) - \alpha D^{eq}\right] = 0$$

Then, since $\Psi > 0$ and $\dot{\Psi} \leq 0$, the asymptotic stability of the internal dynamics is proved. ∎

Proposition 4. *Since the AD model (2) is a minimum-phase system and has a well-defined relative degree r under NOC. Then, the following input-output linearizing controller will make converge exponentially the total concentration of organic substrate S_T to a desired value S_T^* for all $t > 0$*

$$u = \frac{1}{(S_{T,in} - S_T)}\left[k'\mu(.)X_T - \omega(x)\right] \tag{6}$$

where $\omega(x) = Kc(S_T - S_T^*)$ and Kc is a positive constant.

Proof. By implementing the input-output linearizing controller (6) to the AD model in the normal form (5), it is obtained that

$$\dot{z}_1 = -Kc(S_T - S_T^*)$$

$$\dot{z}_2 = q(z) = \mu(.)\phi_j \left[1 - \frac{k'\alpha X_T}{(S_{T,in} - S_T)} \right]$$

This system clearly appears decomposed into a *linear subsystem* $\dot{z}_1$, which is the only responsible for the input-output behavior, and a *non linear subsystem* $\dot{z}_2$, whose behavior is stable. Then, by defining the control error as $e = (S_T - S_T^*)$, it is clear to observe that the solution of the linear subsystem is given by $e = exp(-Kc * t)$, which implies that $S_t \to S_T^*$ as $t \to \infty$. ∎

Remark 1. From Proposition 4, the existence of an input-output linearizing control law capable to regulate exponentially the total concentration of organic substrate S_T in a desired value S_T^* was demonstrated. However, in order to implement this controller in practice, a perfect knowledge of the process dynamics is required. In other words, this implies that either the influent composition $S_{T,in}$ or the process kinetics $k', \mu(.)$ must be perfectly known. Nevertheless, this condition is difficult to satisfy in practice limiting its application. But what about if the uncertain terms can be estimated from available measurements and a control scheme with a similar structure to that of the input-output linearizing controller (6) is used. In the next section, a robust approach is proposed based in this fact.

Robust Geometric Approach

Here, a geometric-based approach capable to regulate the total amount of organic substrate S_T in spite of uncertainties in either the influent composition or the process kinetics is proposed. In order to emulate real operating conditions, the following assumptions are done.

A1: The total amount of organic substrate S_T is available from on-line measurements [32].

A2: The kinetic function $\mu(.)$ is unknown in the sense that no analytical expression of this function is available. Based on biological evidence, it is non restrictive to assume that: i) $\mu(.)$ is at least a function of S_T and X_T and, ii) $\mu(.)$ is a continuous bounded positive-defined function.

A3: It is well-known that, if an advanced sensor is available and capable to measure a chemical component within the medium, it is easier to implement the sensor at the output of the process rater than at the input. This fact is justified by several causes such that: i) the measuring range of the sensor, the presence of suspended solids and the sensor cost [3]. Then, the total concentration of organic substrate in the influent $S_{T,in}$ is assumed to be unknown, piecewise constant and bounded (i.e., $S_{T,in}^{min} \leq S_{T,in} \leq S_{T,in}^{max}$).

A4: In practice, the feeding flow Q is restricted in order to avoid the washout condition due to the process overload or the drag of biomass [21]. Since the control input given by the dilution rate is defined as $D = Q/V$, where V is the digester volume. Then, the control input D is constrained; i.e., $D^{min} \leq D \leq D^{max}$ where the limits D^{min} and D^{max} are well-known.

By the moment, let us consider that only assumptions $A1$-$A3$ hold. Also, without lost of generality, assume that the total concentration of organic substrate in the influent can be described as follows: $S_{T,in} = \tilde{S}_{T,in} + \Delta_{S_T}$, where Δ_{S_T} is an uncertain and bounded function related to the variation of the influent composition around a well-known nominal value $\tilde{S}_{T,in}$. In practice, the nominal value $\tilde{S}_{T,in}$ can be computed by a single off-line measurement of the wastewater to be treated.

Now, let us define a function

$$\eta = \sigma(z, u) = L_f h(\Phi^{-1}(x)) + \Delta_{S_T} u = -k' \mu(.) X_T + \Delta_{S_T} u$$

where all the uncertain terms associated to S_T are lumped. Then, it is possible to rewrite the normal form (5) in the following extended state-space representation

$$\dot{z}_1 = \eta + a'(z)u$$
$$\dot{\eta} = \Xi(z, u) \tag{7}$$
$$\dot{z}_2 = q(z)$$

where $a'(z) = (\tilde{S}_{T,in} - z_1)$. Then, two important properties can be highlighted from system (7) (Femat et al., 1999): (a) it can be proved that the solution of system (5) is a projection of the solution of the extended state-space (7) and, (b) a feature of system (7) is that the uncertainties have been lumped into an uncertain function $\sigma(z, u)$ which can be estimated by a not measured but observable state η. As a consequence, a control law capable to achieve the control objective over the system (7) will be also capable to do it over the system (5) and, consequently, over the AD model (2).

Nowadays, it is possible to find in the literature some works focused in the robust stabilization of nonlinear systems by output feedback [4, 16, 35]. The main idea in these works is the proposal of high-gain observers used as uncertainty estimators, such that the output feedback stabilization can be guaranteed in uncertain nonlinear systems. By following this idea, a high-gain Luenberger observer is constructed to estimate the states z_1, η. This observer was selected due to its simple and linear structure. In addition, it has demonstrated a good performance in the estimation of uncertain terms related to biological and chemical processes [8, 20, 21].

Lemma 2. *The dynamics of the states z_1, η can be reconstructed from measurements of the output $y = z_1 = S_T$ by means of the following high-gain Luenberger observer*

$$\dot{\hat{z}}_1 = \hat{\eta} + (\tilde{S}_{T,in} - \hat{z}_1)u + \Gamma g_1(z_1 - \hat{z}_1) \tag{8}$$
$$\dot{\hat{\eta}} = \Gamma^2 g_2(z_1 - \hat{z}_1)$$

where $u = D$, Γ is a tuning parameter (observer high-gain) and the coefficients g_1, g_2 are selected in such a way that the polynomial $s^2 + g_2 s + g_1 = 0$ be Hurwitz (i.e., eigenvalues with negative real part).

Proof. See Teel and Praly, 1995 page 1477.   ∎

Corollary 1. *From the results shown in Lemma 2, it is possible to guarantee that the estimation error vector $\Omega = [z_1 - \hat{z}_1, \eta - \hat{\eta}]^T \to \varepsilon$ as $t \to \infty$, where ε is an arbitrary small neighborhood of the origin. Then, the control law*

$$u = \frac{1}{(\tilde{S}_{T,in} - z_1)} [-\hat{\eta} - Kc(\hat{z}_1 - S_T^*)] \tag{9}$$

will be capable to achieve the practical asymptotic regulation of the output S_T around a desired value S_T^* for all the pairs X_T, S_T contained in the set of NOC (i.e., $\forall\ X_T, S_T \in U \subset \mathbb{R}^2$).   ∎

Remark 2. From Corollary 1 we claim that, under NOC and assumptions $A1$-$A3$, the robust geometric approach (8)-(9) is capable to regulate S_T around a desired value S_T^* in spite of uncertainties in either the feeding composition or the process kinetics. However, constraints in the manipulated variable ($A4$) have not been considered, which may lead to a significant deterioration of the controller performance and even cause the closed-loop instability [29]. Such performance/stability degradation is usually due to the so-called *windup* phenomenon [34]. In addition, since the robust geometric approach (8)-(9) is based on a high-gain observer, undesired dynamic effects such as the so-called *peaking* phenomenon can be induced [18]. This phenomenon produces large overshoots, which can lead to the saturation of the control input. In such a situation, the feedback is broken and the plant behaves as an open-loop with a constant input, allowing the possible degradation of the closed-loop performance. On the other hand, note that the performance of the high-gain observer (8) is directly related to the computed control input (9). Then, constraints in the control input create an error between the computed control signal (CCS) and the control action acting (CAA) on the process, which can induces a deterioration of the nominal performance. Recently, Méndez-Acosta et al., (2004) have demonstrated that an observer-based antiwindup scheme can be obtained by feeding back the CAA to the high-gain observer (8), instead of the CCS. In this way, the robust geometric approach with antiwindup (RGA-AW) action can be written as follows

$$\dot{\hat{z}}_1 = \hat{\eta} + (\tilde{S}_{T,in} - \hat{z}_1)u_{sat} + \Gamma g_1(z_1 - \hat{z}_1)$$
$$\dot{\hat{\eta}} = \Gamma^2 g_2(z_1 - \hat{z}_1) \tag{10}$$
$$u = \frac{1}{(\tilde{S}_{T,in} - z_1)} [-\hat{\eta} - Kc(\hat{z}_1 - S_T^*)]$$

where $u_{sat} = D_{sat}$ is the dilution rate that is being applied to the AD process. The block diagram of controller (10) is depicted in Figure 4. Note that the

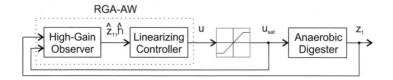

Fig. 4. Block diagram of the robust approach with antiwindup action (10)

RGA-AW presents a simple structure, which can be easy to tune since only the estimation and control gains Γ, Kc can be adjusted to this end.

3.2   Experimental Validation

In this section, the experimental validation of the robust geometric approach is presented. The experimental validation were carried out in a fully-instrumented AD pilot-scale plant used to treat distillery vinasses, which is located in the Laboratoire de Biotechnologie de l'Environement (INRA-LBE) in Narbonne, France.

The AD Pilot-Scale Plant

The AD pilot-scale plant consists of a fixed-bed up-flow reactor operating in a continuous regime which is sketched in Figure 5. First, water is added to rough vinasses (i.e., neither homogenous nor sterile) in the dilution tank, where the pH of the solution is neutralized by adding NaOH with a peristaltic pump controlled by a local PID-controller. Connected to the dilution tank is a remotely controllable peristaltic pump that ensures the desired influent flow rate Q, which allows to disturb the process in a controlled way. After, fresh substrate is mixed with the recycled liquid just before entering the heat exchanger (which regulates the process temperature to 35°C using a local PID-controller). The heated liquid is then introduced at the bottom of the reactor where it is homogenized by the mixing pump. Then, the liquid passes through a PVC multichanel tubular support commercially called Cloisonyl (specific surface of 180 m²/m³) until it reaches the top. The Cloisonyl structure fills 0.0337m3 of the reactor volume leaving 0.9483m³ of effective volume. The liquid from the top of the reactor is collected by overflow in a receiving vessel. Some of this liquid is recycled and the rest is sent to the sewer. At the top of the process, an ultra-filtration module has been installed. This module is composed of a high-pressure pump and a ceramic membrane with a pore diameter of 0.14μm, which provides of suitable samples to the on-line sensors of chemical and physicochemical variables (e.g., volatile fatty acids (VFA), total organic carbon (TOC), bicarbonates (B), total (TA) and partial (PA) alkalinities). At the liquid output of the reactor, there is a degassing system and a gas evacuating system. Finally, the produced flow of biogas and its percentage composition of CO_2 and CH_4 are measured before the biogas is

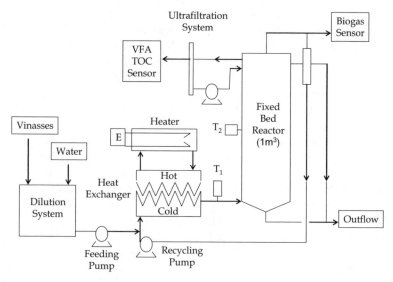

Fig. 5. Fully instrumented AD pilot-scale plant

released to the atmosphere. Readers interested in a more detailed description of the AD pilot-scale plant are referenced to [31, 32].

Experimental Protocol and Results

All the sensors and actuators are connected to an input-output device that allows the acquisition, treatment and storing of data in a personal computer by means of a free software developed in the INRA-LBE which is called Modular MSPC Software©. The implementation of the robust geometric approach was made by using the Modular MSPC Software© in conjunction with the Matlab Software© where the differential and algebraic equations that represents the robust approach were computed. The interval of reception and sending of information between the software and the process was fixed at 2 minutes, as it was considered a good solution to be fast enough for control and supervision purposes, however generating reasonable size data files. The total concentration of organic substrate was on-line measured by means of a Zellweger Analitycs TOC-1950 sensor and a modified Fourier Transform Infrared (FT-IR) spectrometer developed in the INRA-LBE (MIR). Particularly, three experiments were performed in order to test the robust geometric approach under the following conditions.

- (E1) *Effect of the control parameters in the servo-control problem.* In this experiment, the Zellweger Analitycs sensor was used (sampling time 2min). Diluted vinasses (50% H_2O and 50% vinasses) were fed to the process, whose nominal value $\tilde{S}_{T,in}$ was determined by a single TOC off-line measurement, obtaining that $\tilde{S}_{T,in} = 15,900$mg/l. After, four set-point changes of the same magnitude 250 mg/l were performed. The control parameters used in the first two set-point changes were $g_1 = 2h^{-1}$, $g_2 = 1h^{-1}$, $\Gamma = 0.7$ and $Kc = 0.4h^{-1}$; whereas in the rest of the experiment, only the control and estimation gains were increased to $\Gamma = 1.0$ and $Kc = 0.7h^{-1}$. The initial conditions used in the high-gain observer were $z_1(0) = 491$mg/l and $\eta(0) = -0.05$.

- (E2) *Effect of the sampling time in the controller performance.* The procedure followed in this experiment as well as the control parameters were the same of that described in E1. Only the TOC on-line sensor was changed by the infrared MIR sensor, whose sampling time is 30min. This means that the control scheme was calculated every 30min, the same time that the control action remains constant. In this experiment, the initial conditions used in the high-gain observer were $z_1(0) = 911$mg/l and $\eta(0) = -0.02$.
- (E3) *Effect of the long-time saturation of the control input.* Here, the Zellweger Analitycs sensor was used again. Particularly, the experiment was conducted in three steps by following a predetermined feeding policy. First, diluted vinasses were fed to the digester. Then, the TOC concentration was regulated around a desired set-point $S_T^* = 750$mg/l. After, pure vinasses were fed and the set-point was changed to $S_T^* = 1500$mg/l. At the same time, the antiwindup structure of the robust approach was unplugged by feeding back to the high-gain observer the feeding flow calculated by the control scheme (CCS) unlike to the feeding flow measured at the entrance of the process (CAA). Finally, the antiwindup structure was reestablished. It is important to remark that either the nominal value $\tilde{S}_{T,in} = 15900$mg/l or the control parameters used along the experiment were always the same. Particularly, the control parameters were the same of those in the first part of experiment E1 and the initial conditions used in the high-gain observer were $z_1(0) = 761$mg/l and $\eta(0) = -0.034$.

Next, the results of experiment E1 are presented. Figure 6a shows the dynamic behavior of the TOC concentration, S_T when the RGA-AW is applied. In general, the controller performance in the servo-control problem is quite acceptable, since the RGA-AW is capable to regulate the TOC concentration, S_T around the set-point values, S_T^*. However, a small off-set less than 50 mg/l is present. Note that the RGA-AW presents a robust margin in spite of noisy measurements. This phenomenon is explained by the double integrator in the high-gain Luenberger observer, which acts as a low-pass filter [21]. At the end of the second set-point change, see how the controller performance is not deteriorated when a disturbance is induced due to the cleaning procedure of the sensor. Once the control and estimation gains (Kc, Γ) are increased, the controller response becomes faster and the off-set is diminished (see third and fourth set-point changes in Figure 6a).

Figure 6b shows that there is a significant difference between the calculated flow by the RGA-AW (black line) and the flow fed by the peristaltic pump (gray line), which can explain the off-set along experiment E1. An evidence of

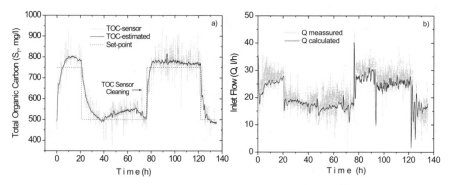

Fig. 6. (a) Set-point tracking performance of the TOC concentration when the RGA-AW is applied (S_T^*: dashed line, S_T: gray line and $\hat{S}_T$: black line). (b) Dynamical response of the manipulated variable (black line: Q calculated by the control law and gray line: Q measured at the entrance of the digester).

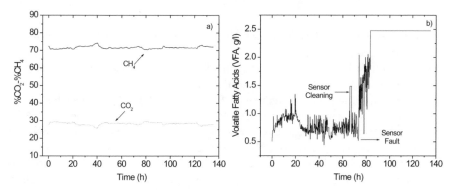

Fig. 7. (a) Percentage production of CO_2 and CH_4. (b) Dynamic behavior of the VFA concentration during experiment E1.

this conjecture is in the second set-point change, when the smaller off-set is achieved (see Figure 6a). From Figure 6b, it can be observed that at this point, the flow calculated by the RGA-AW is almost the average of the flow fed by the peristaltic pump. On the other hand, it can be seen that the dynamical response of the manipulated variable is smooth along the experiment, because it gets saturated neither in the lower nor the upper limit, which are given by $Q^{min} = 5l/h$ and $Q^{max} = 50l/h$. This condition is desired in practice, since it helps to preserve the actuator life. Figure 7a shows that a good percentage production of CH_4 is obtained when the RGA-AW is applied. Note that the percentage of methane is greater than 70% along experiment E1. This result suggests that the process stability is preserved when the AD process is in closed-loop with the RGA-AW. This fact is corroborated in Figure 7b, where

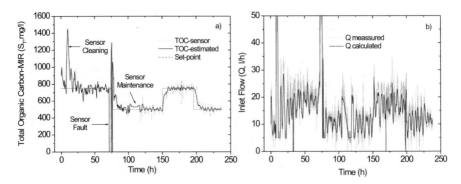

Fig. 8. (a) Set-point tracking performance when the sampling time is increasing from 2 to 30min by using the MIR-sensor (S_T^*: dashed line, S_T: gray line and $\hat{S}_T$: black line). (b) Dynamical response of the manipulated variable when the MIR-sensor is used (black line: Q calculated by the control law and gray line: Q measured at the entrance of the digester).

it can be observed that there is not a signal of VFA accumulation before the fault of the MIR-sensor at $t = 75$h.

Now, the results obtained in experiment E2 are discussed. Figure 8a shows that the RGA-AW is capable to regulate the TOC concentration, S_T even when the sampling time is increased from 2 to 30min by using the MIR-sensor. Also, note that the disturbing rejection performance of the RGA-AW is quite acceptable, because it is capable to reestablish the dynamic behavior of the TOC concentration, S_T in the presence of different disturbances due to the cleaning, fault and maintenance of the sensor (see first and second set-point changes in Figure 8a). In the same way than in E1, the controller response becomes faster and the controlled variable remains closer to the set-point, when the control and estimation gains (Kc, Γ) are increased (see third and fourth set-point changes in Figure 8a). An important fact observed in this experiment is that the off-set disappears, since the TOC concentration moves around the set-point with a non constant error of ±50mg/l. Remember that in E1 the off-set was attributed to the difference between the calculated flow by the RGA-AW (solid line) and the flow fed by the peristaltic pump (dash-dot line). This conjecture is corroborated in Figure 8b where it can be seen that effectively, such a difference is smaller than in experiment E1. Unlike experiment E1, it is observed in Figure8b that the inlet flow, Q gets saturated several times either in the upper and lower limit due to the disturbances induced by the on-line sensor. However, the controller performance is not deteriorated. Figure 9a shows that the percentage production of methane during experiment E2 is also acceptable (i.e., higher than 65%). From 9b it is observed that the VFA accumulation is avoided contributing to assure the operational stability of the process.

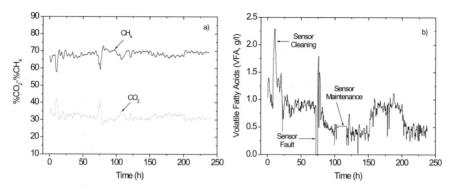

Fig. 9. (a) Percentage production of CO_2 and CH_4. (b) Dynamic behavior of the VFA when the MIR is used as on-line sensor.

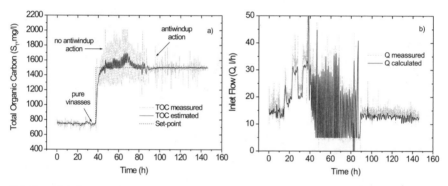

Fig. 10. (a) Set-point tracking performance in experiment E3 (S_T^*: dashed line, S_T: gray line and $\hat{S}_T$: black line). (b) Dynamical response of the manipulated variable when the AD digester is fed with pure vinasses (black line: Q calculated by the control law and gray line: Q measured at the entrance of the digester).

Finally, the results of experiment E3 are presented. First, the TOC concentration, S_T is regulated at 750mg/l by using the Zellweger Analitycs sensor and diluted vinasses. Figure 10a shows that the controller performance in this point is quite acceptable (i.e., no off-set is present and a small non constant error of ±25mg/l is obtained). Then, at $t = 37h$ the dilution tank was filled with pure vinasses and the antiwindup structure was unplugged by feeding back to the high-gain observer the feeding flow calculated by the control scheme (CCS) unlike to the feeding flow measured at the entrance of the process (CAA). After, the set-point was changed to $S_T^* = 1500mg/l$ at $t = 38h$. Note that the controller performance is deteriorated just after the set-point change is induced. This deterioration is evidenced by oscillations of high amplitude and frequency which are characteristics of the windup phenomenon. See in Figure 10b how the manipulated variable begins to wind from the lower to a high value. At $t = 86h$, the windup structure is reestablished and the controller performance is recovered as it is depicted in Figure 10a. Observe that the response of the manipulated variable in this point becomes of lower amplitude and frequency compared to that obtained at the beginning of the experiment (see Figure 10b). This difference is explained because of the big influence of pure vinasses on the system. So that, little changes in the feeding flow are enough to keep the TOC concentration close to the set-point. From Figure 11a, it can be seen that in the first part of the experiment, the percentage production of methane is around 70%. Note that after the controller performance is deteriorated, the percentage of methane diminishes to 60%.

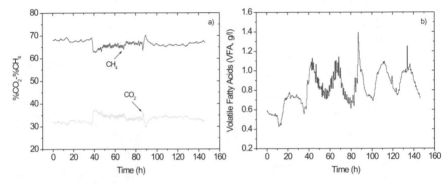

Fig. 11. (a) Percentage production of CO_2 and CH_4. (b) Response of the VFA concentration during E3.

However, once the controller performance is reestablished, the percentage of methane produced is recovered. On the other hand, the response of the VFA concentration along E3 is illustrated in Figure 11b. Observe that the response of the VFA at the beginning of E3 is similar to that obtained in E1 and E2; whereas the response becomes oscillating after pure vinasses are feed to the process. Nevertheless, there are not signs of a maintained accumulation of VFA during the experiment.

4 Regulation of VFA in an AD Process

As was pointed out in previous sections, the main control objective in AD processes is related to guarantee the stable operation of the process, which is also called operational stability [14]. The variables frequently used to monitoring the digester-stability are the biogas production rate and the COD, VFA and alkalinity effluent concentrations. Particularly, VFA are one of the most important intermediaries in the AD process, because their accumulation (generally induced by organic and toxic overloads) may lead to process failure due to the pH-drop they induce and their inhibitory effects in acid form [14, 27]. Thus, the importance of design efficient control systems capable to regulate the VFA concentration is of no doubt. However, this is not an easy task because several factors must be considered in the controller design: i) uncertain load disturbances, since the monitoring of substrate and metabolite is difficult and expensive; ii) modelling errors (uncertain kinetics), due the highly non linear nature of the process; iii) constraints in the control input due to practical operating conditions. As a consequence, few contributions concerning to the VFA regulation in AD processes can be found in the current literature [1, 22, 28], where only the fuzzy approach developed in [28] has been experimentally validated. This fact motivates and justifies the present section, where the main contribution is the design and the experimental validation of a RGA-AW capable to achieve the VFA regulation in spite of factors (i)-(iii), by using the input flow rate as manipulated variable.

4.1   Controller Design

Model Geometric Properties
From AD model (1), it is straightforward to observe that the dynamic behavior of the VFA concentration is given by the following set of equations

$$\begin{aligned}
\dot{X}_1 &= (\mu_1(.) - \alpha D)X_1 \\
\dot{X}_2 &= (\mu_2(.) - \alpha D)X_2 \\
\dot{S}_1 &= (S_{1,in} - S_1)D - k_1\mu_1(.)X_1 \\
\dot{S}_2 &= (S_{2,in} - S_2)D + k_2\mu_1(.)X_1 - k_3\mu_2(.)X_2
\end{aligned} \tag{11}$$

where $\mu_1(.)$, $\mu_2(.)$ are respectively the growth rates associated to the acidogenic and methanogenic bacteria within the digester. Therefore, it can be seen that AD model (11) is also an affine system, because it can be rewritten as it follows

$$\begin{bmatrix} \dot{X}_1 \\ \dot{X}_2 \\ \dot{X}_3 \\ \dot{X}_4 \end{bmatrix} = \begin{bmatrix} \mu_1(.)X_1 \\ \mu_2(.)X_2 \\ -k_1\mu_1(.)X_1 \\ k_2\mu_1(.)X_1 - k_3\mu_2(.)X_2 \end{bmatrix} + \begin{bmatrix} -\alpha X_1 \\ -\alpha X_2 \\ (S_{1,in} - S_1) \\ (S_{2,in} - S_2) \end{bmatrix} D$$

Next, the geometric properties of model (11) are analyzed assuming that the digester is operating under NOC.

Lemma 3. *Let $y = h(x) = S_2$ and $u = D$ respectively the output and input of the system. Then, AD model (11) has a well-defined relative degree $r = 1$ under NOC.*

Proof. By calculating the Lie derivative of the output along the vector fields $f(x)$ and $g(x)$, it is obtained that $L_g L_f^0 h(x) = (S_{2,in} - S_2)$. Then, since $(S_{2,in} - S_2) \neq 0$ the relative degree $r = 1$ is well-defined under NOC. ∎

Since the relative degree is less than the system order (i.e., $r = 1 < n = 4$), AD model (11) is partially linearizable. Then, it is necessary to define ϕ_j complementary functions for $j = r+1, r+2, \dots n$ in order to complete the mapping $\Phi(x)$ that allows the transformation of AD model (11) into the canonical form (4).

Proposition 5. *Let $\phi_2 = X_1/(S_{2,in} - S_2)^\alpha$, $\phi_3 = X_2/(S_{2,in} - S_2)^\alpha$ and $\phi_4 = X_2/(S_{1,in} - S_1)^\alpha$ smooth functions defined in the space of NOC, U. Then, the ϕ_j-functions for $j = r+1, r+2, \dots n$ are complementary functions of system (11) and the mapping*

$$\Phi(x) = \begin{bmatrix} S_2 \\ X_1/(S_{2,in} - S_2)^\alpha \\ X_2/(S_{2,in} - S_2)^\alpha \\ X_2/(S_{1,in} - S_1)^\alpha \end{bmatrix}$$

defines a local coordinate change for all $x \in U \subset \mathbb{R}^4$.

Proof. First, it is possible to verify that the ϕ_j-functions are solutions of the Lie derivative $L_g\phi_j(x) = 0$ given by the following partial differential equation

$$(-\alpha X_1)\frac{\partial \phi_j}{\partial X_1} + (-\alpha X_2)\frac{\partial \phi_j}{\partial X_2} + (S_{1,in} - S_1)\frac{\partial \phi_j}{\partial S_1} + (S_{2,in} - S_2)\frac{\partial \phi_j}{\partial S_2} = 0$$

Now, let

$$J_{\Phi(x)} = \left[\begin{pmatrix} 0 & 0 & 0 & 1 \\ \frac{1}{(S_{2,in}-S_2)^\alpha} & 0 & 0 & \frac{\alpha X_1}{(S_{2,in}-S_2)^{\alpha+1}} \\ 0 & \frac{1}{(S_{2,in}-S_2)^\alpha} & 0 & \frac{\alpha X_2}{(S_{2,in}-S_2)^{\alpha+1}} \\ 0 & \frac{1}{(S_{1,in}-S_1)^\alpha} & \frac{\alpha X_2}{(S_{1,in}-S_1)^{\alpha+1}} & 0 \end{pmatrix} \right]$$

be the jacobian matrix of mapping $\Phi(x)$, whose determinant is given by

$$\det\left(J_{\Phi(x)}\right) = \frac{\alpha X_2}{(S_{2,in} - S_2)^{2\alpha}(S_{1,in} - S_1)^{\alpha+1}}$$

Then, since the determinant is nonzero under NOC, the jacobian matrix is nonsingular and therefore the existence of an inverse

$$\Phi^{-1}(x) = \begin{bmatrix} z_1 \\ z_2(S_{2,in} - z_1)^\alpha \\ z_3/(S_{2,in} - z_1)^\alpha \\ S_{1,in} - (z_2/z_3)^{1/\alpha}(S_{2,in} - z_1) \end{bmatrix}$$

such as $\Phi^{-1}(\Phi(x)) = x$ for all $x \in U \subset \mathbb{R}^4$ is guaranteed, completing the proof.   ∎

The change of coordinates defined in Proposition 5 allow us to rewrite AD model (11) in the normal form (4) as it follows

$$
\begin{aligned}
\dot{z}_1 &= (S_{2,in} - z_1)D + k_2\mu_1(.)X_1 - k_3\mu_2(.)X_2 \\
\dot{z}_2 &= \frac{X_1}{(S_{2,in} - z_1)^\alpha}\left[\mu_1(.) - \alpha\frac{k_3\mu_2(.)X_2 - k_2\mu_1(.)X_1}{(S_{2,in} - z_1)}\right] \\
\dot{z}_3 &= \frac{X_2}{(S_{2,in} - z_1)^\alpha}\left[\mu_2(.) - \alpha\frac{k_3\mu_2(.)X_2 - k_2\mu_1(.)X_1}{(S_{2,in} - z_1)}\right] \\
\dot{z}_4 &= \frac{X_2}{(S_{1,in} - S_1)^\alpha}\left[\mu_2(.) - \frac{\alpha k_1\mu_1(.)X_1}{(S_{2,in} - z_1)}\right]
\end{aligned}
\tag{12}
$$

where $z_1 \in \mathbb{R}_+$ represents the controllable and observable part of the system when the dilution rate, D is used as the control input, whereas the remaining states $z_2, z_3, z_4 \in \mathbb{R}_+^3$ represent the uncontrollable and unobservable part also called internal dynamics [15].

Table 1. Equilibrium points of AD model (11)

	X_1^{eq}	X_2^{eq}	S_1^{eq}	S_2^{eq}
P1	0	0	$S_1^{eq} \neq S_{1,in}$	$S_2^{eq} \neq S_{2,in}$
P2	$-\frac{(S_{2,in} - S_2^{eq})}{\alpha k_2}$	0	$S_1^{eq} \neq S_{1,in}$	$S_2^{eq} \neq S_{2,in}$
P3	x_1^{eq}	0	0	0
P4	$\frac{(S_{1,in} - S_1 eq)}{\alpha k_1}$	$\frac{(S_{2,in} - S_1^{eq}) + \alpha k_2 X_1^{eq}}{\alpha k_3}$	$\mu_1^{eq}(.) = \mu_2^{eq}(.) = \alpha D$	$\mu_1^{eq}(.) = \mu_2^{eq}(.) = \alpha D$

Proposition 6. *AD model (11) is a minimum-phase system under NOC.*

Proof. By analyzing the steady-state behavior of AD model (11), four possible equilibrium points are obtained (see Table 1). However, it is observed that only the point P4 has physical meaning under NOC. This means that under NOC, AD model (11) has a single equilibrium point P4, which depends on the process kinetics and the influent composition. Now, in order to evaluate the stability of the internal dynamics of AD model (11), the following candidate Lyapunov function (CLF) is proposed

$$
\Psi = \frac{\phi_2\phi_4}{\phi_3} = \frac{X_1}{(S_{1,in} - S_1)^\alpha}
$$

which under NOC is positive defined. Therefore, by differentiating the CLF with respect to time and evaluating around the equilibrium point P4, it is obtained that

$$
\dot{\Psi} = \frac{\mu_1^{eq}(.)X_1^{eq}}{(S_{1,in} - S_1^{eq})^\alpha}\left[1 - \frac{\alpha k_1 X_1^{eq}}{(S_{1,in} - S_1^{eq})}\right] = \Psi\left[\mu_1^{eq}(.) - \alpha D^{eq}\right] = 0
$$

Then, the asymptotic stability of the internal dynamics is proved, which completes the proof. ∎

Remark 3. From Proposition 4, it is well-known that if the AD model (11) is a minimum-phase system and it has a well-defined relative degree under NOC, the following input-output linearizing controller guarantee the exponential convergence of the VFA concentration, S_2 to a desired value, S_2^* for all $t > 0$

$$D = \frac{1}{(S_{1,in} - S_1^{eq})} [-k_2\mu_1(.)X_1 + k_3\mu_2(.)X_2 - \omega(x)] \tag{13}$$

where $\omega(x) = Kc(S_2 - S_2^*)$ and Kc is any positive constant (see the proof of Proposition 4). However, as it was established, the linearizing controller (13) cannot be applied directly in practice due to the uncertainties in either the kinetic terms or the influent composition. Then, to overcome this drawback a RGA-AW is proposed in the next section.

Robust Geometric Approach

Here, a RGA-AW is proposed to regulate the VFA concentration, S_2 in spite of uncertainties in either the influent composition or the process kinetics. Therefore, the controller design is carried out under assumptions (A1)-(A4) defined in Section 3.1. Without lost of generality, let us assume that the influent VFA concentration is described as follows: $S_{2,in} = \tilde{S}_{2,in} + \Delta_{S_2}$, where Δ_{S_2} is an uncertain and bounded function related to the variation of the influent composition around a well-known nominal value $\tilde{S}_{2,in}$, which can be determined in practice by a single VFA off-line measurement of the wastewater to be treated.

Now, by defining the function

$$\eta = \sigma(z, u) = L_f h(\Phi^{-1}(x)) + \Delta_{S_2} u = k_2\mu_1(.)X_1 - k_3\mu_2(.)X_2 + \Delta_{S_2} u$$

the AD model in the normal form (12) can be rewritten in the extended state-space representation (7). Then, by following the same ideas described in Section 3, the following RGA-AW is proposed

$$\dot{\hat{z}}_1 = \hat{\eta} + (\tilde{S}_{2,in} - \hat{z}_1)u_{sat} + \Gamma g_1(z_1 - \hat{z}_1)$$
$$\dot{\hat{\eta}} = \Gamma^2 g_2(z_1 - \hat{z}_1) \tag{14}$$
$$u = \frac{1}{(\tilde{S}_{2,in} - z_1)} [-\hat{\eta} - Kc(\hat{z}_1 - S_2^*)]$$

In what follows, the performance and robustness of the RGA-AW (14) is experimentally evaluated for the VFA regulation in the INRA-LBE fully instrumented pilot-scale plant (see Figure 5).

4.2   Experimental Validation

In this section, the results obtained from the experimental validation of the RGA-AW (14) are presented. It is important to remark that the experimental validation was carried out one month after the AD process was re-started

Table 2. Set-point changes induced along the experiment

Time (h)	0	127	192	242	533	600	868
S_2^* (mgVFA/l)	Open-loop	1400	1000	1800	2500	3500	1500

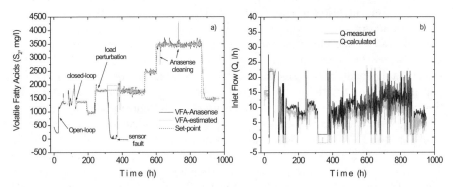

Fig. 12. (a) Response of the AGV concentration and (b) the manipulated variable when the control law (14) is experimentally implemented

after seven months of inactivity, which implies highly uncertain conditions. Some months before the experimental validation, the digester volume was reduced because of local security regulations. Therefore, the effective volume of the digester was diminished from 0.9483m^3 to 0.528m^3, whereas the influent restrictions were fixed at $Q^{min} = 1$ l/h and $Q^{max} = 22$ l/h. The controller algorithm was implemented by using a free software developed by the laboratories INRIA and INRA-LBE which is called ODIN. In addition, a commercial automatic titrimetric analyzer called ANASENSE was used to measure online the VFA concentration with a sampling time of 30min [24], which is fast enough compared to the hydraulic residence time of the process (between 10 to 12 h) allowing the assumption of continuous control. Therefore, the RGA-AW (14) was calculated each 30min being the same time that the control input remains constant. The controller performance was tested for one month under different set-point values and load disturbances. Particularly, six set-point changes were induced (see Table 2), by using the following control parameters: $Kc = 0.4$, $g_1 = 0.7$, $g_2 = 0.7$ and $\tilde{S}_{2,in} = 7500\text{mgVFA/l}$, which corresponds to a single off-line measurement of the VFA concentration of a sample of pure vinasses. This value was selected in order to induce an important error in the actual nominal value $S_{2,in} \approx 5200\text{mgVFA/l}$ to test the robustness of the RGA-AW (14), because diluted vinasses were used to feed the digester.

The response of the VFA concentration under the RGA-AW (14) is shown in Figure 12a. In general, it can be observed that the performance of the control law along the experiment is quite acceptable even when several disturbances were induced. For example, at $t = 260$ a load perturbation was induced by changing the dilution relationship from 11l H_2O-9l vinasses to 8l H_2O-12l vinasses, which implies that the nominal value was varied from $S_{2,in} \approx 5200\text{mgVFA/l}$ to $S_{2,in} \approx 6700\text{mgVFA/l}$. Also, three more disturbances were induced due to the fault and cleaning of the sensor; respectively (see Figure 12a). Note that the proposed scheme (14) is capable to regulate the VFA concentration in spite of these disturbances by using the same control parameters and knowing neither the inlet VFA concentration, $S_{2,in}$ nor the behavior of the kinetic terms $k_2\mu_1(.)X_1$, $k_3\mu_2(.)X_2$. The response of the inlet flow used as manipulated variable is depicted in Figure 12b. In general, the dynamic behavior of the manipulated variable is smooth along the experiment. Note that there is a difference between the inlet flow calculated by the control scheme (14) and that measured at the entrance of the digester. Observe that this error becomes smaller at high flows avoiding the presence of off-set (see Figure 12a). Note also that even when the manipulated variable gets saturated in either the upper and lower limits, the controller performance is never deteriorated On the other hand, the biogas composition is depicted in Figure 13. Remember that this information is used in the monitoring of AD processes, since an increase in the CO_2 concentration is an indication of instability in the process. However, note that the methane production is in average around 70% along the experiment avoiding this possibility. Observe also that the gaseous flow composition is influenced by the VFA concentration.

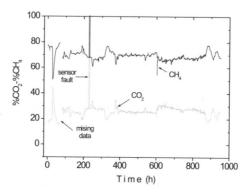

Fig. 13. Percentage of methane and carbon dioxide when the RGA-AW (14) is applied

5   Concluding Remarks

Recently, anaerobic digestion (AD) has recovered the attention of the scientific community, since it represents a suitable and promising solution to reduce the organic matter and pathogenic organisms from wastewater [12]. Nevertheless, AD process is not widely applied at the industrial scale [36], because it is known as an unstable process in the sense that large variations of the dilution rate and influent organic composition may lead to the crash of the digester. This is why actually the scientific efforts are focused not only to extend the number of AD applications, but also in the proposal of control schemes that increase the robustness of the process against disturbances [31]. In this chapter, a robust nonlinear approach was developed and experimentally validated to regulate (i) the total amount of decomposable organic pollutant agents measured as total organic carbon (TOC) and, (ii) the concentration of volatile fatty acids (VFA); by using in both cases the input flow rate as manipulated variable. The proposed approach is a model-based controller obtained from geometric control tools and the definition of an uncertain but observable function that lumps the uncertain terms associated to the dynamics of the controlled variables TOC and VFA (e.g., feeding composition, kinetic growth functions and parameters). The robust scheme is composed of a feedback linearizing control law and a high-gain Luenberger observer which adapts the linearizing control law from estimates of the uncertain function. The robust approach was experimentally validated in a pilot-scale fixed-bed AD process used in the wastewater treatment of industrial distillery vinasses. Several experiments were performed to evaluate the controller performance and robustness under different set-points, feeding concentrations and sampling times (2 and 30min). Results show that both, the regulation of COT and VFA can be addressed in spite of the full ignorance of the kinetic growth functions, noisy measurements and unknown feeding composition.

Acknowledgments. This work was supported by PROMEP/103.5/05/1705 and CONACyT J50282-Y. Also, the authors would like to thanks the anonymous reviewers for their suggestions to improve this paper.

References

[1] V. Alcaraz-González, J. Harmand, A. Rapaport, J.P. Steyer, V. González-Álvarez, and C. Pelayo-Ortíz. Robust interval-based SISO and SIMO regulation for a class of highly uncertain bioreactors: Application to the anaerobic digestion. In *IEEE Conf. Decision and Control (CDC)*, Sydney, Australia, December 2000.

[2] J. Andrews. A mathematical model for the continous culture of microorganisms utilizing inhibitory substrates. *Biotechnol. Bioeng.*, 10:707–723, 1968.

[3] R. Antonelli, J. Harmand, J.P. Steyer, and A. Astolfi. Set-point regulation of an anaerobic digestion process with bounded output feedback. *IEEE Trans. Contr. Syst. Technol.*, 11(4):495–504, 2003.

[4] A.N. Atassi and H.K. Khalil. Separation results for the stabilization of nonlinear systems using different high-gain observer designs. *Systems Control Lett.*, 39:183–191, 2000.

[5] G. Bastin and D. Dochain. *On-Line Estimation and Adaptive Control of Bioreactors.* Elsevier, Amsterdam, 1990.

[6] D.J. Batstone, J. Keller, R.B. Newell, and M. Newland. Model development and full scale validation for anaerobic treatment of protein and fat based wastewater. *Wat. Sci. Technol.*, 36:423–431, 1997.

[7] O. Bernard, Z. Hadj-Sadok, D. Dochain, A. Genovesi, and J.P. Steyer. Dynamical model development and parameter identification for anaerobic wastewater treatment process. *Biotechnol. Bioeng.*, 75(4):424–438, 2001.

[8] R. Femat, J. Alvarez-Ramírez, and M. Rosales-Torres. Robust asymptotic linearization via uncertainty estimation: Regulation of temperature in a fluidized bed reactor. *Comput. Chem. Eng.*, 23:697–708, 1999.

[9] S. Graef and J. Andrews. Mathemathical modelling and control of anaerobic digestion. *Wat. Res.*, 8:261–289, 1974.

[10] E. Heinzle, I.J. Dunn, and G.B. Ryhiner. Modelling and control for anaerobic wastewater treatment. In R. Aarts, M. Aynsley, J.E. Bailey, P.M. Doran, and I.J. Dunn, editors, *Bioprocess Design and Control*, volume 48 of *Advances in Biochemical Engineering/Biotechnology*, pages 79–114. Springer-Verlag, Berlín, 1993.

[11] M.A. Henson and D.E. Seborg. *Nonlinear Process Control*. Prentice Hall, New Jersey, 1997.

[12] M. Henze, P. Harremoes, J. LA Cour Jansen, and E. Arvin. *Wastewater Treatment: Biological and Chemical Processes*. Springer Verlag, 2nd. edition, 1997.

[13] D.T. Hill and C. Barth. A dynamic model for stimulation of animal waste digestion. *Wat. Poll. Cont. Ass.*, 10:2129–2143, 1977.

[14] D.T. Hill, S.A. Cobbs, and J.P. Bolte. Using volatile fatty acid relationships to predict anaerobic digester failure. *Trans. ASAE*, 30:496–501, 1987.

[15] A. Isidori. *Nonlinear Control Systems*. Springer Verlag, third edition, 1995.

[16] A. Isidori, A.R. Teel, and L. Praly. A note on the problem of semiglobal practical stabilization of uncertain nonlinear systems via dynamic output feedback. *Systems Control Lett.*, 39:165–171, 2000.

[17] G. Kiely, G. Tayfur, C. Dolan, and K. Tanji. Physical and mathematical modelling of anaerobic digestion of organic wastes. *Wat. Res.*, 31:534–540, 1997.

[18] M.V. Kothare, P.J. Campo, M. Morari, and C.N. Nett. A unified framework for the study of anti-windup designs. *Automatica*, 30(12):1869–1883, 1994.

[19] J.F. Malina and F.G. Pohland. Design of anaerobic processes for the treatment of industrial and municipal wastes. *CRC Press*, 1992.

[20] H.O. Méndez-Acosta, D.U. Campos-Delgado, R. Femat, and V. González-Álvarez. A robust feedforward/feedback control for an anaerobic digester. *Comput. Chem. Eng.*, 31:1–11, 2005.

[21] H.O. Méndez-Acosta, R. Femat, and D.U. Campos-Delgado. Improving the performance on the COD regulation in anaerobic digestion. *Ind. Eng. Chem. Res.*, 43(1):95–104, 2004.

[22] H.O. Méndez-Acosta, B. Palacios-Ruiz, and V. Alcaraz-González. Robust regulation of volatile fatty acids in an anaerobic digester. In *IFAC-ROCOND*, Toulouse, France, July 2006.

[23] R. Moletta, D. Verrier, and G. Albagnac. Dynamic modeling of anaerobic digestion. *Wat. Res.*, 20:427–434, 1986.

[24] F. Molina, G. Ruiz, E. Roca, and J.M. Lema. Report on full-scale performance of the sensors within telemac. Technical Report D2.7, TELEMAC IST-2000-28156, Spain, 2004.

[25] F. Mosey. Mathematical modeling of the anaerobic digestion process: Regulatory mechanisms for the formation of short-chain volatile acids from glucose. *Wat. Sci. Technol.*, 15:209–232, 1983.

[26] H. Nijmeijer and A.J. van der Schaft. *Nonlinear Dynamical Control Systems*. Springer Verlag, 1991.

[27] G. Olsson, M.K. Nielsen, Z.Yuan, A. Lynggaard-Jensen, and J.P. Steyer. Instrumentation, control and automation in wastewater systems. Technical Report 15, IWA, 2005.

[28] A. Puñal, L. Palazzotto, J.C. Bouvier, T. Conte, and J.P. Steyer. Automatic control of VFA in anaerobic digestion using a fuzzy logic based approach. In *IWA VII Latin America Workshop and Symposium on Anaerobic Digestion*, pages 126–133, Merida, Mexico, 2002.

[29] S. Rönnbäck. *Linear Control of Systems with Actuator Constraints*. PhD thesis, Division of Automatic Control, Lulea University of Technology, May 1993.

[30] A. Rozzi. Alkalinity considerations with respect to anaerobic digester. In *Proceeding 5th Forum Applied Biotechnol.*, volume 56, pages 1499–1514, Med. Fac. Landbouww. Rijksuniv. Gent, 1991.

[31] J.P. Steyer, O. Bernard, D.J. Batstone, and I. Angelidaki. Lessons learnt from 15 years of ICA in anaerobic digesters. *Wat. Sci. Technol.*, 53(4):25–33, 2006.

[32] J.P. Steyer, J.C. Bouvier, T. Conte, P. Gras, J. Harmand, and J.P. Delgenes. On-line measurements of COD, TOC, VFA, total and partial alkalinity in anaerobic digestion process using infra-red spectrometry. *Wat. Sci. Technol.*, 45(10):133–138, 2002.

[33] J.P. Steyer, P. Buffiere, D. Rolland, and R. Moletta. Advanced control of anaerobic digestion processes through disturbances monitoring. *Wat. Res.*, 33(9):2059–2068, 1999.

[34] S. Tarbouriech and G. Garcia. *Stabilization of Linear Discrete-Time Systems with Saturating Controls and Norm-Bounded Time-Varying Uncertainty. Control of uncertain systems with bounded inputs; Lecture Notes in Control and Information Science 227.* Springer-Verlag: Berlin, 1997.

[35] A. Teel and L. Praly. Tools for semiglobal stabilization by partial state and output feedback. *SIAM J. Control Optim.*, 33(5):1443–1488, 1995.

[36] D.E. Totzke. Anaerobic treatment technology overview. Technical report, Applied Technologies Inc., USA, 1999.

[37] J. Van Lier, A. Tilche, B.K. Ahring, H. Macarie, R. Moletta, M. Dohanyos, L.W. Hulshoff Pol, P. Lens, and W. Verstraete. New perspectives in anaerobic digestion. *Wat. Sci. Technol.*, 43(1):1–18, 2001.

Advances in Diagnosis of Biological Anaerobic Wastewater Treatment Plants

L. Lardon and J.P. Steyer

Ingénierie des Procédés, INRA-LBE
{lardonl,steyer}@ensam.inra.fr

Summary. The on-line diagnosis is a key requirement of industrial processes. This is particularly true in the case of biological process due to the composition of media, the requirements of operating conditions and the wide variety of possible disturbances that necessitate careful and constant monitoring of the processes. Moreover, because only partial information is available in an on-line context and because of the technical and biological complexities of the involved processes, specific characteristics are required for diagnosis purposes. Several approaches like quantitative model based, qualitative model based and process history based methods were applied over the years. This chapter presents a methodological framework based on Evidence theory to manage the fault signals generated by conventional approaches (*i.e.*, residuals from hardware and software redundancies, fuzzy logic based modules for process state assessment) and to account for uncertainty. The advantages of using evidence theory like modularity, detection of conflict and doubt in the information sources are illustrated with experimental results from a $1m^3$ fixed bed anaerobic digestion process used for the treatment of industrial distillery wastewater.

1 Introduction

To meet higher standards, industrial processes contain a large number of monitored variables and of actuators managed by automatic controllers. Such an automation contributes to maintain satisfactory operating conditions through dynamic compensation of perturbations and changes occurring on the process. However, some perturbations cannot be compensated and control loops can themselves create a dysfunction of the process *e.g.* in case of fouling of sensors. The role of Fault Detection and Isolation (FDI) is to detect these changes and, if possible, to correct them. In this chapter, after presenting the main methods of FDI, it is shown how an accurate uncertainty representation model – the Evidence theory –, can provide a unifying framework to diagnosis methodologies and improve the performances and the interpretation of diagnosis modules.

H.O. Méndez-Acosta et al. (Eds.): Dyn. & Ctrl. of Chem. & Bio. Proc., LNCIS 361, pp. 201–239, 2007.
springerlink.com   © Springer-Verlag Berlin Heidelberg 2007

1.1   First Definitions

To understand the goals of the particular contributions and to compare the different approaches of diagnosis, the IFAC (*i.e.*, International Federation on Automatic Control) Technical Committee SAFEPROCESS (Fault Detection, Supervision and Safety for Technical Processes) has defined a common terminology. These definitions, detailed in [13] will be adopted throughout the present chapter.

A *fault* is understood as an unpermitted deviation of at least one characteristic property or parameter of the system from the acceptable, usual or standard condition. A fault can stem from several origins as depicted by the Figure 1. It can be caused by an unexpected *perturbation* (*i.e.*, a major deviation from one input acting on the system) or by a *disturbance* (*i.e.*, the action of an unknown and uncontrolled input on the system). Another fault origin can be an *error* of any sensor or actuator, which is a deviation between the measured and the true or specified value.

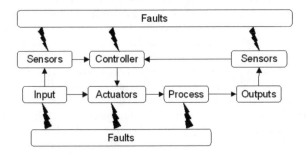

Fig. 1. Possible occurrence of faults

To face these faults, *fault diagnosis* modules are developed. They are based on the observations and interpretations of *symptoms*, *i.e.* the changes of an observable quantity from normal behavior. Fault Diagnosis task can be split in three main steps.

- fault *detection*: determination of the presence of fault(s).
- fault *isolation*: determination of the kind, location and time of a fault.
- fault *identification*: determination of the size and time-variant behavior of the fault.

These diagnosis modules are integrated into monitoring or supervision systems: the *monitoring* of a system is the continuous real-time task of determining the conditions of a physical system, by recording information, recognizing and indicating anomalies in the behavior. The *supervision* of a system consists in monitoring it and taking appropriate actions to maintain the operation in the case of faults, as depicted by Figure 2.

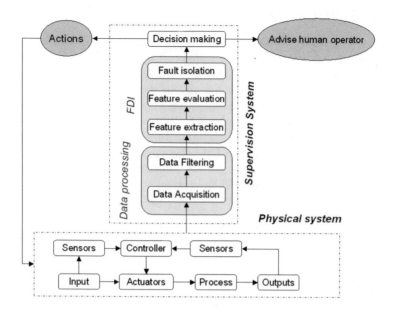

Fig. 2. The different general sub-tasks performed by a supervision system

A great number of approaches have been proposed in the literature to perform the diagnosis of a system. Before briefly detailing them, it is mandatory to define desirable characteristics common to all fault diagnosis systems, as proposed by:

- *quick detection and diagnosis*,
- *isolability*: the ability to distinguish the observed fault between several possible failures,
- *robustness*: providing good result in noisy and uncertain conditions. There is often a trade-off between quick detection and robustness,
- *novelty identifiability*: ability to detect an unknown fault,
- *classification error estimate*: providing to the end user a confidence measure of the performed diagnosis,
- *adaptability-modularity*: process can evolve (*e.g.*, new working states, new sensors, ...) these changes should be included into the scope of diagnosis system,
- *multiple-fault identifiability*: this important requirement is often hard to meet because of the interactions between faults,
- *uncertainty management*: new techniques allow one to generate uncertain data rather than arbitrary values. Such intervals and confidence index have to be managed by the diagnosis system.

1.2   Main FDI Methods

In practice, the most frequently used diagnosis method is to monitor the level (or trend) of a particular variable, and to take a decision when the signal has reached a given threshold. This method of limit checking, even though very simple to implement, has some serious drawbacks. The first one is the possibility of false alarms in the event of noise, input variations and changes of operating point. The second drawback is firstly that a single fault could cause many system signals to exceed their limits and appear as multiple faults, and secondly that different faults can have the same effect on a specific variable. As a consequence, fault isolation is very difficult. A correct interpretation of these observed system signals will thus lie on a model describing the system. Models used for diagnosis can be:

- *quantitative models*: a set of static and dynamic relations linking system variables and parameters in order to describe a system's behavior in quantitative mathematical terms,
- *qualitative models*: a set of static and dynamic relations linking system variables and parameters in order to describe a system's behavior in qualitative terms such as causalities or "if-then" rules,
- *diagnostic models*: a set of static or dynamic relations linking symptoms to faults.

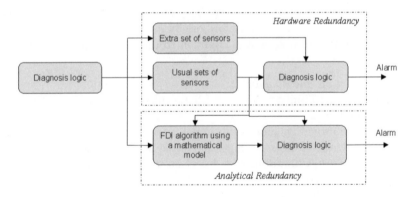

Fig. 3. Hardware vs. analytical redundancy

A traditional approach to fault diagnosis in the wider application context is based on "hardware (*i.e.* physical) redundancy" methods which use multiple lines of sensors, actuators, computers and software to measure and/or control a particular variable. Typically, a voting scheme is applied to the hardware redundant system to decide if and when a fault has occurred and its likely location amongst redundant system components. The use of multiple redundancy in this way is common, for example with digital fly-by-wire flight control systems in the aeronautics field and in other applications such as in nuclear reactors. The major problems encountered with hardware redundancy are the extra equipment and maintenance costs, which limit the applicability of this solution in the wastewater treatment field.

In view of the conflict between the reliability and the cost of adding more hardware, it is sensible to attempt to use the dissimilar measured values together to "cross check"[1] each other, rather than replicating each hardware individually. This is the concept of "analytical (*i.e.* functional) redundancy" which uses redundant analytical (or functional) relationships between various measured variables of the monitored process (*e.g.*, inputs/outputs, outputs/outputs and inputs/inputs). Figure 3 illustrates the hardware and analytical redundancy concepts.

In analytical redundancy schemes, the resulting difference generated from the consistency checking of different variables is called a *residual* signal. The residual should be by convention zero-valued when the system is normal and should diverge from zero when a fault occurs. This zero and non-zero property of the residual is used to determine whether or not a fault has occurred. Analytical redundancy makes use of a quantitative model of the monitored process and is therefore often referred to as the "model-based approach" to fault diagnosis.

Several model-based techniques have been proposed to detect and to isolate faults:

- **Diagnostic observers** consist in the definition of a set of observers from which it is possible to define residuals specific of only one failure [8].
- **Parity relations** are relations derived from an input-output model or a state-space model [11] checking the consistency of process outputs and known process inputs.
- Another way is to build a **state estimator** from a Kalman filter based on the normal operating model [2],[36].
- Finally, on-line **parameter estimation** allows one to diagnose parameter drifts [12].

The models used can be either fixed or adaptive and parametric or nonparametric models. These methods have different performances depending on the kind of fault to be treated (*i.e.*, additive or multiplicative faults). Analytical model-based approaches require knowledge to be expressed in terms of input-output models or first principles quantitative models based on mass and energy balance equations. These methodologies give a consistent base to perform fault detection and isolation. The cost of these advantages relies on the modeling and computational efforts and on the restriction that one places on the class of acceptable models.

Other methods, the *signal-model-based methods*, only use the available online measurements. They can detect changes from the normal behavior using

---

[1] This procedure is sometimes referred as "data reconciliation".

some characteristics of the signal (*e.g.*, the mean values and variance). They also use some methods from the signal-processing field (*e.g.*, segmentation methods). Pattern recognition-based methods are either statistical (*i.e.*, parametric and non-parametric classifiers) or non-statistical (*e.g.*, neural-network classifiers). They are based on the process history data and their main advantages rely on their real-time performances, their facility for knowledge acquisition and their applicability to a wide variety of systems. On the other hand, they are limited in their generalization capability outside the training data and they have difficulties in identifying multiple faults.

Knowledge-based diagnosis uses qualitative models in combination with different search techniques. The hypothesis-and-test using causal models, the finite state-space search of fault trees and the search in malfunction hypothesis hierarchy are among the different methods used. They generally rely on explicit knowledge represented by causal pathways to link the available measurements with the fault origin or by a hierarchy of malfunction hypothesis. They can be used when abundant experience of the process operation is available but not enough detailed to be used in accurate quantitative models. On the other hand, they suffer from the resolution problems related to the ambiguity in qualitative reasoning.

In the last decade, fuzzy logic has gained in applications in several fields. It has been for example applied to signal processing, process modeling and non-linear control. However, despite its popularity, less attention has been paid to its application on fault detection, fault isolation and supervision even though the ability of fuzzy logic to handle vague and imprecise information can be of large interest for advanced supervision systems. Few demonstrations of the advantages of fuzzy-based *FDI* systems (particularly concerning the residual evaluation) can be found in the literature; see for example [10],[21],[33].

Furthermore, general techniques such as *data mining* and *case-based reasoning* have been applied to different fields of application. Data mining allows the automatic detection of features and trends in data. Time-based data mining for monitoring and analyzing evolving trends has developed relatively recently but is now a better-understood technology [4]. It is a valuable technique for detecting and responding to drifts in sensor performance over time. Integrated with data mining, case-based reasoning are other valuable techniques for abstracting from the data to higher-level representations of conditions of the plant. Case-based reasoning involves matching the actual situation against a library of prior cases, modifying them if necessary, using the important characteristics of each case to perform the matching. Moreover, regarding decision support, it is also important to equip the human decision-makers – whether at the local site or the remote monitoring center – with the information they need to make informed decisions [18].

As a conclusion, it is important to state that the use of only one fault detection, isolation and diagnosis method is not suitable for real industrial implementation. Furthermore, a mixture of shallow and deep knowledge is generally available on the real processes and, in some cases, analytical models of subsystems of the process do exist. This is why integration of these different methods has been encouraged over the recent years. Among the desirable properties of a diagnosis system, some are mainly consequences of the tuning of the fault signal generation method. But diagnosis stability, multiple-fault detection, modularity and uncertainty management are directly determined by the method used to combine all these signals. Often fault signals are compared to a threshold and the list of possible faults is determined by a boolean isolation. This technique has several drawbacks like diagnosis instability [16] and is unadapted to manage the uncertainty of information sources (imprecise or conflicting information). Using approximate reasoning techniques rather than boolean can improve performances of the fault isolation.

As it is shown in the following, among all the uncertain reasoning methods, the Dempster-Shafer theory offers a theoretical framework particularly suited to tackle the objectives of the state manager (modularity, uncertainty management, robustness). In addition, its application is independent of the method used to determine the fault signals.

1.3   Specificity of FDI for WWTP

In the recent years, the development and improvements of high performance bioreactors, of on-line facilities and the application of automatic control have brought evidence that bioprocesses could be optimized and, when concerned with biological wastewater treatment, that efficient pollutants removal could be achieved. However, more than their control, the diagnosis of faults occurring in biological processes has become nowadays a very challenging research area. Indeed, even in normal operational conditions, several types of disturbances can be present (*e.g.*, influence of the *inocula*, contamination of the media, presence of toxic, fouling of sensors,...) and they can largely affect the operating conditions. Moreover, these disturbances can be either sudden or slow, and they can be related to normal or faulty process operation provoking real or apparent deviations from the normal operation. Hence, there is a clear need for advanced supervisory control (*i.e.*, gathering on-line control and diagnosis) in order to keep the system performance as close as possible to optimal. This is particularly true for biological processes with environmental purposes (*e.g.*, biological WasteWater Treatment Plants or WWTPs) where the state of "the living" part of the system must be closely monitored together with large possible disturbances occurring on any part of the systems.

In the present chapter, anaerobic digestion (AD) is chosen as an illustrative example of biological WWTPs. It is a set of biological processes that take place in the absence of oxygen and by which organic matter is decomposed and converted on one hand into biogas (*i.e.*, a mixture of mainly carbon dioxide and methane) and, on the other hand, into microbial biomass and residual organic matter. AD is naturally present in many ecosystems such as the digestive tract of animals as well as in natural and cultivated ecosystems like wetlands, marine sediments and rice fields where it is actively involved in biogeochemical cycles of matter. In parallel, AD can be considered as one of the oldest technologies for waste and wastewater treatment.

AD can be described as a two step reaction, transforming a complex organic substrate, characterized by its Chemical Oxygen Demand (COD), into Volatile Fatty Acids (VFA) and then transforming VFAs to CH_4 and CO_2. Both reactions are operated by micro-organisms grouped for modeling reasons, into two groups X_1 and X_2, that respectively denote acidogenous and methanogenous micro-organisms.

$$COD \xrightarrow{X_1} X_1 + VFA + CO_2 \qquad VFA \xrightarrow{X_2} X_2 + CH_4 + CO_2$$

Several advantages are recognized to AD processes when used as WWTPs: high capacity to treat slowly degradable substrates at high concentrations like wine vinasses or aerobic sludges, very low sludge production, potentiality for valuable intermediate metabolites production, low energy requirements and possibility for energy recovery through methane combustion. However AD presents is highly sensitive to overloads of the process because of the inhibition of the methanogenous biomass by high concentrations of its own substrate. This instability explains probably that despite large interests and more than 1,400 commercial installations referenced world-wide in 1999 [32], many industrials are still reluctant to use AD processes, probably because of the counterpart of their efficiency: they can become unstable under some circumstances. This is why actual research aims not only to extend the potentialities of anaerobic digestion [35], but also to optimize anaerobic processes and increase their robustness towards perturbations [34].

Most of FDI approaches previously presented were developed in the field of safety-critical control (*i.e.*, to avoid incidents such as the *Three Mile Island* accident or the *Chernobyl* disaster) but only very few studies were performed for biological processes in general – mostly at the laboratory scale – and wastewater treatment plants in particular. In addition, with only fewer exceptions [15],[31], there is no *industrial* application of diagnosis for biological processes. This can be explained by the fact that biological systems – because they involve living organisms – require specific developments for diagnosis in general. They are indeed very complex non-linear and time-changing systems (maybe the most complicated systems after the human body!) with only very few on-line measurements to monitor them. As a consequence, trying to apply diagnosis strategies developed in other fields generally leads to failures. The approach presented in this chapter tries to afford biological process complexity by grounding the diagnosis of theses process on an uncertainty representation formalism, the Evidence theory.

2 The Dempster and Shafer's Evidence Theory

2.1 Basic Elements of the Evidence Theory

Evidence theory also known as the Dempster-Shafer theory, has been first introduced by Dempster [5], then formalized by [25] and finally axiomized later into the framework of the Transferable Belief Model (Evidence theory) by Smets [28]. The Evidence theory can be understood as an alternative to probability theory for the representation of uncertainty. It allows one to manipulate non-necessarily exclusive events and thus to represent explicitly information uncertainty.

This theory assumes the definition of (i) the *frame of discernment* Ω consisting of the exhaustive and exclusive hypothesis and (ii) the *reference set* 2^{Ω} of all the disjunctions of the elements of Ω. The Evidence theory defines the *basic belief assignment* (*bba*) function as an elementary mass function $m : 2^{\Omega} \rightarrow [0, 1]$ verifying for all elements A of 2^{Ω}:

$$\sum_{\{A \in 2^{\Omega} / A \subseteq \Omega \text{ and } A \neq \emptyset\}} m^{\Omega}(A) \leq 1 \tag{1}$$

The elements of 2^{Ω} whose mass is non-zero are called the *focal* elements of m and constitute the core N_m of the belief assignment.

A *bba* is a distribution of a unit mass of evidence among all the elements of 2^{Ω} (*i.e.*, between all subsets of Ω rather than between the singletons of Ω as it is the case in probability theory). The mass of evidence attributed to a disjunction of singletons represents the amount of knowledge which cannot be more precisely allocated without hypothesis; as a consequence, a *bba* represents the exact knowledge of an information source. Moreover, if all focal elements of a belief structure are singletons, this belief structure is similar to a probability distribution.

A *bba* is also characterized by two functions: the *credibility* function denoted *Cred* and the *plausibility* function denoted *Pl*. The credibility of a subset A is the sum of all pieces of evidence that support A and the plausibility of A the sum of pieces of evidence not supporting $\neg A$.

$$\forall \; A \in 2^{\Omega} \begin{cases} Cred_m^{\Omega}(A) = \sum_{\{B \in 2^{\Omega} / B \subseteq A\}} m^{\Omega}(B) \\ \\ Pl_m^{\Omega}(A) = \sum_{\{B \in 2^{\Omega} / B \not\subseteq \bar{A}\}} m^{\Omega}(B) = 1 - Cred_m^{\Omega}(\bar{A}) \end{cases} \tag{2}$$

The interval $[Cred(A), Pl(A)]$ can thus be considered as the bounds of the not exactly known probability of A.

2.2 Combination Rules

Based on these definitions, combination operators can be defined. For example, it is possible to build a unique elementary mass function m from two elementary mass functions m_1 and m_2, arising from two distinct and independent sources but defined on the same set, such that $m = m_1 \oplus m2$ where $\oplus$ denotes the combination operator. The so-called Dempster's rule consists of calculating:

$$\forall\, C \in 2^\Omega / \{\emptyset\}\ m^\Omega(C) = \tfrac{1}{1-K} \sum_{\{(A,B)\in(2^\Omega)^2 / A\cap B=C\}}\ m_1^\Omega(A) \cdot m_2^\Omega(B)$$
(3)

$$K = \sum_{\{(A,B)\in(2^\Omega)^2 / A\cap B=\emptyset\}}\ m_1^\Omega(A) \cdot m_2^\Omega(B)$$

Let us here take a simple example that, even though being not related diagnosis, illustrates these notions and can help to understand the Evidence theory. Let us suppose that a student has to be evaluated by two professors P_1 and P_2. Each professor has an evaluation grid (*Bad, Average, Good*). Now L_1 will design *Bad*, L_2 *Average* and L_3 *Good*. The problem for the president of the group of professors is to decide the "true" level of the student. Represented into the formalism of Evidence theory the problem defines the following frame of discernment and reference set:

$$\Omega = \{L_1, L_2, L_3\}$$
$$2^\Omega = \{L_1, L_2, L_3, L_1 \cup L_2, L_1 \cup L_3, L_2 \cup L_3, L_1 \cup L_2 \cup L_3\}$$

Let us assume that the first professor says "the student is Bad with a probability of 10%, Bad or Average with a probability of 80% and Good with a probability of 10%", and the second one says "the student is Bad or Average with a probability of 20% and Average or Good with a probability of 80%".

These two independent sources of information can be described by the following belief functions:

$$\begin{vmatrix} m_1^\Omega(L_1) = 0.1 \\ m_1^\Omega(L_1 \cup L_2) = 0.8 \\ m_1^\Omega(L_3) = 0.1 \end{vmatrix} \quad \begin{vmatrix} m_2^\Omega(L_1 \cup L_2) = 0.2 \\ m_2^\Omega(L_2 \cup L_3) = 0.8 \end{vmatrix}$$

The resulting Credibility and Plausibility functions are:

$$\begin{vmatrix} Cred_1^\Omega(L_1) = 0.1 & Pl_1^\Omega(L_1) = 0.9 \\ Cred_2^\Omega(L_2) = 0 & Pl_2^\Omega(L_2) = 0.8 \\ Cred_3^\Omega(L_3) = 0.1 & Pl_3^\Omega(L_3) = 0.1 \end{vmatrix} \quad \begin{vmatrix} Cred_1^\Omega(L_1) = 0 & Pl_1^\Omega(L_1) = 0.2 \\ Cred_2^\Omega(L_2) = 0 & Pl_2^\Omega(L_2) = 1 \\ Cred_3^\Omega(L_3) = 0 & Pl_3^\Omega(L_3) = 0.8 \end{vmatrix}$$

The combination of m_1^Ω and m_2^Ω, illustrated by the Figure 4, gives the following new belief function:

$$m_{1,2}^\Omega = m_1^\Omega \oplus m_2^\Omega \quad \begin{vmatrix} m_{1,2}^\Omega(L_1) = 0.02 & m_{1,2}^\Omega(L_2) = 0.64 \\ m_{1,2}^\Omega(L_3) = 0.08 & m_{1,2}^\Omega(L_1 \cup L_2) = 0.16 \\ m_{1,2}^\Omega(L_3 \cap (L_1 \cup L_2)) + m_{1,2}^\Omega(L_1 \cap (L_2 \cup L_3)) = \\ m_{1,2}^\Omega(\emptyset) = 0.1 \end{vmatrix}$$

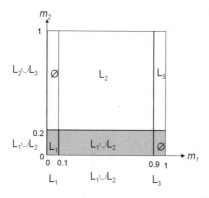

Fig. 4. Illustration of information combination with the student assessment example

In such situations where empty conjunctions of focal elements are present ($e.g.$, $L_3 \cap (L_1 \cup L_2) = \emptyset$), a renormalization is carried over the complete total mass not assigned to the empty set. The normalized belief function of previous example is:

$$m_{1,2}^{\Omega} = m_1^{\Omega} \oplus m_2^{\Omega} \quad \begin{vmatrix} m_{1,2}^{\Omega}(L_1) = 0.0222 & m_{1,2}^{\Omega}(L_2) = 0.7111 \\ m_{1,2}^{\Omega}(L_3) = 0.0889 & m_{1,2}^{\Omega}(L_1 \cup L_2) = 0.1778 \end{vmatrix}$$

The computation of Credibility and Plausibility of all singletons shows that the combination of both information sources has reduced the size of the interval of probability.

$$\begin{vmatrix} Cred_{1,2}^{\Omega}(L_1) = 0.022 & Pl_{1,2}^{\Omega}(L_1) = 0.2 \\ Cred_{1,2}^{\Omega}(L_2) = 0.71 & Pl_{1,2}^{\Omega}(L_2) = 0.888 \\ Cred_{1,2}^{\Omega}(L_3) = 0.09 & Pl_{1,2}^{\Omega}(L_3) = 0.09 \end{vmatrix}$$

The Dempster's rule of combination is often open to criticism since it entails two drawbacks. On the one hand, it has the effect of masking the aspect of conflict of the sources in question. Hence, there is a loss of information. On the other hand, when the conflict is great, renormalization may lead to counter-intuitive results [20]. To solve this problem, several other combination rules have been defined and they often differ by the way the mass of evidence of an empty intersection is allocated. The choice of the combination rule reflects the interpretation of the mass allocated to the empty set.

For example, the Smets' combination rule assumes that the sources are reliable and that the conflict between them can stem only from one or more hypotheses not having been taken into account in the frame of discernment [27]. In other words, this combination rule consists to assign the conflicting mass to the empty set, which is interpreted as a reject class.

$$\forall\, C \in 2^{\Omega} \quad m^{\Omega}(C) = \sum_{\{(A,B) \in (2^{\Omega})^2 / A \cap B = C\}} m_1^{\Omega}(A) m_2^{\Omega}(B) \tag{4}$$

Another view is to see the frame of discernment as certain, so the conflict can stem only from a wrong information source. This is the case of the Yager's method. This combination rule performs a partial disjunctive combination where the conflicting mass is allocated to Ω [37].

$$\begin{cases} \forall\, C \in 2^{\Omega}\backslash\{\Omega\} \quad m(C) = \sum_{\{A\cap B=C\}} m_1^{\Omega}(A)m_2^{\Omega}(B) \\ \\ m(\{\Omega\}) = m_1(\Omega) + m_2(\Omega) + \sum_{\{A\cap B=\emptyset\}} m_1^{\Omega}(A)m_2^{\Omega}(B) \end{cases} \tag{5}$$

Compared to the Dempster's rule, there is no risk of non-linear comportment caused by the normalization factor and the conflict is explicitly represented by the resulting mass of Ω.

A similar rule, proposed by Dubois and Prade [7], considers that in case of conflict between two sources, one of them is wrong and the other one is right. Then, the conflicting mass is allocated to the conjunction of both elements.

$$\forall\, C \in 2^{\Omega} \quad m(C) = \sum_{A\cap B=C} m_1^{\Omega}(A)\cdot m_2^{\Omega}(B) + \sum_{A\cup B=C} m_1^{\Omega}(A)\cdot m_2^{\Omega}(B) \tag{6}$$

A drawback of these two last methods is the loss of associativity. It is then necessary to define a combination strategy or a simultaneous combination, which can become heavy to compute for high numbers of belief assignments.

2.3   Discounting

The discounting is an operation, which consists to put a piece of evidence from the initial evidence distribution to the focal element representing uncertainty, it means Ω. The new belief assignment m' is then for a discounting of α:

$$\begin{cases} \forall\, A \in 2^{\Omega}\backslash\{\Omega\}m'(A) = \alpha m(A) \\ \\ m'(\Omega) = (1-\alpha)\sum_{\{A\in 2^{\Omega}\backslash\{\Omega\}\}} m(A) + m(\Omega) \end{cases} \tag{7}$$

This operation allows one to reduce the information brought by a belief structure according to the confidence into this information. In practice, it is often difficult to estimate this discounting coefficient which may be a compilation of several coefficients. For example, the various coefficients to be taken into account when determining the overall reliability of a sensor may be the representatively of the learning in respect of the parameters exploited by a particular sensor, the reliability of the sensor used to build the belief, etc [14].

2.4   Refinement and Coarsening of Belief

Moreover, in practice, it is possible to have sources of information whose frames of discernment Ω_1 and Ω_2 are different but compatible. To combine and merge these sources, relationships between the frames of discernment have to be defined. To this end, two operations – refinement and coarsening – express the correspondence in the form of compatibility rules. A refinement associates to each element of Ω_1 a set of compatible elements of Ω_2 while the coarsening is the antagonist relation. A mapping $\rho : 2^{\Omega_1} \to 2^{\Omega_2}$ is a refinement if it verifies the following properties:

$$\{\rho(\{\omega\})/\omega \in \Omega_1\} \subseteq 2^{\Omega_2} \text{ is a partition of } \Omega_2$$

$$\forall\, A \subseteq \Omega_1 \rho(A) = \bigcup_{\{\omega \in A\}} \rho(\{\omega\}) \tag{8}$$

2.5   The Decision Step

Finally, the last step is the decision making process which is supported by the results provided by the combination rules. Indeed, as previously highlighted, the combination of the available sources of information provides us with a new belief function which represents the most reliable and complete information. However, if the choice of the most likely hypothesis is straightforward in the probabilistic framework, it can become quite complex in the Evidence theory.

The decision can be based on the maximum of credibility or on the maximum of plausibility, which are respectively pessimistic and optimistic decision rules. Another method proposed in [28] consists to share equally the mass of a proposition between all singletons of which it is composed. The new function is a probability repartition called **pignistic probability** and defined as:

$$\forall\, A \in O, \quad Pig(A) = \sum_{\{B \in 2^\Omega / A \subseteq B\}} \frac{m(B)}{card(B)} \tag{9}$$

where $card(B)$ is the cardinal of the element B. In this case, the decision will be based on the singleton with the maximal pignistic probability.

The application of pignistic probability to previous examples gives the following results:

- using Dempster's rule :
$$\begin{vmatrix} m_{1,2}^{\Omega}(L_1) = 0.022 & m_{1,2}^{\Omega}(L_2) = 0.71 \\ m_{1,2}^{\Omega}(L_3) = 0.09 & m_{1,2}^{\Omega}(L_1 \cup L_2) = 0.178 \end{vmatrix}$$
$$Pig(L_1) = 0.111 \; Pig(L_2) = 0.799 \; Pig(L_3) = 0.09$$

- using Smets' rule :
$$\begin{vmatrix} m_{1,2}^{\Omega}(L_1) = 0.02 & m_{1,2}^{\Omega}(L_2) = 0.64 \\ m_{1,2}^{\Omega}(L_3) = 0.08 & m_{1,2}^{\Omega}(L_1 \cup L_2) = 0.16 \\ m_{1,2}^{\Omega}(\emptyset) = 0.1 \end{vmatrix}$$
$$Pig(L_1) = 0.1 \; Pig(L_2) = 0.72 \; Pig(L_3) = 0.08$$

- using Yager's rule :
$$\begin{vmatrix} m_{1,2}^{\Omega}(L_1) = 0.02 & m_{1,2}^{\Omega}(L_2) = 0.64 \\ m_{1,2}^{\Omega}(L_3) = 0.08 & m_{1,2}^{\Omega}(L_1 \cup L_2) = 0.16 \\ m_{1,2}^{\Omega}(\Omega) = 0.1 \end{vmatrix}$$
$$Pig(L_1) = 0.13 \; Pig(L_2) = 0.723 \; Pig(L_3) = 0.083$$

In this case, the Dubois's rule gives the same result than the Yager's rule because the union of focal elements whose intersection is empty, is equal to Ω.

In this example, three different results are obtained but a decision based on the maximum of pignistic probability is the same in all cases (the level of the student is average, *i.e.*, L_2). However, it is to be noticed that for more complex cases and/or with higher conflict levels, this is not always the case and the final result can be sensitive to the chosen combination method.

Given the various ideas presented above, the Evidence theory has emerged as a suitable formalism within a multisource context. Evidence theory is indeed appropriated for modeling and combining information sources so as to gain access to more reliable knowledge. We also saw that it enabled the reliability of a source to be incorporated through the weakening operation. The Evidence theory thus makes it possible to overcome various constraints in a multisource approach. Nevertheless it should be noticed that for some applications combinations operations can be computation expensive operations and that in these cases approximation or computing strategies have to be applied. Without constituting an exhaustive list, we may mention below a number of applications that use Evidence theory: on-shore exploration in Canada [1], autonomous navigation system for craft [6], battlefield recognition [9], and medical diagnosis system [26]. Recently, application of Evidence theory to diagnosis has been proposed [19],[24] and is further discussed in the following before applying it to anaerobic digestion processes.

3   Application of Evidence Theory to Diagnosis

3.1   Diagnosis Problem Representation

Because it offers a framework to manage uncertain and conflicting information, the Evidence theory can be relevant to combine and to cross check fault signals. In the context of fault diagnosis, the frame of discernment Ω will be the set of all possible states of the system, *i.e.* all the faults that can occur on the supervised process. In other terms, we have:

$$\Omega = \{f_0, f_1, f_2, \ldots, f_n\}$$

where f_0 is the fault free state and f_i is a possible faulty state.

Each residual[2] is considered as an independent source of information on the state of the process. The signature table that describes the relation between the faults and residual is then used to determine the core of the belief assignment corresponding to each residual. For example, let us suppose a system with 3 possible faults and 3 residuals, described by the following table:

---

[2] Let us recall that a residual is a numerical signal equal to 0 when no fault is present and to 1 when a fault is active.

Table 1. Fault signature table

	r_1	r_2	r_3
f_0	0	0	0
f_1	0	1	1
f_2	1	1	0
f_3	1	0	1

None of the residual is specific of one fault but each fault has a specific signature in any space defined the combination of two among the three possible residuals. It means that the diagnosticability is guaranteed and that two residuals are theoretically enough to isolate the fault.

In the Evidence theory framework, the *bba* corresponding to residuals r_1, r_2 and r_3 will have respectively the cores:

$$N_1 = \{f_0 \cup f_1, f_2 \cup f_3\}, \ N_2 = \{f_0 \cup f_3, f_1 \cup f_2\}, \ N_3 = \{f_0 \cup f_2, f_1 \cup f_3\}$$

The value applied to each mass is derived from the value of the associated residual. In a boolean framework, a threshold would be defined such that some values are declared as fault-free (*i.e.*, when the residual is lower than the threshold) and other ones as faulty (*i.e.*, when the residual is higher than the threshold). This method produces for each situation a vector of 0 and 1 that can be compared to the known signatures to isolate the fault. As already pointed out one main drawback is that if the value of the residual oscillates around the threshold, the state associated to the residual oscillates too. On the contrary, in the Evidence theory framework, an infinity of values are possible for each focal element. It is then possible to use smooth functions to produce a *bba* from the residual. For example, the following function has been proposed in [23]:

$$\rho \left|
\begin{aligned}
&[0\ 1] \rightarrow [0\ 1] \\
&\rho(r) = 1/\left(1 + \frac{1-a}{a}\left(\frac{\tau}{r}\right)^{2\gamma}\right)
\end{aligned}
\right. \tag{10}$$

where a is the value assigned to the smoothed residual $\rho(r)$ when the original residual r is equal to τ, and the parameter γ is a smoothing parameter. Compared to a crisp threshold, it gives a progressive response, as illustrated by Figure 5.

3.2 Illustration and Comparison to Boolean Representation

Let us consider the system described by the previous signature table (see Table 1) and let us assume three faults occurring successively:

- f_1 from $t = 10$ h to $t = 20$ h,
- f_2 from $t = 25$ h to $t = 35$ h and
- f_3 from $t = 45$ h to $t = 55$ h.

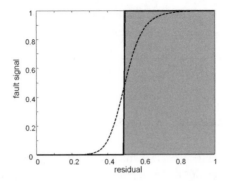

Fig. 5. Belie f assignment function (boolean for the continuous line, smooth for the dashed one)

Figure 6.a shows an example of possible evolutions of the three residuals r_1 to r_3 during the occurrence of the three faults. These residuals are artificially generated on the basis of the signature table, the way they have been generated is not considered here. Figure 6.b presents the isolation results using the boolean method (the real fault occurrences are indicated in grey) whereas Figure 6.c shows the results obtained using the Evidence theory based method.

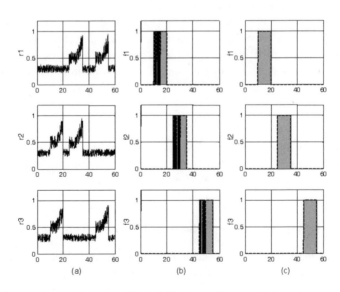

Fig. 6. Comparison of boolean (b) and Evidence theory (c) based approaches with coherent residuals (a). Abscissa dimension is hours.

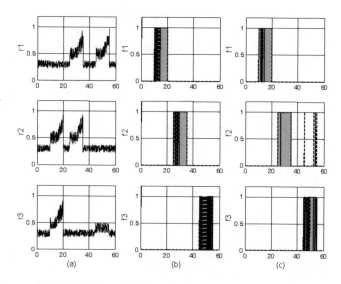

Fig. 7. Comparison of boolean (b) and Evidence theory (c) based approaches with conflicting residuals (a). Abscissa dimension as in Figure 6.

The beliefs are derived from the residuals with the relation (15) with the same threshold than for boolean evolution, *i.e.*, $\tau = 0.4$ for $a = 0.5$. Clearly, the boolean combination leads to an unstable isolation of the faults because the value of r_1 is often oscillating around 0.4. On the contrary, the approach based on the Evidence theory isolates perfectly the faults and does not induce any oscillation. In another situation where the value of residual r_3 is not affected as it should be when the fault f_3 occurs, the boolean combination does not succeed again to isolate the faults (see in particular f_3 in Figure 7.b) whereas the Evidence theory combination methods correctly the faults most of the time (*Cf.* Figure 7.c).

These two simple examples highlight how an uncertainty managing framework like Evidence theory can improve fault isolation performances.

3.3   Multiple Fault Diagnosticability

In the following, we will show that using the Evidence theory can improve even more the diagnosis task because the exact representation of the available knowledge is introduced in a *bba* whereas it is only roughly represented in a classical residual. One of the key advantages of using the Evidence theory for diagnosis lies indeed in the knowledge modeling brought by an information source and coarsening and refinement operations allow one to have a multi-fault detection system.

As it has been explained in section 2.2, only belief functions defined on a same frame of discernment can be combined. This implies that all the possible states of the system should be listed in Ω. This can be long to write for large numbers of sensors. Moreover, because some faults cannot be isolated, it can also be useless. However, it is possible to define a frame of discernment Ω_i for each sensor, each actuator and each element to be diagnosed. These information sources will then be defined on the corresponding Cartesian product of the frames of discernment. For example, let us suppose two sensors S_1 and S_2, each of them having two possible states: fault-free or faulty. These two states will respectively be noted ok and ko and the frames of discernment $\Omega_i = \{ok_i, ko_i\}$ are defined for each sensor. The verification of a relation linking these two sensors (a physical redundancy relation by example) will produce a belief structure defined on the Cartesian product of Ω_1 and Ω_2.

Moreover, a relevant modeling of an information source lies in the correct definition of its frame of discernment and of its focal elements. If this task is correctly performed, we have potentially a set of belief assignments defined on different frames of discernment. However, it is needed to refine them to combine them. In fact, a refinement associates a set of compatible elements of another frame of discernment Ω' to an element of Ω and a coarsening is the antagonist relation. The projection of a belief assignment defined on a set Ω_i into a set $\Omega = \Omega_i \times \Omega_j$ is then a particular case of a refinement. The following rule of correspondence Ψ shows how elements of the initial target reference set 2^{Ω_i} are expressed in the target reference set $2^{\Omega_i \times \Omega_j}$:

$$\forall\, A \in \Omega_i \quad m_{i,j}(\Psi(A)) = m_i(A)$$

$$\text{where} \quad \left| \begin{array}{l} 2^{\Omega_i} \rightarrow 2^{\Omega_i \times \Omega_j} \\ \forall\, A \in 2^{\Omega_i} \quad \Psi(A) = \bigcup_{\{B \in \Omega_j\}} \{A, B\} \end{array} \right. \tag{11}$$

Applying this relation to the Cartesian product of all frames of discernment of each input (*i.e.*, $\Omega = \Omega_1 \times \Omega_2 \times \ldots \times \Omega_n$ if there are n sensors), it is possible to refine all available belief functions into a common reference set and then to combine them to produce a unique belief structure that, in addition, is able to indicate the conflict between information sources.

From this function, it is possible to produce belief assignments defined on several mono-dimensional reference sets corresponding to each sensor states frame of discernment Ω_i. This is done using the following rule of correspondence based on the Cartesian projection $p_{\Omega \rightarrow \Omega_j}$ of Ω on Ω_j:

$$\forall\, A \in 2^{\Omega_i} \quad m_i(A) = \sum_{\{B \in 2^{\Omega} | p_{\Omega \rightarrow \Omega_j}(B)=A\}} m_{i,j}(B) \tag{12}$$

If we denote Ψ_k the refinement of $m_{i,j}$ from $\Omega_i \times \Omega_j$ to $\Omega_i \times \Omega_j \times \Omega_k$, the combination of $m_{i,j}$ and $m_{i,k}$ will check the following intersection:

$$\forall\, A, B \in \left(2^{\Omega_i \times \Omega_j}, 2^{\Omega_i \times \Omega_k}\right)$$

$$\Psi_k(A) \cap \Psi_j(B) = \left(\bigcup_{\{X \in 2^{\Omega_k}\}} (\{A, X\})\right) \cap \left(\bigcup_{\{X \in 2^{\Omega_j}\}} (\{B, X\})\right) \tag{13}$$

The existence of a cross-checking between two information sources is expressed by the existence of a non-empty intersection between $\Psi_k(A)$ and $\Psi_j(B)$ where A, B are respectively elements of $\Omega_i \times \Omega_j$ and $\Omega_i \times \Omega_k$. In other terms, there is a cross-checking for the sensor S_i if and only if:

$$
\begin{aligned}
&\exists\, C \in \{ok_i, ko_i\} \\
&\exists\, (A, B) \in (N_{i,j}, N_{i,k}) \quad p_{\Omega i,j,k \to \Omega_i}((\Psi_k(A) \cap \Psi_j(B))) = C
\end{aligned}
\tag{14}
$$

where $N_{i,j}$ and $N_{i,k}$ are the core of $m_{i,j}$ and $m_{i,k}$.

Since the combination operation performs all intersections between the focal elements of each refined belief assignment, this method of refinement/ combination/ coarsening guarantees that all the possible cross-checks have been done. The focal elements resulting of the projection have the highest level of details allowed by the whole information put in the decision system. This means that if there was no crosschecking relation for a sensor, then the only focal element would be the one of total doubt, i.e. $\{\Omega_i\}$.

4   From Residuals to Belief Assignments – Application to a Sensors Network

Despite all the advantages of the Evidence theory, diagnosis performances will deeply rely on the modeling of the information supported by the residual. This means that a valid list of focal elements and coherent numerical values have to be provided. In the literature, there is a great number of methods used for mass affectation: conversion of probabilities, conversion of distances, expert rules, neural networks, etc. In the following, some methods of residuals generation and conversions are evaluated for the diagnosis of a sensors network and their relevance to represent information provided by the fault signals is discussed. Experimental illustration is provided from using data available from a fully instrumented anaerobic digestion process, available at LBE-INRA, France.

4.1   Presentation of the Anaerobic Digestion Process

The anaerobic digestion (AD) process is a pilot-scale up-flow fixed bed reactor made of a circular column of 3.5 m height, 0.6 m diameter and an originally useful volume of 0.984 m^3 (*Cf.* Figure 8). It performs the biological anaerobic fermentation of distillery wastewater, transforming organic carbon into methane and carbon dioxide.

This process has a classical on-line instrumentation gathering measurements of liquid flow rates (at the input of the reactor and in the recirculation loop), temperature and pH in the reactor and biogas flow rate and composition (*i.e.*, CO_2, CH_4 and H_2 content in the biogas) [30]. In addition, the following sensors were installed over the years: a TOC analyzer, a titrimetric sensor [3] and a FT-IR spectrometer [29]. Since 1998, this instrumentation provides us with the following on-line measurements in the liquid phase every

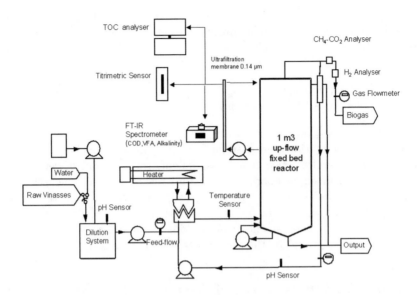

Fig. 8. Schematic view of the INRA anaerobic digestion process and the associated on-line instrumentation

2 to 30 minutes: soluble chemical oxygen demand (COD), total organic carbon (TOC), total volatile fatty acids (VFA), acetate (Ac), dissolved CO_2 (CO_{2d}), and bicarbonate concentrations and total and partial alkalinity, some of them being measured twice or even three times. However, because of the hard environmental conditions and the heterogeneity of influent, sensors failure occurs (fouling, deviation, jumps, hieratic behavior ...) often and can have great consequences if they are used for a closed-loop control. Restarting this kind of biological process in case of major faults can indeed take up to few months.

4.2 Belief Based on Hardware Redundancy

This relation is the ground relation of diagnosis and the easiest to use. It consists to check that two sensors providing measurements of the same variable give the same values. Given two sensors S_1 and S_2, providing values V_1 and V_2 of the variable, the residual r is defined by the normalized distance d between V_1 and V_2:

$$
d \left| \begin{array}{l} \mathbb{R}^2 \to [0,\ 1] \\ d(V_1, V_2) = min\left(\sqrt{\left(\frac{V_1 - V_2}{\sigma_V}\right)^2},\ 1 \right) \end{array} \right.
\tag{15}
$$

where σ_V is the standard deviation of the measure of the variable on the process.

The transformation of the residual into a *bba* follows the method defined in section 3.1 with the operator ρ described by equation (15):

$$m_{1,2}(\{ok_1, ok_2\}) = 1 - \rho(d) \quad m_{1,2}(\{ko_1, ok_2\}, \{ok_1, ko_2\}, \{ko_1, ko_2\}) = \rho(d)$$

For each redundancy, a belief is then defined on a two-dimensional state-universe. When dealing with the AD process previously presented, this relation is used for volatile fatty acids, total and partial alkalinity provided by the titrimetric sensor and the FT-IR spectrometer.

4.3   Belief Based on Software Redundancy

It is also possible to create a redundancy between measurements with models linking several variables. Static and robust models like acid-base equilibrium and gas-liquid equilibrium can be used easily. However, models with strong assumptions and/or with unknown validity should be used with caution. In this last case, we also have to take into account that the model itself can be faulty and a dimension of the state-universe of the corresponding *bba* is the faulty state of the model.

The Henry's law provides the following relation between the CO_2 flow-rate, the dissolved CO_2 concentration and the partial pressure of CO_2 in the gas:

$$q_{CO_2} = k_L a([CO_2] - K_H \cdot p_{CO_2} \cdot P_T) \tag{16}$$

which leads to:

$$[CO_2] = p_{CO_2} \left(\frac{Q_{gas}}{V_{Dig} \cdot k_L a \cdot V_{mol}} + K_H \cdot P_T \right) \tag{17}$$

By combining the relation (17) with the acid base equilibrium, the relation becomes:

$$[Bic] = \frac{K_b}{[H^+]} p_{CO_2} \left(\frac{Q_{gas}}{V_{Dig} \cdot k_{La} \cdot V_{mol}} + K_H \cdot P_T \right) \tag{18}$$

where K_b is the acidity constant of the couple CO_2/Bicarbonate.

Equations (17) and (18) allow one to link respectively (i) the dissolved CO_2 concentration with the gas composition and the gas flow rate and (ii) the bicarbonate concentration with the gas composition, the gas flow rate and the pH. The estimated value can be then compared to the measured one to build a belief assignment with the same methodology and parameters than in the case of the strict physical redundancy.

However, in this case, where several different variables can be faulty, the residuals do not have the same sensitivity to each sensor fault. For example, let us consider the distance dCO_2 between the measured and the computed value of $[CO_2]$ and let us assume that the faults are additive for each sensor (*i.e.*, a fault term d is added to each measurement). Then, we obtain:

$$V_{fault-free} - V_{faulty} =$$
$$p_{CO_2} \left(\frac{\delta Q_{gas}}{V_{Dig} \cdot k_{La} \cdot V_{mol}} \right) + \delta p_{CO_2} \left(\frac{Q_{gas} + \delta Q_{gas}}{V_{Dig} \cdot k_{La} \cdot V_{mol}} + K_H \cdot P_T \right) \tag{19}$$

From (19), we can determine the sensitivity of the residual to each fault:

$$\begin{cases} \dfrac{d(r_{CO_2})}{d(\delta_{p_{CO_2}})} = \left(\dfrac{Q_{gas} + \delta_{Q_{gas}}}{V_{Dig} \cdot k_{La} \cdot V_{mol}} + K_H \cdot P_T \right) \dfrac{1}{\sigma_{CO_2}} \\[3mm] \dfrac{d(r_{CO_2})}{d(\delta_{Q_{gas}})} = \left(\dfrac{p_{CO_2} + \delta_{p_{CO_2}}}{V_{Dig} \cdot k_{La} \cdot V_{mol}} \right) \dfrac{1}{\sigma_{CO_2}} \end{cases} \tag{20}$$

The comparison of these two terms for faults of comparable intensity (*i.e.*, a deviation of 10% from the true value) shows that for nominal values of Q_{gas} the residual built from the estimation of $[CO_2]$ is much more sensitive to a fault of p_{CO_2} than to a fault of Q_{gas}. Two options are possible to manage this: (i) select a low threshold to detect faults even for the less sensitive sensor with the risk to have a great number of false detections, (ii) select a threshold so that the most sensitive fault is correctly detected and include faults of the second sensor into this residual.

The first choice does not change the structure of the belief assignment but it affects the value of the term d. On the contrary, the second one leads to the following belief function:

$$\begin{cases} m_{CO_2}(\{ok_{CO_2}, ok_{Q_{gas}}, ok_{p_{CO_2}}\} \cup \{ok_{CO_2}, ok_{Q_{gas}}, ok_{p_{CO_2}}\}) = 1 - d \\[3mm] m_{CO_2}(\{\Omega_{CO_2} \times \Omega_{Q_{gas}} \times \Omega_{p_{CO_2}}\} \setminus \{ok_{CO_2}, ok_{Q_{gas}}, ok_{p_{CO_2}}\}) = d \end{cases} \tag{21}$$

4.4   Consistency Checking Based on Dynamical Modeling

In some cases, it is not possible to estimate directly a variable from other ones. However, dynamic models can express a relationship between variables. Moreover, some of the needed variables can be partially known, for example when only an interval on the values is provided.

Alkalinity Mass-Balance
The total alkalinity Z is the sum of the concentrations of all cations involved in occurring reactions. Its dynamic is described by:

$$\dot{Z} = D(Z_{in} - Z) \tag{22}$$

where the dilution rate D is the ratio between the feed-flow and the digester volume.

The integration of this relation between t and $t - T$ leads to:

$$\phi_Z(t - T, t, Z_{in}) = Z(t) - Z(t - T) - \int_{t-T}^{t} D(Z_{in} - Z) d\tau = 0 \tag{23}$$

Because in the most cases, the input concentration is imprecisely known, the value Z_{in} can be approximated by an interval. Providing minimal and maximal bounds to the value Z_{in} is equivalent to say that:

$$\exists\, \beta \in [0, 1] \quad Z_{in} = \beta Z_{in}^{min} + (1 - \beta) Z_{in}^{max} \tag{24}$$

By combination of relations (23) and (24), we obtain:

$$\exists\, \beta \in [0,\ 1] \quad \beta\phi_Z(t-T,t,Z_{in}^{min}) + (1-\beta)\phi_Z(t-T,t,Z_{in}^{max}) = 0 \qquad (25)$$

$$\exists\, \beta \in [0,\ 1] \quad \beta = \frac{\phi_Z(t-T,t,Z_{in}^{max})}{\phi_Z(t-T,t,Z_{in}^{max}) - \phi_Z(t-T,t,Z_{in}^{min})} \qquad (26)$$

COD Mass Balance

In an anaerobic process, the *COD* follows the relation:

$$\dot{COD} = D(COD_{in} - COD) - \zeta q_{CH_4} \qquad (27)$$

where ζ is the *COD* equivalent of 1 mole of CH$_4$. The same approach than for alkalinity leads to:

$$\exists\, \gamma \in [0\ 1] \quad \gamma = \frac{\phi_{COD}(t-T,t,COD_{in}^{max})}{\phi_{COD}(t-T,t,COD_{in}^{max}) - \phi_{COD}(t-T,t,COD_{in}^{min})}$$

where $\phi_{COD}(t_1, t_2, COD_{in}) = \qquad (28)$

$$COD(t_2) - COD(t_1) - \int_{t_1}^{t_2} [D(COD_{in} - COD) - \zeta q_{CH_4}]\, d\tau$$

A value of γ lower than 0 or higher than 1 shows an error from one of the inputs of the relation. The residual built from this relation is then a normalized distance between the computed value of g and the nearest authorized bound (*i.e.*, 0 or 1).

$$d_{[COD_{min}, COD_{max}]} \begin{vmatrix} \mathbb{R} \to \mathbb{R} \\[2mm] d_{[COD_{min}, COD_{max}]}(\gamma) = 0 \\[2mm] \text{if } \gamma \in [COD_{min}, COD_{max}] \\[2mm] d_{[COD_{min}, COD_{max}]}(\gamma) = \\[2mm] min(\gamma - COD_{max}, COD_{min} - \gamma) \text{ otherwise} \end{vmatrix} \qquad (29)$$

For small deviations from the true value of one of the variables D, COD_{in}, COD and q_{CH_4}, this interval-based relation can remain valid. In addition, the number of undetected faults will be large if large intervals are assumed for these variables. This can be expressed in the belief definition by adding a discounting factor proportional to the size of the interval. The resulting belief is:

$$
\begin{vmatrix}
m_{\phi_{COD}}^{\Omega_{COD_{in}} \times \Omega_{COD}}(\{ok_{COD_{in}}, ok_{COD}\}) = \\[2mm]
(1-\alpha)\left(1 - d_{[COD_{in}^{min}, COD_{in}^{max}]}\right) \\[2mm]
m_{\phi_{COD}}^{\Omega_{COD_{in}} \times \Omega_{COD}}((\Omega_{COD_{in}} \times \Omega_{COD}) \setminus \{ok_{COD_{in}}, ok_{COD}\}) = \\[2mm]
(1-\alpha)d_{[COD_{in}^{min}, COD_{in}^{max}]} \\[2mm]
m_{\phi_Z}^{\Omega_{Z_{in}} \times \Omega_Z}(\Omega_{COD_{in}} \times \Omega_{COD}) = \alpha
\end{vmatrix}
\qquad (30)
$$

where α is the value of $\frac{COD_{in}^{max}(t) - COD_{in}^{min}(t)}{max(COD_{in}^{max}) - min(COD_{in}^{min})}$.

It is to be noticed that he same relations applies to β replacing COD_{in} by Z_{in} and COD by Z.

Table 2. Signature table obtained from Figure 9

sensor	r_1	r_2	r_3	r_4	r_5	r_6	r_7	r_8	r_9	r_{10}	r_{11}
pH	0	0	0	1	0	1	0	0	0	0	0
$gasA$	0	0	0	1	1	0	0	1	0	0	0
$Qgas$	0	0	0	1	1	0	0	1	0	0	1
$titri$	1	1	1	1	0	1	0	0	1	0	0
$spectro$	1	1	1	0	1	1	1	1	0	1	0
COD_{in}	0	0	0	0	0	0	0	1	0	0	0
COD_{in}	0	0	0	0	0	0	0	0	1	1	0
COD_{in}	0	0	0	0	0	0	1	0	0	0	0

4.5 Experimental Illustration

Diagnosticability Analysis

The presented approach has been applied on the process illustrated in Figure 8. In this case, several physical redundancy relations are available but the same method can be successfully applied with a standard instrumentation fulfilled by software sensors [17]. r_1, r_2, r_3, r_6, r_7 and r_11 are based on physical redundancy, r_4 and r_5 on functional redundancy, r_8, r_9 and r_10 are derived from dynamical modeling. As expressed in Table 2, there is not one single residual that is representative of a specific fault. A diagnosticability analysis of the signature table shows that among the 256 different states of the system:

- 16 states and among them all the single-fault states are characterized by one specific residual set,
- the 240 remaining states are only partially isolable and can be represented by 36 signatures.

Section 3.2 has shown that Evidence theory could improve the performances of a FDI system by supporting a non boolean (*i.e.*, not crisp) evaluation of the residuals. In the following, it will be shown how the method performs information crosschecking and exhibits uncertainty.

Clear residuals (*i.e.*, with values close to 0 or 1) corresponding to all possible states of the sensors network are artificially generated and expressed under the form of belief structure. The diagnosis task is performed by the combination of the belief structure in respect with the previously presented method. Isolation performances are detailed by the Table 3.

All single faults, and more generally all well characterized states, are correctly isolated and no points is wrongly labelled. The most interesting and original property is that when the states cannot be distinguished, the performed isolation provides the best possible results, labelling precisely the sensors when it is possible and attributing a total uncertainty about the state of others. This illustrates that, structurally, a FDI system based on the Evidence theory will give the best possible results and will not perform unachieved assumptions for the isolation.

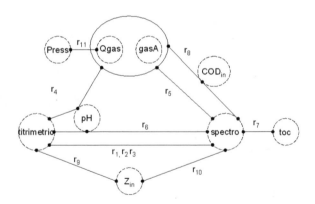

Fig. 9. Relations network on the instrumented AD process

Table 3. Fault isolation performances of the Evidence theory based FDI system

Fault isolation	Number of cases
good	16
wrong	0
partial	240

Application

The core of the Evidence theory lies in the combination of belief assignments. As already described, several belief combination rules exist, each of them corresponding to an interpretation of the conflict between *bba*. As a consequence, a rule should be first chosen according to the interpretation of conflict appropriate to the final objective. In addition, some mathematical considerations should be also accounted for. Indeed, only the Dempster's and Smets' rules are associative. This means that for other combination rules (*e.g.*, Yager's or Dubois and Prade's rules), we have (i) either to perform a simultaneous combination of all available *bba* or (ii) to determine the sequence of combinations appropriate to the final objective. The first choice is satisfactory since it does not need to perform any assumption on the combination order. However, it implies to compute a great number of intersections of focal elements and is thus difficult to apply for more than seven *bba* (due to the computation time required).

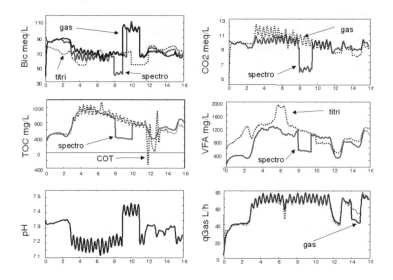

Fig. 10. The data set (abscissa dimension is days)

When analyzing the sensors network just presented above, the second choice was done due to the high number of bba obtained from Table 3. The Smets's rule could be preferred for its associativity property but one of its main drawback is that the resulting belief structure converges toward total uncertainty when a great number of *bba* are combined. The Dempster's rule cannot be accepted because in case of high conflict between sources, the normalizing factor is inducing instability of final results [19]. However, sub-systems can be defined and combined with the Dempster's rule when a small conflict is expected and attributed to noise on the residual generation: this is the case of residuals r_1, r_2 and r_3 which express the physical redundancy between the titrimetric sensor and the FT-IR spectrometer for three different variables. For other combinations, the Yager's rule has been preferred.

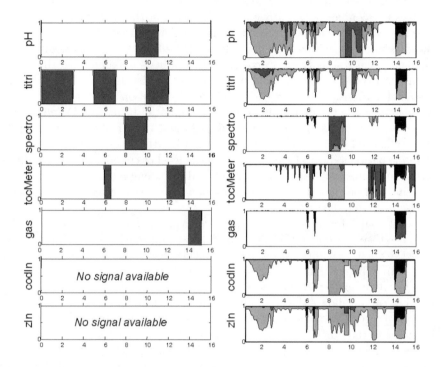

Fig. 11. Manual expert assignment of the state and on-line evolutions of the beliefs for each sensor (white = *ok*, light grey = *doubt*, dark grey = *ko* and black = conflict)

In the following, the diagnosis method has been validated using a data set of 2 weeks, with a serie of on-line measurements every 30 minutes (see Figure 10). During this period, dynamical changes were imposed on the process through Q_{in}, the feed flow rate, while the input concentrations (*i.e.*, COD_{in} and Z_{in}) were supposed to be constant. However, no on-line measurements were performed on these variables so only intervals for the values of COD_{in} and Z_{in} are known. Moreover, several faults occurred on different on-line sensors:

- on the pH probe between days 9 and 11,
- on the titrimetric sensor (*titri*) between days 0 and 3, 5 and 7, and finally between days 10 and 12,
- on the spectrometric sensor (*spectro*) between days 8 and 10,
- on the TOCmeter at day 6 and between days 12 and 13,
- on the gas sensor between days 14 and 15.

These faults are depicted in Figure 11 after an expert analysis of the data curves, and compared to the on-line evolution of the *bba* for each sensor. For each fault the belief in the *OK* state decreases and those of *KO* or doubt (*i.e.* Ω) increase. Moreover, when there is no uncertainty in the information sources nor conflict between residuals, the belief in the *KO* state is maximal (see for example the spectrometric sensor belief between days 8 and 9 and the TOCmeter belief between days 12 and 13) while, if the fault isolability is poor or if the residuals are in conflict, the doubt is maximal (see the belief of pH between 8 and 9). In addition, due to the structure and the order of the combinations, more doubt is sometimes assumed on the related variables compared to the original fault (see for example the belief in TOCmeter between days 8 and 9 and *titri* between days 14 and 16). Finally, when residuals are conflicting, a piece of evidence is measuring the level of conflict (see by example between days 14 and 15), giving the level of confidence into the diagnosis. Some mismatch occur and are caused by wrong residuals but they are rare and can be caused either by a great number of wrong residuals either by a false modeling of the residual.

Additional information can be withdrawn from these curves. Indeed, the evidence theory is one of the very few diagnosis approaches that allows one to detect explicitly conflicts in the measurements signals. As a consequence, one can easily imagine that, in case of persisting doubt, the supervisory software should call an expert (if present on site). Also, such an analysis of the sensors network can invalidate the assumptions made on the influent composition (see the beliefs on COD_{in} and Z_{in} between days 8 and 13 in Figure 4). The person in charge of the process has here an indication on the needed manual analysis to perform for better characterization of the wastewater to be treated. These two aspects are key requirements for telemonitoring issues as detailed in [18].

5   Diagnosis of the Overall Biological State of an AD Process

The previous section concentrated on the management of a hard and soft sensors network. This is an important step since the information sources must be carefully checked before being further used. This section will be devoted to the diagnosis of the overall biological state of the process. In particular, it will illustrate that the use of the Evidence theory approach improves the fault diagnosis system in terms of modularity and dynamical adaptation.

5.1   Fault Detection Based on Fuzzy Inference Systems

A global diagnosis system (*i.e.*, accounting for all the available information sources in a single box) could have been developed. However, modularity would then be very low since adding or removing one information source (*e.g.*, one sensor) would break down the overall structure and lead to false alarms or wrong diagnosis. This is why, as presented in Figure 12, the diagnosis system was chosen to be split into several parts: after selection of the usable information sources (using the approach presented in the previous section), the fault signals are generated by several fuzzy logic based modules, each of them manipulating two or three process variables at the most for simplicity reason. When manipulating only two different sources, the fuzzy modules also use the trends of the variables, computed by a linear regression on a moving window. The outputs of the fuzzy modules (*i.e.*, the fault signals) are then sent to a state manager where they are analyzed and combined using the Evidence theory.

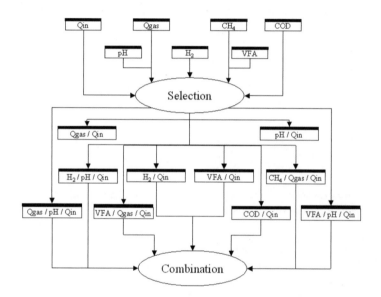

Fig. 12. Modular fault detection system

As an illustration of the different modules, Figure 13 describes the rules associated to Q_{in} and pH (*i.e.*, the basic instrumentation at the industrial scale). In these figures, the following nomenclature was adopted:

- *HO*: hydraulic overload,
- *OO*: organic overload,
- *UL*: underload,
- *Tox*: presence of a toxic,

while "- -" to "+ +" is used for the fuzzy qualification of the variables from "very low" to "very high". It is to be noticed that in some occasions, the module cannot distinguish between two states (*e.g.*, an organic overload and the presence of a toxic). This doubt is normal and is due to the low number of sensors that are managed to assess the state of the AD process. The same

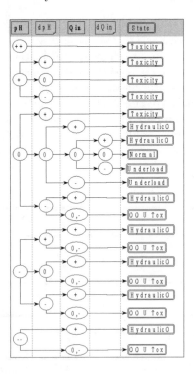

Fig. 13. Rules associated to the fault detection module analyzing Q_{in} and pH, d_{pH} and $d_{Q_{in}}$ indicate here the trends of the pH and Q_{in} signal

impossibility to distinguish between two (or more) situations is present in all the modules which makes relevant the choice of using Evidence theory within the state manager.

Of course, neither all the modules of Figure 12 – nor all the sensors in Figure 9 – are needed to supervise an AD process and once again, modularity is central in this diagnosis architecture. A set of sensors able to use from 4 to 6 boxes among the 10 presented, can be considered as enough rich to supervise AD.

In order to combine the conclusions of several such Fuzzy Inference Systems, they will be translated into belief structures, according to the method proposed in [22]. Each Fuzzy Inference System presented Figure 12 is the association of fuzzification functions (providing a numerical evaluation of the membership of a variable to fuzzy sets) and of rules linking these observations to different classes which can be states or disjunctions of states of the process.

The rules of a Fuzzy Inference System link observations on the space of observed variables to singletons or disjunctions of singletons from the frame of discernment Ω. It means that a rule can be expressed by:

$$R_i : (V_1 \in C_{1,R_i}, \dots, V_n \in C_{n,R_i}) \Rightarrow E_{R_i} \tag{31}$$

where $C_{i,j}$ is the jst class of the variable V_i and E is an element of 2^{Ω}.

It is possible to write, from the fuzzification function of each input variable, a belief structure defined on the size x_i of the variable V_i

$$\begin{cases} m(x_i \in C_{i,1}) = \mu_{C_{i,1}}(x_i) \\ m(x_i \in C_{i,2}) = \mu_{C_{i,2}}(x_i) \\ \dots \end{cases} \tag{32}$$

where $\mu_{C_{i,j}}(x)$ is the membership function of the jst class of the variable V_i.

It is then possible to combine these n belief structures to produce a belief structure defined on 2^Ω according to the combination rule:

$$\forall E \in 2^\Omega \ m(E) = \sum_{C_{j_1} \cap \dots \cap C_{j_n} \Rightarrow E} m_1(x_1 \in C_{1,j_1}) \times \dots \times m_n(x_n \in C_{n,j_n}) \tag{33}$$

This belief structure represents the belief into the state of the system of the set of rules associated to the fuzzification functions.

5.2 Experimental Illustration of Diagnosis of the Overall Biological State

During the 12 days experimental data presented in Figure 14, several faulty situations were encountered:

- from $t = 0$ to $t = 10$ hours, the situation is normal (*i.e.*, Q_{in}, the input liquid flow rate is around 20 l/h, VFA, the volatile fatty acids concentration in the reactor, is low (less than 0.5 g/l), pH in the process is around neutrality and CH$_4$ in the biogas is between 60 and 70%)
- from 10 to 25 hours, the input flow rate shows a step increase and the process response is normal for this small hydraulic overload (even though pH in the influent is lower than usual, *i.e.*, 4 instead of 6)
- Starting at $t = 40$ h, a large increase of the input flow rate is applied, inducing a high accumulation of VFA in the reactor (more than 6 g/l), thus lowering the pH in the process (down to almost 5). This leads to a large inhibition of the biological reaction scheme (the gas flow rate decreases sharply while it should have increased if no inhibition was present).
- To solve this problem, a first solution was applied through the increase of the pH in the influent at $t = 60$ h.
- However, this was not enough and thus, the input feed flow was decreased to 10 l/h after 120 hours and the process was underloaded.
- In addition, at about $t = 150$h, a toxicant was added to the process (*Cf.* the spike in the feed flow) inducing an increase of the pH in the reactor. Fortunately, the problem lasted for only few hours and the process could slowly recover until $t = 300$h (the fast step change of Q_{in} at $t = 200$h was done on purpose to evaluate the process performance during the recovery).

As an illustration of this statement, Figure 15 to Figure 17 detail the faults detected by fuzzy module Q_{in}-pH, module Q_{in}-Q_{gas}-pH and module Q_{in}-Q_{gas}-CH$_4$ when analyzing the experimental data described in Figure 14. These

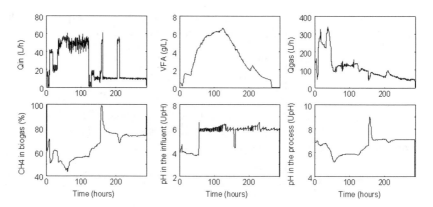

Fig. 14. A 12 day experiment showing inhibition of the overall AD process

different sensors are chosen because they are the most widely used at the industrial scale. In these different plots, a signal close to 0 means that the associated process state is not present whereas when the signal reaches 1, the associated state is present.

As it can be seen, the fuzzy module Q_{in}-pH detects well the hydraulic overloads (*i.e.*, HO) from 10 to 25 and from 40 to 120 hours (*Cf.* Figure 15.c). Then, it cannot differentiate among organic overload and toxicity (*i.e.*, $OO \cup Tox$) between 120 and 150 hours (*Cf.* Figure 15.e). Addition of the toxic is well detected between 150 and 155 hours (*Cf.* Figure 15.d) while after that, underload (*i.e.*, UL) is the main fault detected (*Cf.* Figure 15.b).

Module Q_{in}-Q_{gas}-pH, despite managing more sensors than module Q_{in}-pH, provides also more uncertainty. Indeed, as can be seen in Figure 16.f, some unknown faults are detected after 155 hours while it cannot distinguish between a hydraulic overload and the presence of a toxic from 50 to 120 hours and from 200 and 210 hours (*Cf.* Figure 16.g). A similar doubt can be noticed between an organic overload and the presence of a toxic from 120 to 150 hours, from 165 to 190 hours and from 210 hours until the end (*Cf.* Figure 16.h). Other process states are correctly assessed by this fuzzy module, even though in some cases, uncertainty is still present (see for example Figure 16.b where the fault signal never reaches 1).

This uncertainty is lowered when using the composition of methane in the biogas instead of pH as it is the case in module Q_{in}-Q_{gas}-CH$_4$. However, despite very good and accurate process state assessments for underload (*Cf.* Figure 17.b), organic overload (*Cf.* Figure 17.c), hydraulic overload (*Cf.* Figure 17.d), unknown situations are still present after 155 hours (*Cf.* Figure 17.e) and distinction cannot be made in some cases between organic overload and toxicity (*Cf.* Figure 17.f).

As illustrated in Figure 15 to Figure 17, each of these three modules, when taken separately, cannot provide clear and definitive process state assessment

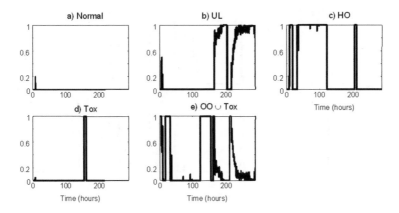

Fig. 15. Fault detection when analyzing only Q_{in} and pH

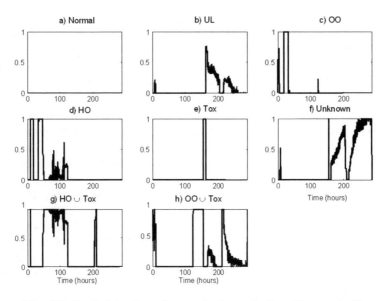

Fig. 16. Fault detection when analyzing only Q_{in}, Q_{gas} and pH

and they need to be combined within the state manager to reinforce the diagnosis. To show the benefits of such a combination using the Evidence theory, results obtained from the application of the Yager's rule for the combination of these three modules (*i.e.*, modules Q_{in}-pH, Q_{in}-Q_{gas}-pH and Q_{in}-Q_{gas}-CH$_4$) are presented in Figure 18. Clearly, the different faults are well analyzed until 155 hours, more belief being put afterwards on the underload of the process while unknown situations are detected at a low level.

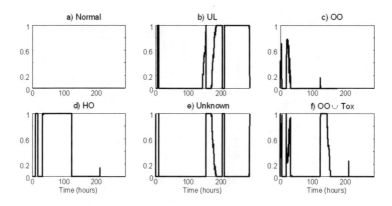

Fig. 17. Fault detection when analyzing only Q_{in}, Q_{gas} and CH_4

In fact, these results are in very good agreement with our expectations in such situations. Indeed, during these 12 days, none of the sensors provided erratic signals and the fuzzy fault detection modules were developed using a deep expertise on the process. As a consequence, each module reinforced the others ones in their belief about the encountered events. Another main advantage of the Evidence theory is that it can also be used to detect conflicts in the expert rules associated to each module. Indeed, a badly tuned rule or an inconsistency in the rule base could be detected following again the same approach. This aspect is of particular interest during the development of the overall diagnosis system and speeds up the tuning stage.

Last but not least, the benefits of this diagnosis structure also lie in the minimality of the needed modules. For example, fault detection results only using the fuzzy module Q_{in}-Q_{gas}-*VFA* are presented in Figure 19 and they are very similar to those presented in Figure 18. One could thus think that module Q_{in}-Q_{gas}-*VFA* alone could manage the overall AD process and, as a consequence, she/he could choose to implement only a gas flow meter and a *VFA* on-line sensor. However, in this situation, robustness of the diagnosis would be very low. Indeed, if the *VFA* sensor delivers wrong measurements (which is likely to appear in practice due to interference of chemical species, to sensor fouling,...), the whole supervisory scheme breaks down. In critical safety industries (*e.g.*, nuclear and aeronautic), the solution would be to implement two *VFA* sensors. However, in WWTPs, this solution is not viable economically. And maybe more importantly, it is felt that this would not be an appropriate solution. Indeed, if pH and gas composition sensors are implemented in addition to only one *VFA* sensors (instead of one gas flow meter and two *VFA* sensors), many more faulty situations can be handled (*e.g.*, presence of different toxic compounds, technical problems,...) while validating the obtained measurements by cross-checking each of them using the Evidence theory.

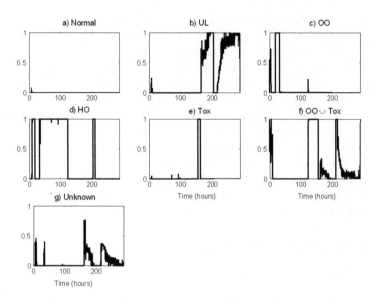

Fig. 18. Results from the state manager when receiving faults detection signals from module Q_{in}-pH, module Q_{in}-Q_{gas}-pH and module Q_{in}-Q_{gas}-CH$_4$

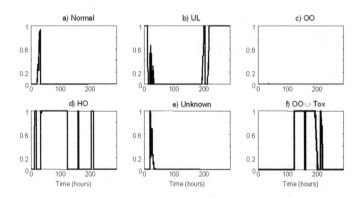

Fig. 19. Fault detection when analyzing only Q_{in}, Q_{gas} and *VFA*

6   Conclusion

Biological process in general, and WWTP in particular, are very complex and dynamical systems that require careful and efficient monitoring. However, in addition to their specific characteristics induced by the living aspects of the material involved (*i.e.* the micro-organisms), these processes present large uncertainty that classical control and/or diagnosis approaches cannot always manage.

In the present chapter, it is demonstrated that the Evidence theory offers a very well adapted methodological framework to handle this uncertainty. This ability was experimentally demonstrated for the management of hard/soft sensors network and for the assessment of the biological state. Moreover the Evidence theory provides a framework able to model exactly the information brought by any diagnosis method (which faults are detected, how precisely they are isolated,...) and then to use benefits from the combination of different methods (historical data based, analytical model based or knowledge based) to process FDI. Globally this approach provides a unifying and integrating framework for different diagnosis methodologies.

For all the reasons detailed throughout the chapter (*i.e.*, modularity, robustness, simplicity) it is our strong belief that Evidence theory will play a major role in the next future for advanced diagnosis of bioprocesses.

List of Abbreviations and Symbols

Related to bioprocesses:

X	biomass $(g \cdot L^{-1})$
COD	chemical oxygen demand $(g \cdot L^{-1})$
VFA	volatile fatty acids $(mmol \cdot L^{-1})$
Z	total alkalinity $(meq \cdot L^{-1})$
TOC	total organic carbon $(mg \cdot L^{-1})$
q_A	molar flow-rate of the molecule A $(mol \cdot L^{-1} \cdot h^{-1})$
D	dilution rate (day^{-1})
Q_{gas}	total biogas flow-rate $(L \cdot h^{-1})$
$[A]$	concentration of the molecule A $(mmol \cdot L^{-1})$
V_{Dig}	liquid volume in the digester (L)
V_{mol}	molar volume $(L \cdot mol^{-1})$
P_t	total pressure of the gas phase (bar)
K_H	Henry constant $(mmol \cdot L^{-1} \cdot bar^{-1})$
K_b	acidity constant of the couple $(CO_2$-Bicarbonate)
p_A	composition of gas in A (%)
$k_L a$	liquid-gas exchange constant (day^{-1})
σ_A	standard deviation of the variable A

Subscripts:

in	in the influent
gas	in the gas phase

Related to Evidence theory:

Ω	frame of discernment
2^Ω	reference set
bba	basic belief assignment
m^Ω	belief structure defined on Ω

N_m core of the belief structure m
C_{ij} j^{st} class of the i^{st} variable

Related to fault detection and isolation:
$Cred$ credibility function
f fault
FDI fault detection and isolation
ko any faulty state
ok good functioning state
Pig pignistic probability
Pl plausibility function
r residual

References

[1] P. An. *Spatial reasoning and integration techniques for geophysical and geological exploration data.* PhD thesis, University of Manitoba, Canada, 1992.

[2] M. Basseville. Detecting changes in signals and systems–A survey. *Automatica*, 24(3):309–326, 1988.

[3] J.C. Bouvier, J.P. Steyer, T. Conte, P. Gras, and J.P. Delgenès. On-line titrimetric sensor for the control of anaerobic digestion processes. In *Latin American Workshop and Symposium on Anaerobic Digestion*, pages 65–68, Merida, Mexico, 2002.

[4] X. Chen and I. Petrounias. *A development framework for temporal data mining.* Bramer, M.A., 1999.

[5] A.P. Dempster. Upper and lower probabilities induced by a multivalued mapping. *Annals. of Mathematical Statistics*, 38:325–339, 1967.

[6] J. Dezert. *Towards a New Concept of Autonomous Craft Navigation. A Link Between Probabilistic Data Association Filtering and the Theory of Evidence.* PhD thesis, Université d'Orsay, France, 1990.

[7] D. Dubois and H. Prade. Representation and combination of uncertainty with belief functions and possibility measures. *Computational Intelligence*, 4:244–264, 1998.

[8] P.M. Frank. Fault diagnosis in dynamic systems using analytical and knowledge-based redundancy: A survey and some new results. *Automatica*, 26(3):459–474, 1990.

[9] T.G. Garvey. Evidential reasoning for geographic evaluation for helicopter route planning. *IEEE Trans. Geoscience and Remote Sensing*, 25(3):294–304, 1987.

[10] A. Genovesi, J. Harmand, and J.P. Steyer. Integrated fault detection and isolation: Application to a winery's wastewater treatment plant. *Applied Intelligence Journal (APIN)*, 13:207–224, 2000.

[11] J. Gertler and D. Singer. A new structural framework for parity equations-based failure detection and isolation. *Automatica*, 26:381–388, 1990.

[12] R. Isermann. Process fault detection based on modeling and estimation methods–A survey. *Automatica*, 20(4):387–404, 1984.

[13] R. Isermann and P. Balle. Trends in the application of model-based fault detection and diagnosis of technical processes. *Control Eng. Pract.*, 5(5):709–719, 1997.

[14] F. Janez. *Fusion of informations sources defined on different nonexhaustive reference sets.* PhD thesis, Université d'Angers, France, 1996.

[15] K. Kipling, M. Willis, J. Glassey, and G. Montague. Bioprocess improvement through knowledge elicitation. In *International Symposium on Computer Applications in Biotechnology (CAB)*, pages 56–61, Nancy, France, 2004.

[16] M.A. Kramer. Malfunction diagnosis using quantitative models with non-boolean reasonning in expert systems. *A.I.Ch.E. Journal*, 33(1):130–140, 1987.

[17] L. Lardon, O. Bernard, and J.P. Steyer. Application du modèle des croyances transférables pour le diagnostic d'un réseau de capteurs et d'observateurs: application a un procédé de traitement des eaux. In *CIFA*, Douz, Tunisie, 2004.

[18] L. Lardon, J.P. Steyer E. Roca, J. Lema, S. Lambert, P. Ratini, S. Frattesi, and O. Bernard. Specifications of modular internet-based remote supervision systems for wastewater treatment plants. In *Eur. Conf. Artificial Intell. (ECAI)*, pages 5.1–5.5, Lyon, France, 2002.

[19] L. Lardon and J.P. Steyer. Using evidence theory for diagnosis of sensors networks: application to a wastewater treatment process. In *Int. Joint Conf. Artificial Intell. (IJCAI)*, pages 29–36, Acapulco, Mexico, 2003.

[20] E. Lefevre, O. Colot, and P. Vannoorenberghe. Belief function combination and conflict management. *Information Fusion*, 3(2):149–162, 2002.

[21] J. Montmain and S. Gentil. Decision-making in fault detection: A fuzzy approach. In *Int. Conf. TOOLDIAG*, Toulouse, France, 1993.

[22] J.M. Nigro and M. Rombaut. IDRES: A rule-based system for driving situation recognition with uncertainty management. *Information Fusion*, 4(4):309–317, 2003.

[23] A. Rakar and D. Juricic. Diagnostic reasoning under conflicting data: the application of the transferable belief model. *J. Process Control*, 12(1):55–67, 2002.

[24] A. Rakar, D. Juricic, and P. Balle. Diagnostic reasoning under conflicting data: the application of the transferable belief model. *Eng. Applic. Artificial Intell.*, 12(5):555–567, 1999.

[25] G. Shafer. *A mathematical theory of evidence.* Princeton University Press, 1976.

[26] P. Smets. *A mathematico-statistical model simulating the process of medical diagnosis.* PhD thesis, Université Libre de Bruxelles, Faculty of Medicine, 1978.

[27] P. Smets. The combination of evidence in the transferable belief model. *IEEE Trans. Pattern Analysis and Machine Intelligence*, 12(5):447–458, 1990.

[28] P. Smets and R. Kennes. The transferable belief model. 66(2):191–234, 1994.

[29] J.P. Steyer, J.C. Bouvier, T. Conte, P. Gras, J. Harmand, and J.P. Delgenes. On-line measurements of COD, TOC, VFA, total and partial alkalinity in anaerobic digestion process using infra-red spectrometry. *Wat. Sci. Technol.*, 45(10):133–138, 2002.

[30] J.P. Steyer, J.C. Bouvier, T. Conte, P. Gras, and P. Sousbie. Evaluation of a four year experience with a fully instrumented anaerobic digestion process. *Wat. Sci. Technol.*, 45(4-5):495–502, 2002.

[31] J.P. Steyer, I. Queinnec, and D. Simoes. BIOTECH: A real time application of artificial intelligence for fermentation processes. *Control Eng. Pract.*, 1(2):315–321, 1993.

[32] D.E. Totzke. Anaerobic treatment technology overview. Technical report, Applied Technologies Inc., USA, 1999.

[33] M. Ulieru and R. Isermann. Design of a fuzzy-logic based diagnostic model for technical processes. *Fuzzy Sets and Systems*, 58:249–271, 1993.

[34] J. Van Lier, A. Tilche, B.K. Ahring, H. Macarie, R. Moletta, M. Dohanyos, L.W. Hulshoff Pol, P. Lens, and W. Verstraete. New perspectives in anaerobic digestion. *Wat. Sci. Technol.*, 43(1):1–18, 2001.

[35] W. Verstraete and P. Vandevivere. New and broader applications of anaerobic digestion. *Critical Reviews in Environmental Science and Technology*, 29(2):151–165, 1999.

[36] A.S. Willsky. A survey of design methods for failure detection in dynamic systems. *Automatica*, 12(6):601–611, 1976.

[37] R.R. Yager. On the Dempster-Shafer framework and new combination rules. *Information Sciences*, 41(2):93–137, 1987.

Part III

Dynamics of Controlled Reactors

Nonisothermal Stirred-Tank Reactor with Irreversible Exothermic Reaction $A \rightarrow B$: 2. Nonlinear Phenomena

M. Pérez-Polo[1] and P. Albertos[2]

[1] Department of Physics, System Engineering and Signal Theory, EPS
`manolo@dfists.ua.es`
[2] Department of Systems Engineering and Control, ETSII
`pedro@aii.upv.es`

Summary. In this chapter, the non-linear dynamics of a nonisothermal CSTR where a simple first order reaction takes place is considered. From the mathematical model of the reactor without any control system, and by using dimensionless variables, it has been corroborated that an external periodic disturbance, either inlet stream temperature and coolant flow rate, can lead to chaotic dynamics. The chaotic behavior is analyzed from the sensitivity to initial conditions, the Lyapunov exponents and the power spectrum of reactant concentration. Another interesting case is the one researched from the self-oscillating regime, showing that a periodic variation of coolant flow rate can also produce chaotic behavior. Finally, steady-state, self-oscillating and chaotic behavior with two PI controllers have been investigated. From different parameters of the PI controllers, it has been verified that a new self-oscillating regime can appear. In this case, the saturation in a control valve gives a Shilnikov type homoclinic orbit, which implies chaotic dynamics. The existence of a new set of strange attractors with PI control has been analyzed from the initial conditions sensitivity. An Appendix to show a computationally simple form to implement the calculation of Lyapunov exponents is also presented.

1 Introduction

It is well known that self-oscillation theory concerns the branching of periodic solutions of a system of differential equations at an equilibrium point. From Poincaré, Andronov [4] up to the classical paper by Hopf [12], [18], non-linear oscillators have been considered in many contexts. An example of the classical electrical non-oscillator of van der Pol can be found in the paper of Cartwright [7]. Poore and later Uppal [32] were the first researchers who applied the theory of nonlinear oscillators to an irreversible exothermic reaction $A \rightarrow B$ in a CSTR. Afterwards, several examples of self-oscillation (Andronov-Poincaré-Hopf bifurcation) have been studied in CSTR and tubular reactors. Another

H.O. Méndez-Acosta et al. (Eds.): Dyn. & Ctrl. of Chem. & Bio. Proc., LNCIS 361, pp. 243–279, 2007.
springerlink.com

example taken from mechanics and electronics can be found in [11], [15], [24], [31], [34].

The experiments and the simulation of CSTR models have revealed a complex dynamic behavior that can be predicted by the classical Andronov-Poincaré-Hopf theory, including limit cycles, multiple limit cycles, quasi-periodic oscillations, transitions to chaotic dynamic and chaotic behavior. Examples of self-oscillation for reacting systems can be found in [4], [17], [18], [22], [23], [29], [30], [32], [33], [36]. The paper of Mankin and Hudson [17] where a CSTR with a simple reaction $A \to B$ takes place, shows that it is possible to drive the reactor to chaos by perturbing the cooling temperature. In the paper by Pérez, Font and Montava [22], it has been shown that a CSTR can be driven to chaos by perturbing the coolant flow rate. It has been also deduced, by means of numerical simulation, that periodic, quasi-periodic and chaotic behaviors can appear.

More recently, the problem of self-oscillation and chaotic behavior of a CSTR with a control system has been considered in others papers and books [2], [3], [8], [9], [13], [14], [20], [21], [27]. In the previously cited papers, the control strategy varies from simple PID to robust asymptotic stabilization. In these papers, the transition from self-oscillating to chaotic behavior is investigated, showing that there are different routes to chaos from period doubling to the existence of a Shilnikov homoclinic orbit [25], [26]. It is interesting to remark that in an uncontrolled CSTR with a simple irreversible reaction $A \to B$ it does not appear any homoclinic orbit with a saddle point. Consequently, Melnikov method cannot be applied to corroborate the existence of chaotic dynamic [34].

In the present chapter, steady state, self-oscillating and chaotic behavior of an exothermic CSTR without control and with PI control is considered. The mathematical models have been explained in part one, so it is possible to use a simplified model and a more complex model taking into account the presence of inert. When the reactor works without any control system, and with a simple first order irreversible reaction, it will be shown that there are intervals of the inlet flow temperature and concentration from which a small region or lobe can appears. This lobe is not a basin of attraction or a strange attractor. It represents a zone in the parameters-plane inlet stream flow temperature-concentration where the reactor has self-oscillating behavior, without any periodic external disturbance.

In this situation, a periodic variation of coolant flow rate into the reactor jacket, depending on the values of the amplitude and frequency, may drive to reactor to chaotic dynamics. With PI control, and taking into account that the reaction is carried out without excess of inert (see [1]), it will be shown that it the existence of a homoclinic Shilnikov orbit is possible. This orbit appears as a result of saturation of the control valve, and is responsible for the chaotic dynamics. The chaotic dynamics is investigated by means of the eigenvalues of the linearized system, bifurcation diagram, divergence of nearby trajectories, Fourier power spectra, and Lyapunov's exponents.

2   CSTR Models Equations for a Simple Reaction $A \to B$

Let us assume that an irreversible exothermic reaction $A \to B$ takes place in a CSTR, as shown in Figure 1. The cooling jacket surrounding the reactor removes the reaction heat. Perfectly mixed and negligible heat losses are assumed. The jacket is assumed to be perfectly mixed and the mass of the metal walls is considered negligible.

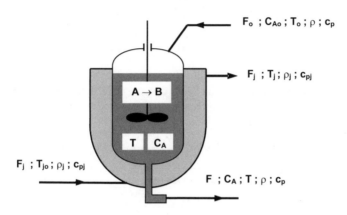

Fig. 1. Perfectly mixed CSTR

The reactor model without control includes the following equations (see [1]):

$$
\begin{aligned}
\frac{dC_A}{dt} &= \frac{F_0}{V}(C_{A0} - C_A) - \alpha C_A e^{-E/RT} \\
\frac{dT}{dt} &= \frac{F_0}{V}(T_0 - T) + \frac{(-\Delta H_R)\alpha}{\rho c_p} C_A e^{-E/RT} - \frac{UA}{\rho c_p V}(T - T_j) \qquad (1) \\
\frac{dT_j}{dt} &= \frac{F_j}{V_j}(T_{j0} - T_j) + \frac{UA}{\rho_j c_{pj} V_j}(T - T_j)
\end{aligned}
$$

In order to simplify the mathematical model of the reactor, the following dimensionless variables are introduced

$$
\tau = \frac{F_0 t}{V}; \quad x = \frac{C_A}{C'_{A0}}; \quad x_0 = \frac{C_{A0}}{C'_{A0}} \qquad (2)
$$

where C'_{A0} is the initial concentration of reactant A in the reactor. The temperatures are transformed to dimensionless variables from the following equations:

$$
y = \frac{RT}{E}; \quad z_j = \frac{RT_j}{E}; \quad y_0 = \frac{RT_0}{E}; \quad z_{j0} = \frac{RT_{j0}}{E} \qquad (3)
$$

By substituting the values of $t, C_A, C_{A0}, T, T_j, T_0, T_{j0}$ from Eq. (2) and (3) into Eq. (1), the mathematical model of the reactor in dimensionless variables can be written as:

$$\frac{dx}{d\tau} = x_0 - x - c_0 x e^{-1/y}$$

$$\frac{dy}{d\tau} = y_0 - y + c_1 x e^{-1/y} - c_2(y - z_j) \qquad (4)$$

$$\frac{dz}{d\tau} = c_3(z_{j0} - z_j) + c_4(y - z_j)$$

where the parameters c_0, c_1, c_2, c_3, c_4 are defined by the following equations:

$$c_0 = \frac{V}{F_0}\,\alpha; \quad c_1 = \frac{VR(-\Delta H_R)C'_{A0}}{F_0\,E\,\rho\,c_p}\,\alpha; \quad c_2 = \frac{UA}{\rho\,c_p\,F_0}$$

$$c_3 = \frac{V\,F_j}{V_j F_0}; \quad c_4 = \frac{\rho\,c_p\,V}{\rho_j\,c_{pj}\,V_j}\,c_2 \qquad (5)$$

Typical values of the reactor parameters, to be used in the following discussion, are shown in Table 1. These values are different respect to those used in [1] because they represent another first order reaction.

Table 1. Parameter values of the reactor for the reaction $A \to B$

Variable	Description	Value
V	Reactor volume (m^3)	1.3592
V_j	Jacket volume (m^3)	0.0849
C_{A0}	Reactant concentration inlet stream (kmol A/m^3)	8
F_0	Volumetric flow rate for the inlet stream (m^3/h)	1.1326
F_j	Volumetric flow rate of cooling water (m^3/h)	1.4130
α	Preexponential factor Arrhenius law (h^{-1})	7.08×10^{10}
E	Activation energy (kJ/kmol)	69,631
A	Heat transfer area (m^2)	23.255
U	Overall heat transfer in the jacket (kJ/h·m^2·°C)	3068.5
T_0	Inlet stream temperature (K)	294.4
T_{j0}	Inlet stream cooling water temperature (K)	294.4
R	Perfect-gas constant (kJ/kmol·K)	8.314
$-\Delta H_R$	Enthalpy of reaction (kJ/kmol)	69,828
c_p	Heat capacity inlet and out streams (kJ/kg·K)	3.142
c_{pj}	Heat capacity of cooling water (kJ/kg·K)	4.189
ρ	Density of the inlet and out streams (kg/m^3)	800
ρ_j	Density of cooling water (kg/m^3)	1000

Using these parameters, Eq.(4) can be simplified considering the dimensionless time constant for the reactor τ_y and the dynamics of the jacket τ_z whose values are:

$$\tau_y = \frac{1}{1 + c_2}; \quad \tau_z = \frac{1}{c_3 + c_4} \tag{6}$$

If $\tau_y \gg \tau_z$ the jacket's dynamic can be neglected. Therefore, the jacket's temperature is very close to the reactor temperature, and the dimensionless temperature in the jacket z_j can be removed from Eq.(4) as follows:

$$z_j = \frac{c_3 \, z_{j0} + c_4 \, y}{c_3 + c_4} \tag{7}$$

Remark 1. The assumption $\tau_y \gg \tau_z$ is verified for small reactors. See Figure 7 and Table 2 in [1].

Substituting Eq.(7) into (4) the simplified equations of the reactor are obtained:

$$\begin{aligned}
\frac{dx}{d\tau} &= x_0 - x - c_0 x e^{-1/y} \\
\frac{dy}{d\tau} &= y_0 - y + c_1 x e^{-1/y} - c_5(y - z_{j0})
\end{aligned} \tag{8}$$

where $c_5 = (c_2 c_3)/(c_3 + c_4)$. Eq.(8) are used to investigate the existence of self-oscillating and chaotic behavior.

3 Self-oscillation and Chaotic Behavior of a CSTR Without Feedback Control

Equations (4) and (8) can be used to simulate the reactor at point P_3 of Figure 5 in [1]. Remember that point P_2 is unstable, so if the initial conditions are those corresponding to this point, it is easy to show [16], [28], the reactor evolves to points P_1 or P_3. Then, two forcing actions on the reactor are considered: 1) when the coolant flow rate and the inlet stream temperature are varied as sine waves, and 2) reactor being in self-oscillating mode, an external disturbance in the coolant flow rate can drive it to chaotic behavior.

3.1 Chaotic Behavior with Double External Periodic Disturbance

It is well known that a nonlinear system with an external periodic disturbance can reach chaotic dynamics. In a CSTR, it has been shown that the variation of the coolant temperature, from a basic self-oscillation state makes the reactor to change from periodic behavior to chaotic one [17]. On the other hand, in [22], it has been shown that it is possible to reach chaotic behavior from an external sine wave disturbance of the coolant flow rate. Note that a periodic disturbance can appear, for instance, when the parameters of the PID controller which manipulates the coolant flow rate are being tuned by using the Ziegler-Nichols rules. The chaotic behavior is difficult to obtain from normal

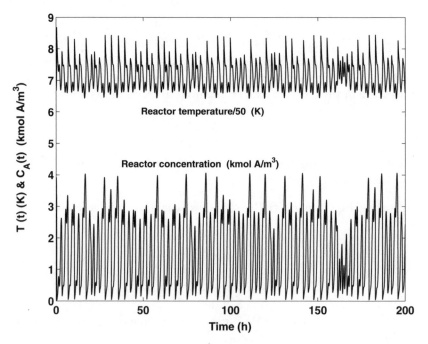

Fig. 2. Chaotic oscillations for three-dimensional model Eq.(1)-(4). $C'_{A0} = 8$ kmol/m^3, $A_t = 111.1$ K, $A_f = 0.849$ m^3/h, $\omega = 5$ rad/h.

operating conditions, but it is possible to obtain at the start and shutdown operations.

In order to investigate this behavior, we consider the mathematical model of the reactor given by Eq.(1) or (4), and assuming that the reactor is at the steady state corresponding at point P_3 of Figure 2 in [1]. The disturbances of the dimensionless inlet stream temperature and the coolant flow rate are the following:

$$y_0 \longrightarrow y_0 + \frac{RA_t}{E} sin(\omega \, \tau)$$
$$F_j \longrightarrow F_j + A_f sin(\omega \, \tau) \tag{9}$$

where A_t and A_f are the amplitudes of the inlet stream temperature and coolant flow rate external disturbances respectively, and ω the dimensionless frequency. Figure 2 shows chaotic oscillations of concentration and reactor temperature, and Figure 3 shows the strange attractor in the $C_A - T$ plane.

Figure 4 shows a pattern of the concentration when the chaotic motion is established as well as the evolution of the deviation from two very close initial conditions. Note that nowadays it is very difficult to prove rigorously that a strange attractor is chaotic. In accordance with [35], a nonlinear system has chaotic dynamics if:

1. It has sensitive dependence on initial conditions, i.e. two very close initial conditions diverge with time.
2. The orbits are dense in a state space region i.e. the orbits fills the phase space zone of the strange attractor Ω.
3. The orbits are topologically transitive in Ω i.e. for any two open sets $\Omega_1, \Omega_2 \subset \Omega$ there is a time from which any orbit starting at Ω_1 ends at Ω_2.

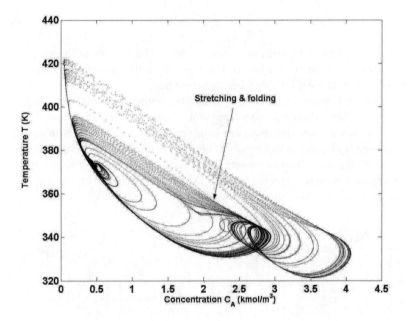

Fig. 3. Strange attractor for model (1)-(4) driven by Eq.(9)

While the conditions 1,2 can be verified approximately by simulation, proving the condition 3 is very difficult. Note that in many studies of chaotic behavior of a CSTR, only the conditions 1,2 are verified, which does not imply chaotic dynamics, from a rigorous point of view. Nevertheless, the fulfillment of conditions 1,2, can be enough to assure the long time chaotic behavior i.e. that the chaotic motion is not transitory. From the global bifurcations and catastrophe theory other chaotic behavior can be considered throughout the disappearance of a saddle-node fixed point [10], [19], [26].

Remark 2. Note that Eq.(1) and the external disturbance (9) can be represented by a four-dimensional space state. Therefore, it is possible to consider an autonomous dynamical system by introducing a new variable $x_4(t)$. Eqs.(1) and (9) can be rewritten as follows:

$$\frac{dC_A}{dt} = \frac{F_0}{V}(C_{A0} - C_A) - \alpha\, C_A\, e^{-E/RT}$$

$$\frac{dT}{dt} = \frac{F_0}{V}(T_0 + A_t sin(x_4) - T) + \frac{(-\Delta H_R)\,\alpha}{\rho\, c_p}\, C_A\, e^{-E/RT}$$

$$- \frac{UA}{\rho\, c_p\, V}\,(T - T_j) \tag{10}$$

$$\frac{dT_j}{dt} = \frac{F_j + A_f sin(x_4)}{V_j}(T_{j0} - T_j) + \frac{UA}{\rho_j\, c_{pj}\, V_j}\,(T - T_j)$$

$$\frac{x_4}{dt} = \omega$$

Eq.(10) represents a four-dimensional model of the reactor with external forcing disturbance, which can be used to investigate the chaotic dynamics. Similar to Eq.(4), the Eq.(10) can be normalized.

The simulation result (Figure 4) shows that when two initial conditions are very close, after a dimensionless time of 40 units the concentration of reactant A and the reactor temperature are completely different. This means that the system has a chaotic behavior and their dynamical states diverge from each other very quickly, i.e. the system has high sensitivity to initial conditions. This separation increases with time and the exponential divergence of adjacent phase points has a very important consequence for the chaotic attractor, i.e.

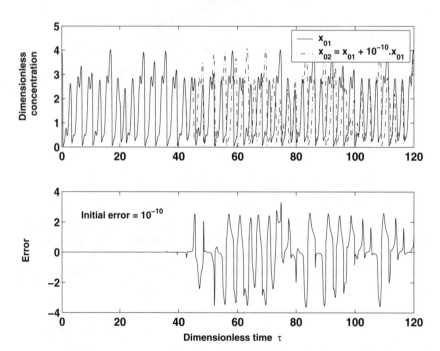

Fig. 4. Sensitivity to initial conditions for the model Eq.(1)-(4) with two very close initial conditions

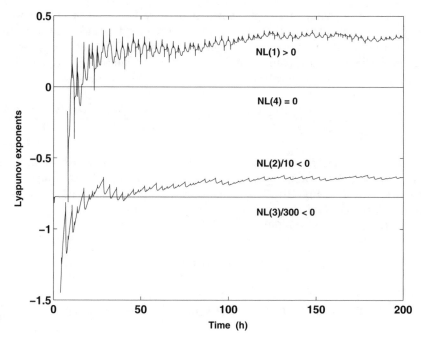

Fig. 5. Lyapunov exponents. $C'_{A0} = 8$ kmol/m^3, $A_t = 111.1$ K, $A_f = 0.849$ m^3/h, $\omega = 5$ rad/h.

the stretching and folding of the phase space. So, when the reactor reaches chaotic behavior, the trajectories of two adjacent phase points remain bounded without intersecting, folding themselves, and producing a four-dimensional chaotic attractor with an infinite number of layers. Consequently, the strange attractor of Figure 3 is a projection of the four-dimensional system (10) on the phase plane $C_A(t) - T(t)$, and the intersection of trajectories is only apparent. This is in accordance with the fact that phase space trajectories can not intersect themselves.

The reactor has four Lyapunov exponents because the model with the external forcing has four state variables, and their sum should be negative since the system is dissipative. The Lyapunov exponents for the system (10) are shown in Figure 5. Note that one exponent (of variable x_4) corresponds to the direction parallel to the trajectory, so it contributes nothing to the expansion or contraction of the phase volume, and therefore the corresponding Lyapunov exponent is zero. The rest of exponents are negative in the appropriate directions of phase volume, whereas there is one exponent positive, indicating divergence of trajectories. The algorithm used in the calculation of Lyapunov exponents can be found in [5],[6] and it is based on the linearized system trough the trajectory, so it must be combined with the simulation of the nonlinear system (10) (see Appendix).

On the other hand, it is well known that there is a relationship between Lyapunov exponents and the divergence of the vector field deduced from the differential equations describing a dynamical system. This relation provides a test on the numerical values obtained from the simulation algorithm. This relationship is, according to the definition of Lyapunov exponents:

$$div\bar{F} = \sum_{i=1}^{4} \lambda_i; \quad div\bar{F} = \frac{\partial \bar{F}_1(C_A, T, T_j)}{\partial C_A} + \frac{\partial \bar{F}_2(C_A, T, T_j)}{\partial T} + \frac{\partial \bar{F}_3(C_A, T, T_j)}{\partial T_j}$$

(11)

where $div\bar{F}$ means the median value of the divergence of vector field, λ_i is the corresponding Lyapunov exponent and $\bar{F}_1, \bar{F}_2, \bar{F}_3$ are the components of the vector field of Eq.(10). The simulation was carried out with fifth order Runge-Kutta-Fehlberg method, with a step interval of T = 0.004 and 0.002 h, and the results are the following:

$$T = 0.004 \Rightarrow div\bar{F} = -241.8684; \quad \sum_{i=1}^{4} \lambda_i = -238.4406$$

$$T = 0.002 \Rightarrow div\bar{F} = -241.8267; \quad \sum_{i=1}^{4} \lambda_i = -241.9691$$

Consequently, we can ensure that the simulation process is correct. Note that if there is at least one positive Lyapunov exponent, trajectories obtained from two very close initial conditions diverge, and when all Lyapunov exponents are negative the same trajectories converge. A practical procedure to numerically determine the Lyapunov exponents is given in the Appendix.

Exercise 1. From the values of Table 1 and Eq.(10), write a computer program using a fourth order Runge-Kutta or fifth order Runge-Kutta-Fehlberg method and reproduce Figures 2, 3, 4, 5. In order to check that the chaotic behavior has been reached, it is necessary to run the program with two initial conditions very close, for example:

$$x_{01} = [C_A = 3.93, T = 333.3, T_j = 330.39, x_4 = 0]$$
$$x_{02} = x_{01} + 10^{-10} x_{01}$$

The dimensionless simulation time must be greater than 100, with a simulation step smaller than 0.004. As the angular variable $x_4(t)$ grows with time, it is convenient to reduce it between $0 - 2\pi$. The function "atan2(x,y)" is very useful. Note that the chaotic behavior is very sensitive to the parameter values, and can be very interesting to investigate the bifurcation diagram in order to obtain the range of values in the parameter space where the reactor behavior is chaotic.

Figure 6 shows the Fourier power spectrum for the concentration of reactant A (see Figure 2). Taking into account that the time series at Figure 2 are

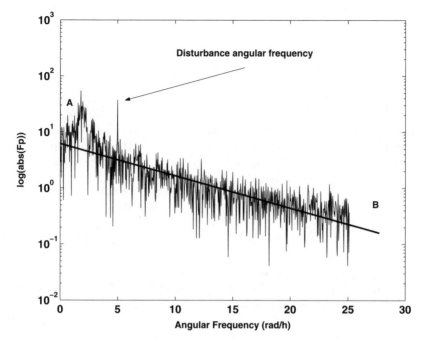

Fig. 6. Power spectrum of reactant concentration A for chaotic behavior. The peak is located at the disturbance angular frequency.

irregular, the corresponding power spectrum is broadband and contains substantial power at low frequencies. Note that there is a sharp component at the forcing frequency 5 rad/h. It is important to point out that although a positive Lyapunov exponent gives sensitivity to initial conditions it does not guarantee chaotic dynamics. However, in practice, a positive Lyapunov exponent and broad spectrum is sometimes used as an indicator of chaos. The straight line AB has been drawn through the peaks of the spectrum of frequencies to remark the character of chaotic dynamic, probably due to a route of period-doubling sequences (see [11], [15], [35]).

Remark 3. Lyapunov exponents and Fourier analysis have been used as standard criterion for distinguishing between chaotic and non-chaotic dynamics in the past. Nevertheless, these criteria should be interpreted cautiously.

3.2   Self-oscillation and Chaotic Behavior

A much more interesting case of chaotic dynamics of the reactor can be obtained from the study of the self-oscillating behavior. Consider the simplified mathematical model (8) and suppose that the reactor is in steady state with a reactant concentration of C'_{A0}. From Eq.(8) the equilibrium point $[x^*, y^*]$ can be deduced as follows:

$$x_0 - x^* - c_0 x^* e^{-1/y^*} = 0 \quad \Rightarrow \quad x^* = \frac{x_0}{1 + c_0 e^{-1/y^*}} \tag{12}$$

$$y_0 - y^* + c_1 x^* e^{-1/y^*} - c_5(y^* - z_{j0}) = 0$$
$$\Rightarrow \quad y_0 = (1 + c_5)y^* - c_5 z_{j0} - \frac{c_1 x_0}{c_0 + e^{1/y^*}} \tag{13}$$

Eq.(13) is transcendent and the variable y^* can not be deduced as a function of the inlet stream dimensionless temperature. Nevertheless it is possible to obtain y_0 as a function of all possible reactor equilibrium temperatures y^*. From Eq.(13) it is deduced that:

$$\text{If } y^* \to 0 \Rightarrow y_0 \approx (1 + c_5)y^* - c_5 z_{j0} = 0$$
$$\text{If } y^* \to \infty \Rightarrow y_0 \approx (1 + c_5)y^* - \left[c_5 z_{j0} + \frac{c_1 x_0}{c_0 + 1}\right] \tag{14}$$

In the plane $y_0 - y^*$ Eqs.(14) are straight lines of slope $(1 + c_5)$, and from intermediate values of y^* it is obtained a curve whose form depends on the value of x_0, as shown in Figure 7.

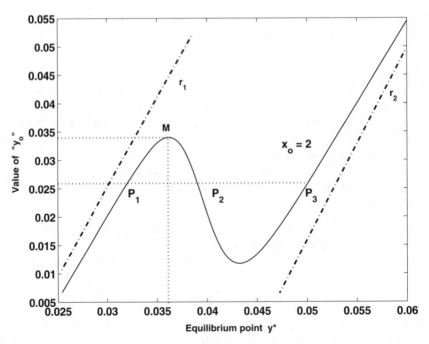

Fig. 7. Inlet stream dimensionless temperature vs. equilibrium temperature y^* at a fixed value of the inlet stream dimensionless concentration x_0. The straight-line r_1 and r_2 are defined by Eqs.(14).

From Figure 7 it is deduced that the number of the equilibrium states depends on the number of points where the straight line $y_0 = $ constant intersects with the curve defined by Eq.(13). With a value of $y_0 \approx 0.025$, there are three equilibrium points P_1, P_2, P_3, being P_1 stable, P_2 unstable and P_3 can be stable or unstable depending on the real part of the eigenvalues of the linearized system at this point. When the line $y_0 = $ constant is tangent to the curve $y_0 = f(y^*)$ (i.e. point M) a new behavior of the reactor appears, which can be characterized from $dy_0/dy^* = 0$ in Eq.(13) as follows:

$$\frac{dy_0}{dy^*} = 0 \quad \Rightarrow \quad 1 + c_5 - \frac{c_1 x_0 e^{1/y^*} 1/(y^*)^2}{(c_0 + e^{1/y^*})^2} = 0 \tag{15}$$

From Eq.(15) the value of x_0 can be deduced as a function of the equilibrium value y^*. By substituting this value in Eq.(13), the following equations are obtained:

$$x_0 = \frac{1 + c_5}{c_1} (y^*)^2 \left(c_0 + e^{-1/y^*} \right)^2 e^{-1/y^*}$$

$$y_0 = (1 + c_5)y^* - c_5 z_{j0} - (1 + c_5)(y^*)^2 \left(1 + c_0 e^{-1/y^*} \right) \tag{16}$$

These equations represent a parametric curve with parameter y^*. From a set of values of the parameter y^* it is possible to draw a curve in the $x_0 - y_0$ plane, so we obtain a bifurcation curve as a function of parameter y^*. This curve with a cusp point can be considered as the border that dividing the plane $x_0 - y_0$ into domains with one and three equilibrium states respectively.

In order to obtain the self-oscillation zone at the plane $x_0 - y_0$ consider the Jacobian of the linearized system (8):

$$J = \begin{bmatrix} a_{11} & a_{12} \\ a_{21} & a_{22} \end{bmatrix} = \begin{bmatrix} -1 - c_0 e^{-1/y^*} & -c_0 \frac{x^*}{(y^*)^2} e^{-1/y^*} \\ c_1 e^{-1/y^*} & -(1 + c_5) + c_1 \frac{x^*}{(y^*)^2} e^{-1/y^*} \end{bmatrix} \tag{17}$$

The eigenvalues of matrix (17) depend on the coefficients a_{ij} from the following equation:

$$\lambda^2 - \sigma \cdot \lambda + \Delta = 0; \quad \sigma = -(a_{11} + a_{22}); \quad \Delta = a_{11} a_{22} - a_{12} a_{21} \tag{18}$$

If the following conditions are imposed:

$$\sigma = 0; \quad \Delta > 0 \tag{19}$$

Eq.(18) has two complex roots with real part equal to zero, and consequently it is possible to deduce a relation between x^* and y^*. By substituting Eq.(18) into Eq.(12) one obtains a parametric equation $x_0 = f_1(y^*)$. Eliminating x_0 between $x_0 = f_1(y^*)$ and Eq.(13), the parametric equations of self-oscillating behavior are deduced:

$$x_0 = \frac{c_0(y^*)^2}{c_1} \left[3 + c_5 + \left(\frac{2 + c_5}{c_0} \right) e^{-1/y^*} + c_0 e^{-1/y^*} \right]$$

$$y_0 = (1 + c_5)y^* - c_5 z_{j0} - (y^*)^2 (2 + c_5 + c_0 e^{-1/y^*}) \tag{20}$$

Eqs.(20) define a closed curve or lobe in the plane $x_0 - y_0$. Eqs. (16) and (20) are plotted in Figure 8. This Figure shows that when the values of the inlet stream concentration and temperature are inside the dash zone (i.e. inside the lobe and outside the curve with cusp point), the reactor has a self-oscillating behavior. For a point outside the lobe and curve with cusp point, there is only one equilibrium point; this means that the values of inlet stream concentration x_0 and the inlet temperature y_0 give a curve such as the straight line $y_0 =$ constant only intersects the bifurcation curve of Figure 7 at one point. Another interesting consideration is that the lobe is usually small, and so it can be very difficult to choose values of temperature and concentration of reactant A in the inlet stream to the reactor to get an oscillating regime. If the reactor is being forced with an external periodic disturbance, and the values of x_0, y_0 are inside the lobe, it is possible that the reactor reaches the chaotic behavior.

Figure 8 shows a point P inside the lobe, and Figure 9 shows the simulation results when the inlet stream concentration and temperature are inside the lobe: a trajectory approaches a closed curve or limit cycle and remains there. This behavior has been corroborated in an industrial environment.

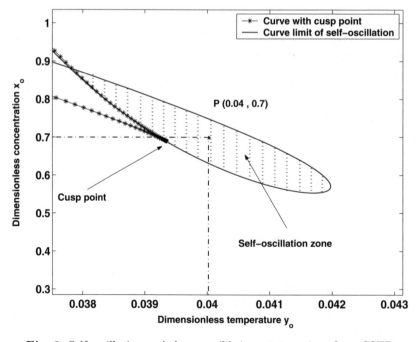

Fig. 8. Self-oscillating and three-equilibrium state regions for a CSTR

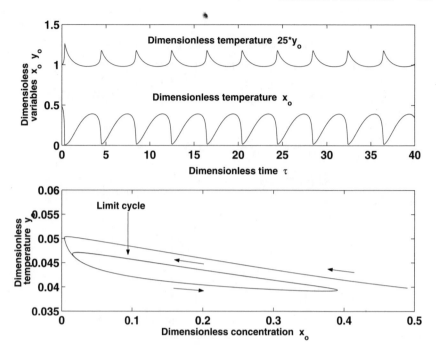

Fig. 9. Self-oscillating behavior with values $x_0 = 0.7$ and $y_0 = 0.04$ inside the lobe

Exercise 2. Using the values from Table 1 and the Eqs.(5), (13) write a computer program to obtain Figure 7 from different values of x_0 and check the Eq.(14). From different values of equilibrium point (x^*, y^*) determine the curve with cusp point and the lobe of Figure 8. These curves can be difficult to visualize if the values of (x^*, y^*) are not appropriate. In order to do it choose values of y^* between the maximum and minimum of the curve at Figure 7. Why? Using the program of exercise 1 and taking $A_t = 0$, $A_f = 0$ and a point (x_0, y_0) inside the lobe check the self-oscillating behavior shown at Figure 9. Make a zoom at Figure 8 to obtain a pair of values (x_0, y_0) inside the zone of curve with cusp point and curve limit of self-oscillation. Can we obtain self-oscillation behavior? Why?.

Exercise 3. From the simulation program of exercise 1 or 2 determine the power spectrum of Figure 6. An easy way is the following one. From the simulation program one obtains the values of $x(t)$, $y(t)$ and $z(t)$, so the fast Fourier transform can be calculated from a standard commercial program such as Maple, Mathematica, Matlab etc. Only the half components of positive frequency must be considered. The absolute values of these components using semilog coordinates give us the power spectra of Figure 6.

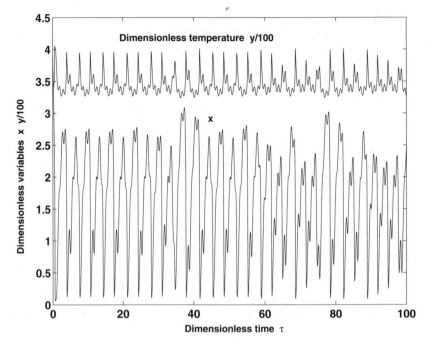

Fig. 10. Chaotic oscillations for the two-dimensional model (8). $A_t = 0$; $A_f = 0.849$ m^3/h; $\omega = 6.5$ rad/h.

Another interesting aspect of the self-oscillating behavior is the following one. If the values of (x_0, y_0) are inside the lobe, an external periodic disturbance of the coolant flow rate can drive the reactor to chaotic behavior.

It is important to remark that this behavior is similar to that previously considered by Eqs.(9), when two external periodic disturbances are applied. Nevertheless, this behavior can be very difficult to obtain, because the lobe of Figure 8 is small. Figures 10 and 11 shows chaotic oscillations and a new strange attractor. By simulation it is possible to obtain plots similar to those in Figures 2, 4, 5 and 6.

4   Regular Self-oscillation and Chaotic Behavior of a CSTR with PI Control

In this section we consider a CSTR with a very simple control system formed by two PI controllers. The first controller manipulates the outlet flow rate as a function of the volume in the tank reactor. A second PI controller manipulates the flow rate of cooling water to the jacket as a function of error in reactor's temperature. The control scheme is shown in Figure 12 where the manipulated variables are the inlet coolant flow rate F_j and the outlet flow rate F respectively.

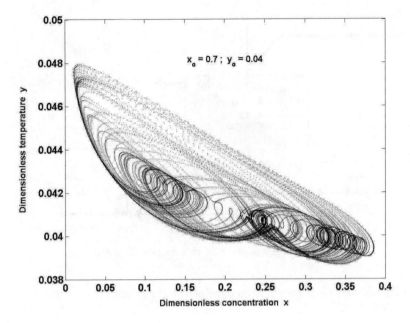

Fig. 11. Strange attractor for the model (8). $C'_{A0} = 8$ kmol/m^3.

Considering the assumptions taken in section 2, the modelling equations of the reactor with two PI controllers are the following:

$$\frac{dV}{dt} = F - F_0$$

$$\frac{dC_A}{dt} = \frac{F_0}{V}(C_{A0} - C_A) - \alpha\, C_A\, e^{-E/RT}$$

$$\frac{dT}{dt} = \frac{F_0}{V}(T_0 - T) + \frac{(-\Delta H_R)\alpha}{\rho\, c_p}\, C_A\, e^{-E/RT} - \frac{UA}{\rho\, c_p\, V}(T - T_j)$$

$$\frac{dT_j}{dt} = \frac{F_j}{V_j}(T_{j0} - T_j) + \frac{UA}{\rho_j\, c_{pj}\, V_j}(T - T_j) \qquad (21)$$

$$\frac{dF}{dt} = K_v\frac{dV}{dt} + \frac{K_v}{t_1}(V - V_s)$$

$$\frac{dF_j}{dt} = K_t\frac{dT}{dt} + \frac{K_t}{t_2}(T - T_{set})$$

where the volume of liquid in the reactor must be considered as variable and the last two equations of (21) represent the equations of the PI controllers. The variables V_s and T_{set} are the steady state volume and set point temperature of the reactor respectively, and K_v, t_1 and K_t, t_2 are the proportional and integral action parameters of each one of the PI controllers. We assume that

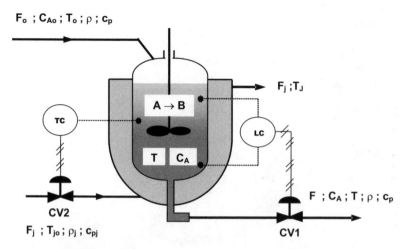

Fig. 12. Perfectly mixed CSTR with two PI controllers

the manipulated variable F_j is constrained due to saturation of the control valve CV2:

$$0 \leq F_j \leq F_{jmax} \tag{22}$$

where F_{jmax} is the maximum value of cooling water through the control valve CV2. The outlet flow through the control valve CV1 is not limited because it is assumed that its size exceeds the nominal one. The variables F_0, C_{A0}, T_0, T_{j0} can be considered as external load disturbances whose effect must be minimized by the PI controllers. Equation (21) and (22) can be difficult to analyze. So, similarly to section 2, in order to simplify the mathematical treatment, it is useful to introduce dimensionless variables as follows:

$$\tau = \frac{F_{0s}t}{V_s}; \ x_1 = \frac{V}{V_s}; \ x_5 = \frac{F}{F_{0s}}; \ x_{50} = \frac{F_0}{F_{0s}}; \ x_2 = \frac{C_A}{C'_{A0}}; \ x_{20} = \frac{C_{A0}}{C'_{A0}} \tag{23}$$

$$x_3 = \frac{RT}{E}; \ x_{30} = \frac{RT_0}{E}; \ x_4 = \frac{RT_j}{E}; \ x_{40} = \frac{RT_{j0}}{E}$$
$$x_s = \frac{RT_{set}}{E}; \ x_6 = \frac{F_j}{F_{js}}; \ x_{6max} = \frac{F_{jmax}}{F_{js}} \tag{24}$$

where F_{0s} is the steady state inlet flow rate, C'_{A0} is the initial concentration of the reactant A in the reactor and F_{js} is the steady state of the coolant flow rate. The dimensionless constants of the PI controllers are the following:

$$K_{vd} = K_v \frac{V_s}{F_{0s}}; \ \tau_{1d} = t_1 \frac{F_{0s}}{V_s}; \ K_{td} = K_t \frac{T_0}{F_{js}}; \ \tau_{2d} = t_2 \frac{F_{0s}}{V_s} \tag{25}$$

Introducing the parameters:

$$c_0 = \frac{V_s}{F_{0s}}\,\alpha; \quad c_1 = \frac{V_s R(-\Delta H_R)C'_{A0}}{F_{0s}\ E\ \rho\ c_p}\,\alpha; \quad c_2 = \frac{U\,A}{\rho\ c_p\ F_{0s}}$$

$$c_3 = \frac{V_s F_{js}}{V_j F_{0s}}; \quad c_4 = \frac{\rho\ c_p\ V_s}{\rho_j\ c_{pj}\ V_j}\,c_2 \tag{26}$$

the equations of the reactor in dimensionless variables can be written as:

$$\frac{dx_1}{d\tau} = x_{50} - x_5$$

$$\frac{dx_2}{d\tau} = \frac{x_{50}}{x_1}(x_{20} - x_2) - c_0 x_2 e^{-1/x_3}$$

$$\frac{dx_3}{d\tau} = \frac{x_{50}}{x_1}(x_{30} - x_3) + c_1 x_2 e^{-1/x_3} - \frac{c_2}{x_1}(x_3 - x_4)$$

$$\frac{dx_4}{d\tau} = c_3 x_6 (x_{40} - x_4) + c_4 (x_3 - x_4) \tag{27}$$

$$\frac{dx_5}{d\tau} = K_{vd}(x_{50} - x_5) + \frac{K_{vd}}{\tau_{1d}}(x_1 - 1)$$

$$\frac{dx_6}{d\tau} = \frac{K_{td}}{x_{30}}\frac{dx_3}{d\tau} + \frac{K_{dt}}{x_{30}\tau_{2d}}(x_3 - x_s)$$

$$0 \le x_6 \le x_{6max}$$

where only the sate variable x_6 is bounded.

From Eqs.(27) it is possible to determine the equilibrium points following a similar procedure to section 3.1 and, as result, Figures such as 7, 8 can be obtained.

Exercise 4. From Eq.(27) prove that the dimensionless coolant temperature can be written as:

$$x_{4e} = \left(\frac{x_{50} + c_2}{c_2}\right) x_5 - \frac{x_{30}x_{50}}{c_2} - \frac{c_1}{c_2}\frac{x_{20}x_{50}}{x_{50}\ e^{1/x_s} + c_0} \tag{28}$$

For different values of x_s draw the function $x_{4e} = f(x_5)$ taking x_{20} as parameter. Show the possibility to obtain one, two or three equilibrium points. Prove that the following inequality is verified:

$$x_{4emin} \le \frac{x_s + x_{6max}x_{40}(c_3/c_4)}{1 + x_{6max}(c_3/c_4)} \le x_5 \tag{29}$$

From different values of x_{20} and by using the values from table 1 write a computer program to obtain a figure similar to Figure 7.

Exercise 5. Taking into account exercise 4 prove that the parametric equations of curve with cusp point are the following:

$$x_{20} = \left(\frac{x_{50} + c_2}{c_1}\right) x_s^2\ e^{-1/x_s}\ (c_0 + e^{1/x_s})^2 \tag{30}$$

$$x_{30} = (x_{50} + c_2)x_s - c_2 x_{4e} - (x_{50} + c_2)x_s^2(1 + c_0\ e^{-1/x_s})$$

Eq.(27) can be simplified taking into account that the first and fifth equations are independent, so eliminating the x_5 variable the following linear differential equation is obtained:

$$\frac{d^2 x_1}{d\tau^2} + K_{vd}\frac{dx_1}{d\tau} + \frac{K_{vd}}{\tau_{1d}}x_1 = \frac{K_{vd}}{\tau_{1d}}; \quad x_1(0) = 1; \quad \left(\frac{dx_1(\tau)}{d\tau}\right)_{\tau=0} = x_{50} - 1 \quad (31)$$

Note that there is no interaction between the volume of the reactor and its temperature.

Eq.(31) can be used to adjust the parameter of the PI controller LC at Figure 12. For example, complex roots of the characteristic equation of (30) give the following dimensionless volume of the reactor:

$$x_1(\tau) = 1 + \frac{x_{50} - 1}{\sqrt{\frac{K_{vd}}{\tau_{1d}}}\sqrt{1 - \frac{\tau_{1d}K_{vd}}{4}}} \, e^{-\frac{K_{vd}}{2}\tau} \, sin\left\{\sqrt{\frac{K_{vd}}{\tau_{1d}}}\sqrt{1 - \frac{\tau_{1d}K_{vd}}{4}} \cdot \tau\right\}$$

$$(32)$$

Eq.(32) can be written as:

$$x_1(\tau) = 1 + f(\tau); \quad (33)$$

$$f(\tau) = \frac{x_{50} - 1}{\sqrt{\frac{K_{vd}}{\tau_{1d}}}\sqrt{1 - \frac{\tau_{1d}K_{vd}}{4}}} \, e^{-\frac{K_{vd}}{2}\tau} \, sin\left\{\sqrt{\frac{K_{vd}}{\tau_{1d}}}\sqrt{1 - \frac{\tau_{1d}K_{vd}}{4}} \cdot \tau\right\}$$

By substituting Eq.(33) into Eq.(27) the dimensionless equations of the reactor are simplified as follows:

$$\frac{dx_2}{d\tau} = x_{50}(x_{20} - x_2) - c_0\, x_2\, e^{-1/x_3} - \frac{x_{50}f(\tau)}{1 + f(\tau)}(x_{20} - x_2)$$

$$\frac{dx_3}{d\tau} = x_{50}(x_{30} - x_3) + c_1\, x_2\, e^{-1/x_3} - c_2(x_3 - x_4)$$

$$- \frac{f(\tau)}{1 + f(\tau)}[x_{50}(x_{30} - x_3) - c_2(x_3 - x_4)] \quad (34)$$

$$\frac{dx_4}{d\tau} = c_3 x_6(x_{40} - x_4) + c_4(x_3 - x_4)$$

$$\frac{dx_6}{d\tau} = \frac{K_{td}}{x_{30}}\frac{dx_3}{d\tau} + \frac{K_{td}}{x_{30}\tau_{2d}}(x_3 - x_s)$$

Eq.(34) are a set of differential equations, which lead a flow in a four dimensional phase space $\mathbb{R}^4$. This flow can be simplified to three dimensional phase space $\mathbb{R}^3$ when the dynamics of the jacket can be considered negligible respect to the reactor's dynamics. Putting $dx4/d\tau = 0$ the dimensionless jacket's temperature x_4 can be eliminated from Eq.(34), and the simplified mathematical model of the reactor can be written as

$$\frac{dx_2}{d\tau} = x_{50}(x_{20} - x_2) - c_0\, x_2\, e^{-1/x_3} - \frac{x_{50}f(\tau)}{1 + f(\tau)}(x_{20} - x_2)$$

$$\frac{dx_3}{d\tau} = x_{50}(x_{30} - x_3) + c_1\, x_2\, e^{-1/x_3} - \frac{c_2 c_3 x_6(x_3 - x_{40})}{c_3 x_6 + c_4}$$

$$- \frac{f(\tau)}{1 + f(\tau)}\left[x_{50}(x_{30} - x_3) - \frac{c_2 c_3 x_6(x_3 - x_{40})}{c_3 x_6 + c_4} \right] \tag{35}$$

$$\frac{dx_6}{d\tau} = \frac{K_{td}}{x_{30}}\frac{dx_3}{d\tau} + \frac{K_{td}}{x_{30}\tau_{2d}}(x_3 - x_s)$$

Note that from Eq.(33) the function $f(\tau) = 0$ if there is not disturbance in the inlet flow to the reactor ($x_{50} = 1$) and consequently, the equilibrium point of the reactor can be obtained from the following equations:

$$\frac{dx_2}{d\tau} = f_1(x_2, x_3, x_6) = x_{50}(x_{20} - x_2) - c_0\, x_2\, e^{-1/x_3}$$

$$\frac{dx_3}{d\tau} = f_2(x_2, x_3, x_6) = x_{50}(x_{30} - x_3) + c_1\, x_2\, e^{-1/x_3} - \frac{c_2 c_3 x_6(x_3 - x_{40})}{c_3 x_6 + c_4} \tag{36}$$

$$\frac{dx_6}{d\tau} = f_3(x_2, x_3, x_6) = \frac{K_{td}}{x_{30}}\left[x_{50}(x_{30} - x_3) + c_1\, x_2\, e^{-1/x_3} - \frac{c_2 c_3 x_6(x_3 - x_{40})}{c_3 x_6 + c_4} \right]$$

$$+ \frac{K_{td}}{x_{30}\tau_{2d}}(x_3 - x_s)$$

The same Eq.(36) can be approximately considered when the dimensionless time τ is high that the exponential factor of $f(\tau)$ in Eq.(33) can be considered negligible.

Exercise 6. Show that the equilibrium point of the model $\mathbb{R}^4$ defined by Eq.(34) and the simplified model $\mathbb{R}^3$ given by Eq.(35), i.e. when the dynamics of the jacket is considered negligible, are the same. Deduce the jacobian of the system (35) at the corresponding equilibrium point. Write a computer program to determine the eigenvalues of the linearized model $\mathbb{R}^3$ at the equilibrium point as a function of the dimensionless inlet flow x_{50}. Values of the dimensionless parameters of the PI controller can be fixed at $K_{td} = 1.52$; $\tau_{2d} = 5$. The set point dimensionless temperature and the inlet coolant flow rate temperature are $x_s = 0.0398$, $x_{40} = 0.0351$ respectively. An appropriate value of dimensionless reference concentration is $C'_{A0} = 0.245$. Does it exist some value of x_{50} for which the eigenvalues of the linearized system $\mathbb{R}^3$ at the equilibrium point are complex with zero real part? Note that it is necessary to vary x_{50} from small to great values. Check the possibility to obtain similar results for the $\mathbb{R}^4$ model.

It is interesting to point out that the regular behavior, where a steady state is reached by the reactor, can be investigated from the models defined by Eqs.(33) or (34), taking into account different values of the PI controllers. The values of K_{vd} and τ_{1d} are chosen from the inequality:

$$\left(\frac{K_{vd}}{2}\right)^2 < \frac{K_{vd}}{\tau_{1d}} \qquad (37)$$

The values of K_{td} and τ_{2d} from Eq.(36) can be obtained from the transfer function of the linearized model at the equilibrium point, applying conventional methods from the linear control theory (see [1]). In order to investigate the self-oscillating behavior, one can determine the linearized system at the equilibrium point, and the corresponding complex eigenvalues with zero real part, when the parameters K_{td} and τ_{2d} of the PI controller are varied. For example, taking into account Eq.(34), the jacobian matrix of the linearized system at dimensionless set point temperature xs is the following:

$$J = \begin{bmatrix} -x_{50} - c_0 e^{-1/x_s} & -c_0 \frac{x_{2e}}{x_s^2} e^{-1/x_s} & 0 & 0 \\ c_1 e^{-1/x_s} & -(x_{50}+c_2) + c_1 \frac{x_{2e}}{x_s^2} e^{-1/x_s} & c_2 & 0 \\ 0 & c_4 & -(c_3 x_{6e}+c_4) c_3(x_{40}-x_{4e}) \\ \frac{K_{td}}{x_{30}} c_1 e^{-1/x_s} & \frac{K_{td}}{x_{30}} \left[-(x_{50}+c_2) + c_1 \frac{x_{2e}}{x_s^2} e^{-1/x_s} \right] & \frac{K_{td}}{x_{30}} c_2 & 0 \\ & + \frac{K_{td}}{x_{30}\tau_{2d}} \end{bmatrix}$$

$$(38)$$

and the characteristic equation of can be written as:

$$|\lambda I - J| = \lambda^4 + S_1\lambda^3 + S_2\lambda^2 + S_3\lambda + S_4; \; S_i = (-1)^i \sum |A_{ik}^{ik}| \qquad (39)$$

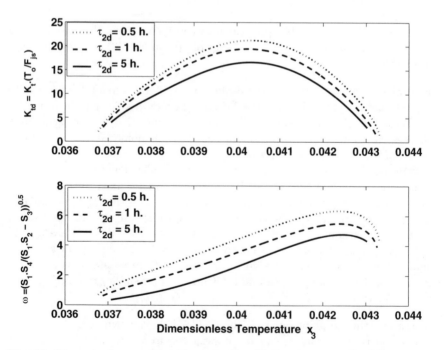

Fig. 13. Dimensionless constant K_{td} and self-oscillating frequency for different constant integral action τ_{2d}

where A_{ik}^{ik} are the minors of determinant (38) and S_i are the sum of the principal minors of J. Taking into account the Routh stability criterion, the Routh array is formed as given below:

$$\begin{vmatrix} 1 & S_2 \; S_4 \\ S_1 & S_3 \\ (S_1 S_2 - S_3)/S_1 & S_4 \\ \frac{(S_1 S_2 S_3 - S_3)^2/S_1 - S_1 S_4}{(S_1 S_2 - S_3)/S_1} \\ S_4 \end{vmatrix}$$

Assuming that $S_1 > 0$, $S_4 > 0$ and $S_1 S_2 - S_3 > 0$, the condition of self-oscillating behavior is given by the equation:

$$S_1 S_2 S_3 - S_3^2 - S_1^2 S_4 = 0 \tag{40}$$

and the frequency of self-oscillations is:

$$\omega = \sqrt{\frac{S_1 S_4}{S_1 S_2 - S_3}} \tag{41}$$

Figure 13 shows the variation of K_{td} for various values of τ_{2d} and the corresponding frequencies of self-oscillation. Figure 14 shows the oscillation behavior of the reactor with the value $x_s = 0.0398$ and $\tau_{2d} = 0.5$, $K_{td} = 19.6$.

Note that Figure 13 can be used to compare the parameters of the controller when they are obtained from the Ziegler-Nichols or Cohen-Coom rules. On the other hand, at Figure 14 it can be observed that the outlet dimensionless flow rate and the reactor volume reaches the steady state whereas the dimensionless reactor temperature remains in self-oscillation. The knowledge of the self-oscillation regime in a CSTR is important, both from theoretical and experimental point of view, because there is experimental evidence that the self-oscillation behavior can be useful in an industrial environment.

In Eqs.(33), (34) and (35) the effect of the constrained coolant flow rate due to control valve saturation is not considered, however this limitation can be introduced in the mathematical model of the reactor as follows:

$$sat(x_6) = \begin{cases} 0 & \text{for} \quad x_6 \leq 0 \\ x_6 & \text{for } 0 \leq x_6 \leq x_{6max} \\ x_{6max} & \text{for} \quad x_6 > x_{6max} \end{cases} \tag{42}$$

The model $\mathbb{R}^3$ of the reactor can be written as:

$$\frac{dx_2}{d\tau} = x_{50}(x_{20} - x_2) - c_0 \, x_2 \, e^{-1/x_3} - \frac{x_{50} f(\tau)}{1 + f(\tau)}(x_{20} - x_2)$$

$$\frac{dx_3}{d\tau} = x_{50}(x_{30} - x_3) + c_1 \, x_2 \, e^{-1/x_3} - \frac{c_2 c_3 sat(x_6)(x_3 - x_{40})}{c_3 sat(x_6) + c_4} \tag{43}$$

$$- \frac{f(\tau)}{1 + f(\tau)} \left[x_{50}(x_{30} - x_3) - \frac{c_2 c_3 sat(x_6)(x_3 - x_{40})}{c_3 sat(x_6) + c_4} \right]$$

$$\frac{dx_6}{d\tau} = \frac{K_{td}}{x_{30}}\frac{dx_3}{d\tau} + \frac{K_{td}}{x_{30}\tau_{2d}}(x_3 - x_s)$$

Eq.(42) and (43) will be used in the analysis of the chaotic behavior.

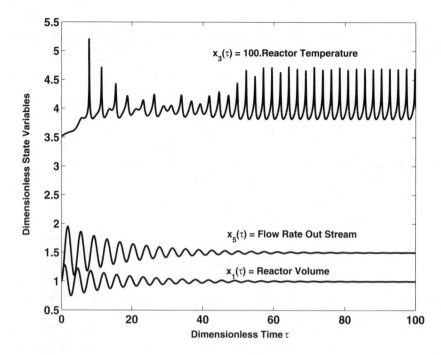

Fig. 14. Self-oscillation behavior with $K_{td} = 19.6$; $\tau_{2d} = 0.5$

4.1   Analysis of the Constrained Coolant Flow Rate

Consider the model's reactor assuming saturation in the control valve and negligible dynamic's jacket as shows in Eq.(43). The steady states of system (43) can be obtained considering that the derivatives of the left hand of Eq.(43) are nil and

$$\lim_{\tau \to \infty} f(\tau) = 0 \tag{44}$$

However, due to limitation in x_6 dimensionless flow rate of cooling water it is necessary to consider two cases, depending on wherever x_6 is constrained or limited by the maximum opening of the control valve.

The Fluid Flow Through the Control Valve Is Not Limited

This means that the value of the flow rate cooling water is enough to cool down the reactor and consequently, the reactor temperature can reach the desired set point. Therefore, the coolant flow rate reaches a certain steady state value x_{6e}, then from the third of Eq.(43) it can be deduced that:

$$\left(\frac{dx_6}{d\tau}\right)_{equilibrium} = \frac{K_{td}}{x_{30}}\left(\frac{dx_3}{d\tau}\right)_{equilibrium} + \frac{K_{td}}{x_{30}\tau_{2d}}(x_{3e} - x_s) \tag{45}$$

where x_{3e} is the equilibrium dimensionless temperature of the reactor. Taking into account that in equilibrium the derivatives of Eq.(45) are nil, it is deduced that the equilibrium dimensionless reactor temperature must be equal to dimensionless set point temperature $x_{3e} = x_s$, thus the control system drives the reactor to the desired set point.

From Eqs.(43) and (44), and considering the derivatives of the left-hand side to be zero, the values of dimensionless variables in steady state are the following:

$$x_{2e} = \frac{x_{50}x_{20}}{x_{50} + c_0 e^{-1/x_s}} \tag{46}$$

$$x_{6e} = \frac{c_4 \left[x_{50}(x_{30} - x_s) + c_1 \, x_{2e} \, e^{-1/x_s}\right]}{c_2 c_3 (x_s - x_{40}) - c_3 x_{50}(x_{30} - x_s) - c_1 c_3 x_{2e} e^{-1/x_s}} \tag{47}$$

where the values of x_{2e} and x_{6e} are the values of steady state. The equilibrium dimensionless jacket's temperature is calculated from the equation:

$$x_{4e} = \frac{c_1 x_{6e} x_{40} + c_4 x_s}{c_3 x_{6e} + c_4} \tag{48}$$

The Dimensionless Cooling Flow Is Constrained

In this case it is not possible to reach any value of equilibrium dimensionless coolant flow rate x_{6e}, because when x_{6e} is greater than x_{6max}, it is constrained to the maximum value x_{6max} due to the flow rate limitation through the control valve. From this moment, the derivative $(dx_6/d\tau)$ is zero and the flow rate cooling x_6 remains constant. Consequently, the coolant flow rate cannot decrease the reactor temperature, which reaches a value greater than the set point, and the corresponding reactant concentration will be smaller. From Eq.(43) the set point temperature must be equal to x_{3e}, and as a result it is impossible that the reactor temperature would be able to reach the set point temperature x_s, an consequently the control system cannot drive the reactor to the desired equilibrium point. The equilibrium values of dimensionless variables are given by the same Eqs.(45), (46) and (47), but making the substitutions:

$$x_{6e} = x_{6max}; \quad x_s = x_{3e} \tag{49}$$

By substituting Eq.(49) into Eqs.(43), it can be deduced that

$$\frac{c_2 c_3 x_{6max}}{c_3 x_{6max} + c_4} = \frac{x_{50}(x_{30} - x_{3e})}{x_{3e} - x_{40}} + \frac{x_{50}x_{20}c_1}{(x_{3e} - x_{40})(c_0 + x_{50}e^{-1/x_{3e}})} \tag{50}$$

Eq.(50) shows the variation of the equilibrium dimensionless temperature as a function of the maximum value of the dimensionless coolant flow rate x_{6max}. Plotting x_{6max} versus x_{3e} a bifurcation curve can be obtained, from which it is possible to determine the value of x_{6max} which gives a different behavior of the reactor in steady state. It is interesting to note that Eq.(50) is equal to Eq.(47) when we make the substitutions of Eq.(49) into Eq.(47).

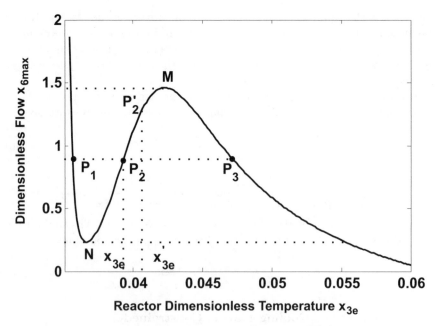

Fig. 15. Dimensionless bifurcation plot when the coolant flow rate is constrained

However, these equations are conceptually different because from a practical viewpoint, the value of x_s is limited to a small value, while x_{3e} is not.

The curve of maximum dimensionless cooling temperature x_{6max} versus the equilibrium dimensionless reactor temperature from Eq.(50) is shown in Figure 15. This plot shows the possibility of three steady state states corresponding to the same value of x_{6max}. The three steady sates are the points of intersection P_1, P_2 and P_3 of the plot with the horizontal line corresponding to a value of x_{6max}. Once the steady state has been found, it is necessary to know how stable they are. Let us assume that in equilibrium the reactor is at point P_2'. At this point the equilibrium temperature is greater than point P_2. This will cause the maximum dimensionless coolant flow rate x_{6max} inlet reactor' jacket to raise, and the raising will continue until the upper maximum point M. It is easy to show, using similar arguments that P_1 and P_3 are steady state and P_2 is an unstable one.

However, more precisely, the stability at the different steady sates can be determined by calculating the eigenvalues of the matrix of the linearized model of the CSTR. If there is an eigenvalue with positive real part, the steady state is unstable, and all eigenvalues with negative real part indicates a stable steady state. Thus, by simulation it can be verified that the steady states P_1 and P_3 are stable and P_2 is unstable. This means that it is impossible to reach the point P_2 when the coolant flow rate is constrained.

The value of x_{6max} corresponding to the point M represents a global bifurcation. Thus, values of this flow above this point give a reactor behavior similar to case 4.1, i.e. when there is not limitation in the flow trough the control valve, and the control system can to drive the reactor to the desired set point. For values of x_{6max} under the corresponding to point N there is only one equilibrium point and this is the case 3.1.2, due to the low flow rate of cooling, the reactor temperature cannot reach the set point. In the next section, the study of chaotic behavior is carried out considering the $\mathbb{R}^4$ and $\mathbb{R}^3$ models of the reactor given by Eqs.(34) and (35) respectively. Note that both $\mathbb{R}^3$ and $\mathbb{R}^4$ models have the same equilibrium points.

4.2   Analysis of the Chaotic Behavior

Consider the $\mathbb{R}^3$ model defined by Eq.(43). In order to analyze the stability at the equilibrium point (x_{2e}, x_{3e}, x_{6e}), it is necessary to introduce the deviation variables given as:

$$x = x_2 - x_{2e}; \quad y = x_3 - x_{3e}; \quad z = x_6 - x_{6e} \tag{51}$$

So, Eq.(43) can be rewritten as:

$$
\begin{aligned}
\frac{dx}{d\tau} &= f_1(x,y,z) = x_{50}(x_{20} - (x + x_{2e})) - c_0(x + x_{2e})\, e^{-1/(y+x_s)} \\
\frac{dy}{d\tau} &= f_2(x,y,z) = x_{50}(x_{30} - (y + x_s)) + c_1(x + x_{2e})\, e^{-1/(y+x_s)} \\
&\quad - \frac{c_2 c_3 \, sat(z + x_{6e})(y + x_s - x_{40})}{c_3 \, sat(z + x_{6e}) + c_4} \\
\frac{dz}{d\tau} &= f_3(x,y,z) = \frac{K_{td}}{x_{30}}\frac{dy}{d\tau} + \frac{K_{td}}{x_{30}\tau_{2d}}\, y
\end{aligned}
\tag{52}
$$

where the $sat()$ term is defined by Eq.(42). Equations $f_i(x,y,z)$ can be expanded in series of Taylor up to first order approximation, so the linearized system at the origin can be written as:

$$
\begin{bmatrix} \dot{x} \\ \dot{y} \\ \dot{z} \end{bmatrix} =
\begin{bmatrix}
-(x_{50} + c_0 e^{-1/x_s}) & -c_0\frac{x_{2e}}{x_s^2} e^{-1/x_s} & 0 \\[2mm]
c_1 e^{-1/x_s} & -x_{50} + c_1\frac{x_{2e}}{x_s^2} e^{-1/x_s} & -\frac{c_2 c_3 (x_s - x_{40})}{(c_3 x_{6e} + c_4)^2} \\
& -\frac{c_2 c_3 x_{6e}}{c_3 x_{6e} + c_4} & \\[2mm]
\frac{K_{td}}{x_{30}} c_1 e^{-1/x_s} & \frac{K_{td}}{x_{30}}\left[-x_{50} + c_1\frac{x_{2e}}{x_s^2} e^{-1/x_s}\right. & \frac{K_{td}}{x_{30}}\left[-\frac{c_2 c_3 (x_s - x_{40})}{(c_3 x_{6e} + c_4)^2}\right] \\
& \left. -\frac{c_2 c_3 x_{6e}}{c_3 x_{6e} + c_4} + \frac{1}{\tau_{2d}}\right] &
\end{bmatrix}
\begin{bmatrix} x \\ y \\ z \end{bmatrix}
\tag{53}
$$

The eigenvalues of the linearized model $\mathbb{R}^3$ given by Eq.(53) can be deduced for a value of the set point temperature x_s, and the values of the PI controller as a function of the inlet dimensionless flow x_{50}. Note that the variables x_{50}, x_{20}, x_{30}, x_{40} can be considered as external disturbances, however only x_{50} will be considered as variable, whereas x_{20}, x_{30}, x_{40} are considered as constant.

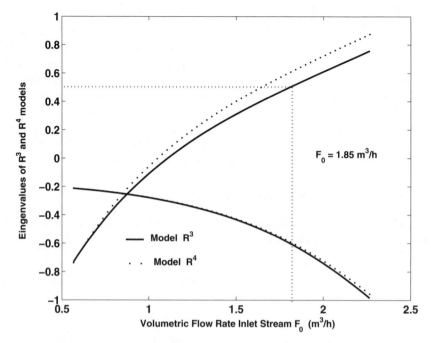

Fig. 16. Real part of eigenvalues of the vector fields $\mathbb{R}^3$ and $\mathbb{R}^4$ vs. volumetric flow rate inlet stream F_0

Remark 4. From Eq.(34), similar equations to (51) and (52) can be obtained for $\mathbb{R}^4$ model. Nevertheless both $\mathbb{R}^3$ and $\mathbb{R}^4$ models have the same equilibrium points.

In Figure 16, the variation of real part of eingenvalues vs. the volumetric flow rate inlet stream for x_{50}, both $\mathbb{R}^3$ and $\mathbb{R}^4$ vector fields, when x_{20}, x_{30}, x_{40} remain constant, are shown. It can be observed that these values do not differ very much. This means that the jacket's dynamics can be considered negligible. It is clear that from a certain value of $x_{50} = F_0/F_{0s}$ there is an eigenvalues with positive real part, and so the equilibrium point will be unstable.

Consider the space state model $\mathbb{R}^3$ defined by Eq.(52), showing an equilibrium point such that the matrix of the linearized system at this point has a real negative eigenvalue λ and a pair of complex eigenvalues $\alpha \pm j\beta$, $(j = \sqrt{-1})$ with positive real parts α. In this situation, the equilibrium point has one-dimensional stable manifold and two-dimensional unstable manifold. If the condition $\lambda < |\alpha|$ is verified, it is possible that an homoclinic orbit appears, which tends to the equilibrium point. This orbit is very singular, and then the Shilnikov theorem asserts that every neighborhood of the homoclinic orbit contains infinite number of unstable periodic orbits.

The existence of the equilibrium point and the conditions for the eigenvalues of the linearized model are not difficult to find, nevertheless the hypothesis

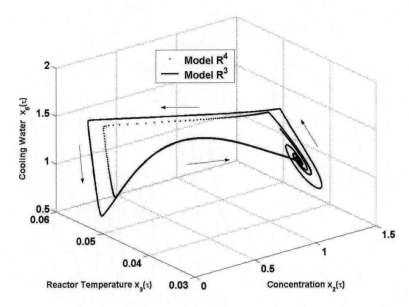

Fig. 17. Homoclinic orbit of Shilnikov type for the vector fields $\mathbb{R}^3$ and $\mathbb{R}^4$. Eigenvalues at the equilibrium point for $\mathbb{R}^3$: $\lambda = -0.56$; $\alpha \pm \beta j = 0.39 \pm 1.92j$.

on the existence of a Shilnikov homoclinic orbit is usually very difficult to establish. For example, with a disturbance inlet flow rate of $F_0 = 1.85$ m^3/h, from Figure 16 the values of eigenvalues are: $0.5 \pm 1.92j$; -0.57, so the Jordan quasi-diagonal form is:

$$\Lambda = \begin{bmatrix} \rho & -\omega & 0 \\ \omega & \rho & 0 \\ 0 & 0 & \lambda \end{bmatrix} ; \quad \lambda_{1,2} = \rho \pm \omega j; \quad \lambda_3 = \lambda \qquad (54)$$

If ρ, $\omega > 0$; $-\lambda > 0$ and $-\lambda/\rho > 1$ the equilibrium point is unstable, and a Shilnikov' orbit may appear. For the reactor, with a value of $x_{50} > 1$ and $x_{6max} \approx (x_{6max})_M$ (see Figure 15), by simulation it is possible to verify the presence of a homoclinic orbit to the equilibrium point. Figure 17 shows the homoclinic orbit for the model $\mathbb{R}^3$ and $\mathbb{R}^4$, when the steady state has been reached. Note that the Shilnikov orbit appear when the coolant flow rate is constrained. If there is no limitation of the coolant flow rate, a limit cycle is obtained both in models $\mathbb{R}^3$ and $\mathbb{R}^4$, by simulation.

According to the Shilnikov's theorem, the reactor presents a chaotic behavior. In order to test the presence of a strange attractor, it is necessary to raise the value of x_{6max} to introduce a perturbation in the vector field around the homoclinic orbit. Taking $x_{6max} = 5$, the results of the simulation are shown in Figure 18, where the sensitive dependence on initial conditions has been corroborated.

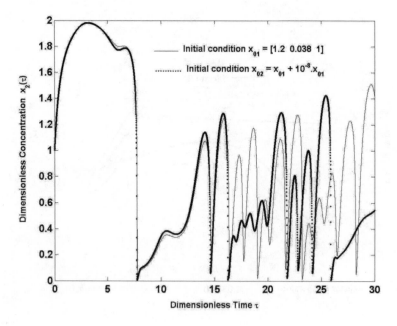

Fig. 18. Sensitive dependence of the dimensionless concentration of $\mathbb{R}^3$ model with two very close initial conditions. $K_t = 0.094 \ \mathrm{m}^3/\mathrm{h}^\circ\mathrm{C}$; $t_2 = 5$ h.

It is interesting to note that in chaotic regime, the flow rate outlet stream, which is manipulated by the control valve CV1 (see Figure 12), and the reactor volume, are driven by the PI controller to the equilibrium point without chaotic oscillations. However, the other variables have a chaotic behavior as shown in Figure 18. So it is possible to obtain a reactor behavior, in which some variables are in steady state and the others are in regime of chaotic oscillations, due to the decoupling or serial connection phenomena. In this case the control system and the volumetric flow limitation of coolant flow rate through the control valve VC2, are the responsible of this behavior. Similar results can be obtained from $\mathbb{R}^4$ model.

5   Conclusion

From the study presented in this chapter, it has been demonstrated that a CSTR in which an exothermic first order irreversible reaction takes place, can work with steady-state, self-oscillating or chaotic dynamic. By using dimensionless variables, and taking into account an external periodic disturbance in the inlet stream temperature and coolant flow rate, it has been shown that chaotic dynamic may appear. This behavior has been analyzed from the Lyapunov exponents and the power spectrum.

Although this behavior is interesting, another more complicated case is researched from the analysis of the eingenvalues of the reactor equations linearized at the equilibrium point. In this case, it has been corroborated that the steady state can be modified throughout the variation of the composition and temperature of inlet stream to the reactor. From the parameters-plane inlet stream temperature-composition, the presence of various zones, i.e. a lobe and a curve with cusp point, with different dynamical behavior has been shown. So, if the dimensionless concentration and temperature of the inlet flow rate to the reactor are inside the lobe, the reactor reaches a self-oscillating behavior. This self-oscillating dynamical behavior can be very difficult to obtain, because the area of the lobe is very small. When the reactor is in self-oscillating mode, applying an external periodic disturbance in the coolant flow rate, chaotic dynamics can appear. This chaotic behavior is very difficult to obtain with only an external periodic disturbance in the coolant flow rate.

The non-linear dynamics of the reactor with two PI controllers that manipulates the outlet stream flow rate and the coolant flow rate are also presented. The more interesting result, from the non-linear dynamic point of view, is the possibility to obtain chaotic behavior without any external periodic forcing. The results for the CSTR show that the non-linearities and the control valve saturation, which manipulates the coolant flow rate, are the cause of this abnormal behavior. By simulation, a homoclinic of Shilnikov type has been found at the equilibrium point. In this case, chaotic behavior appears at and around the parameter values from which the previously cited orbit is generated.

From the results presented in this chapter, more advanced studies from the bifurcation theory can be planed. For example, inside the lobe, the behavior of the reactor is self-oscillating, i.e. an Andronov-Poincare-Hopf bifurcation can be researched from the calculation of the first Lyapunov value, in order to know if a weak focus may appear, or the conditions which give a Bogdanov-Takens bifurcation etc. Finally, it is interesting to remark that the previously analyzed phenomena should be known by the control engineer in order to either avoid them or use them, depending on the process type.

References

[1] P. Albertos and M. Pérez Polo. *Selected Topics in Dynamics and Control of Chemical and Biochemical Processes*, chapter Nonisothermal stirred-tank reactor with irreversible exothermic reaction $A \to B$. 1.Modelling and local control. LNCIS. Springer-Verlag, 2005 (in this volume).

[2] J. Alvarez-Ramírez, R. Femat, and J. González-Trejo. Robust control of a class of uncertain first-order systems with least prior knowledge. *Chem. Eng. Sci.*, 53(15):2701–2710, 1998.

[3] J. Alvarez-Ramírez, J. Suarez, and R. Femat. Robust stabilization of temperature in continuous-stirred tank reactors. *Chem. Eng. Sci.*, 52(14):2223–2230, 1997.

[4] A.A. Andronov, A. Vitt, and S.E. Khaikin. *Theory of Oscillations*. Pergamon Press, Oxford, 1966.

[5] G. Benettin, L. Galgani, and J-M. Strelcyn. Lyapunov characteristic exponents for smooth dynamical systems and for Hamiltonian systems: a method for computing all of them. *Meccanica*, 15:9–20, 1980.

[6] G. Benettin, L. Galgani, and J-M. Strelcyn. Lyapunov characteristic exponents for smooth dynamical systems and for Hamiltonian systems: a method for computing all of them. *Meccanica*, 15:21–30, 1980.

[7] M.L. Cartwright and J.E. Littlewood. On nonlinear differential equations of the second order, I. The equation: $\ddot{y} + k(1 - y^2)\dot{y} + y = b\lambda cos(\lambda t + a)$, k large. *Lond. Math. Soc.*, 20:180–189, 1945.

[8] M. Dolnik, A.S. Banks, and I.R. Epstein. Oscillatory chemical reaction in a CSTR with feedback control of flow rate. *Journal Phys. Chem. A*, 101:5148–5154, 1997.

[9] R. Femat. Chaos in a class of reacting systems induced by robust asymptotic feedback. *Physica D*, 136:193–204, 2000.

[10] P. Glendinning. *Stability, instability and chaos: an introduction to the theory of nonlinear differential equations.* Cambridge University Press, 1994.

[11] J. Guckenheimer and P. Holmes. *Nonlinear Oscillations, Dynamical Systems, and Bifurcations of Vector Fields.* Springer-Verlag, New York, 1983.

[12] E. Hopf. Abzweigung einer periodischen Lösung von einer stationaren Lösung eines Differetiasystems. *Ver Math.-Phys. Kl. Sachs, Acad Wiss*, Leipzig 94:1–22, 1942.

[13] Z. Kubickova, M. Kubicek, and M. Marek. Feed-batch operation of stirred reactors. *Chem. Eng. Sci.*, 42(2):327–333, 1987.

[14] M.J. Kurtz, G. Yan Zhu, and M.A. Henson. Constrained output feedback control of a multivariable polymerization reactor. *IEEE Trans. Contr. Syst. Technol.*, 8:87–97, 2000.

[15] A.J. Lichtenberg and M.A. Lieberman. *Regular and Chaotic Dynamics.* Springer-Verlag, New York, 2nd. edition, 1992.

[16] W.L. Luyben. *Process Modeling, Simulation and Control for Chemical Engineering.* McGraw-Hill, New York, 2nd. edition, 1990.

[17] J.C. Mankin and J.L. Hudson. Oscillatory and chaotic behaviour of a forced exothermic chemical reaction. *Chem. Eng. Sci.*, 39(12):1807–1814, 1984.

[18] J.E. Marsden and M. McCraken. *The Hopf Bifurcation and its Applications.* Springer, New York, 1976.

[19] E. Ott. *Chaos in Dynamical Systems.* Cambridge University Press, 2002.

[20] L. Pellegrini and G. Biardi. Chaotic behaviour of a controlled CSTR. *Comput. Chem. Eng.*, 14:1237–1247, 1990.

[21] M. Pérez and P. Albertos. Self-oscillating and chaotic behaviour of a PI-controlled CSTR with control valve saturation. *J. Process Control*, 14:51–57, 2004.

[22] M. Pérez, R. Font, and M.A. Montava. Regular self-oscillating and chaotic dynamics of a continuos stirred tank reactor. *Comput. Chem. Eng.*, 26:889–901, 2002.

[23] J.B. Planeaux and K.F. Jensen. Bifurcation phenomena in CSTR dynamics: A system with extraneous thermal capacitance. *Chem. Eng. Sci.*, 41(6):1497–1523, 1986.

[24] R. Seydel. *Practical Bifurcation and Stability Analysis From Equilibrium to Chaos.* Springer-Verlag, New York, 2nd. edition, 1994.

[25] L.P. Shilnikov. A case of the existence of a demensurable set of periodic motions. *Sov. Math. Dokl.*, 6:163–166, 1965.

[26] L.P. Shilnikov, A.L. Shilnikov, P.V. Turaev, and O.L. Chua. *Part II. World Scientific*, chapter Methods of qualitative theory in nonlinear dynamics. Series in Nonlinear Science, 2001.

[27] M. Soroush. Nonlinear state-observer design with applications to reactors. *Chem. Eng. Sci.*, 52(3):387–404, 1987.

[28] G. Stephanopoulos. *Chemical Process Control: An introduction to theory and practice*. Prentice Hall, New Jersey, 1984.

[29] F. Teymour. Dynamics of semibatch polymerization reactors: I. Theoretical analysis. *A.I.Ch.E. Journal*, 43(1):145–156, 1997.

[30] F. Teymour and W.H. Ray. The dynamic behavior of continuous solution polymerization reactors-IV. Dynamic stability and bifurcation analysis of an experimental reactor. *Chem. Eng. Sci.*, 44(9):1967–1982, 1989.

[31] K. Tomita. *Periodically forced nonlinear oscillators*. A.V. Holden, Princenton Univ. Press.

[32] Y. Uppal, W.H. Ray, and A.B. Poore. On the dynamic behavior of continuous stirred tank reactors. *Chem. Eng. Sci.*, 29:967–985, 1974.

[33] D.A. Vaganov, N.G.V. Samoilenko, and V.G. Abranov. Periodic regimes of continuous stirred tank reactors. *Chem. Eng. Sci.*, 33:1133–1140, 1978.

[34] S. Wiggins. *Global Bifurcations and Chaos*. Springer, New York, 1988.

[35] S. Wiggins. *Introduction to Applied Nonlinear Dynamical System and Chaos*. Springer, New York, 1990.

[36] A.M. Zhabotinsky and A.B. Rovinskii. *Self-Organization, Autowaves and Structures Far from Equilibrium*. Springer, Berlin, 1984.

Appendix

The Lyapunov exponents provide a computable measure of the sensitivity to initial conditions, i.e. characterize the mean exponential rate of divergence of two nearby trajectories if there is at least one positive Lyapunov exponent, or convergence when all Lyapunov exponents are negative. The Lyapunov exponents are defined for autonomous dynamical systems and can be described by:

$$\frac{dx_i(t)}{dt} = f_i[x(t)]; \; i = 1, 2, \ldots, n \Rightarrow \frac{dx(t)}{dt} = f[x(t)]; \; x(t) \in \mathbb{R}^n \qquad (55)$$

Consider a trajectory in the $\mathbb{R}^n$ n-dimensional space of $x(t)$ and a nearby trajectory $x(t) + \delta x(t)$, where the symbol δ means an infinitesimal variation, i.e. an arbitrary infinitesimal change not tangent to the initial trajectory. Eq.(55) can be linearized throughout the trajectory to obtain

$$\frac{d[x(t) + \delta x(t)]}{dt} = f[x(t) + \delta x(t)] \qquad (56)$$

Now the function $f[x(t) + \delta x(t)]$ can be expanded in series up to first order term to give

$$f[x(t) + \delta x(t)] = f[x(t)] + \left(\frac{\partial f}{\partial x}\right)_{\partial x(t)=0} \delta x(t) \qquad (57)$$

Substituting Eq.(57) into Eq.(56) the following linearized system throughout the trajectory is obtained:

$$\delta \dot{x} = J[x(t)] \cdot \delta x(t) \qquad (58)$$

where the dot indicates differentiation respect to time and $J[x(t)]$ is the jacobiam matrix of system (55).

The Lyapunov exponents of order one are defined by:

$$\lambda_i^1 = \lim_{t \to \infty} \frac{1}{t} \log_2 \left(\frac{\delta(x_0, t)}{\delta(x_0, 0)}\right); \; i = 1, 2, \ldots, n \qquad (59)$$

This value is a measure of the mean exponential rate of divergence (convergence) of two initially very close trajectories, i.e. when $\delta(x_0, 0) \to 0$. The values from (59) are the so called Lyapunov characteristic exponents, which can be ordered by size:

$$\lambda_1^1 \geq \lambda_2^1 \geq \ldots \geq \lambda_n^1 \qquad (60)$$

It is possible to generalize the previous concept to describe the mean rate of exponential growth (decrease) of a m-dimensional volume in the tangent space of $\mathbb{R}^n$, where $m \leq n$. The Lyapunov exponent of order m is defined as

$$\lambda^m = \lim_{t \to \infty} \frac{1}{t} \log_2 \left(\frac{V_m(t)}{V_m(0)} \right); \ m = 1, 2, \ldots, n \qquad (61)$$

where:

$$V_m = \delta \vec{x}_1 \wedge \delta \vec{x}_2 \wedge \ldots \wedge \delta \vec{x}_m \qquad (62)$$

is the volume of an m-dimensional parallelepiped whose edges are $\delta \vec{x}_1$, $\delta \vec{x}_2$, ..., $\delta \vec{x}_m$. Then, the Lyapunov exponent associated to the direction m is defined as:

$$\lambda_m = \lambda^m - \lambda^{m-1} \qquad (63)$$

In order to determine λ_m, the values of Lyapunov exponents or order m must be known and consequently, using Eqs.(61) and (62), it will be necessary to integrate Eq.(58). However, this equation presents numerical problems because the divergence of nearby trajectories can be very strong (In chaotic dynamic this phenomenon is known as stretching and folding), and the linearization process is numerically unstable. This problem is overcome by using the orthonormalization Gram-Schmidt procedure. This is the idea of the method developed in [5].

The calculation of the Lyapunov exponents of order m has been carried out as follows. The following notation is useful:

$$\delta x_{k-1}^j(T)$$

where the superindex means the vector number, the subindex a generic integration step and T is the integration step used in the numerical algorithm. For example, $\delta x_{k-1}^1(T)$ means the vector one which comes from the vector $\delta x_{k-1}^1(0)$ as results of integration process of Eq.(58) when the state of the non-linear system (55) has changed from $x[(k-1)T]$ to $x[KT]$,i.e.

$$x[(k-1)T] \to x[kT] \Rightarrow \delta x_{k-1}^1(0) \to \delta x_{k-1}^1(T) \qquad (64)$$

When all $\delta x_{k-1}^j(T)$; $j = 1, 2, \ldots, n$ are known, the Gram-Schmidt orthonormalization procedure is applied, so a new set of $\delta x_k^j(0)$ is obtained and a new iteration starts. Once $\delta x_{k-1}^j(T)$; $j = 1, 2, \ldots, n$ are known, the following Euclidian distances are calculated

$$d_k^j = \|\delta x_{k-1}^j(T)\|; \ j = 1, 2, \ldots, m \qquad (65)$$

Then, the Lyapunov exponent of order m is calculated from:

$$\lambda^m = \lim_{n \to \infty} \frac{1}{nT} \sum_{i=1}^{n} \log_2(d_i^1 \cdot d_i^2 \cdot \ldots \cdot d_i^m) \qquad (66)$$

Taking into account Eq.(63)the value of the Lyapunov exponent associated to the direction m is deduced as follows:

$$\lambda^m = \lim_{n \to \infty} \frac{1}{nT} \sum_{i=1}^{n} \log_2(d_i^m) \tag{67}$$

The Gram-Schmidt orthonormalization method is applied using the following steps.

i. The first value $\delta x_k^j(0)$ is defined by

$$\delta x_k^1(0) = \frac{\delta x_{k-1}^1(T)}{d_k^1} = \frac{\delta x_{k-1}^1(T)}{\|\delta_{k-1}^1(T)\|} \tag{68}$$

i.e. the normalized value of $\delta x_{k-1}^j(T)$. Now, the values of $\delta x_{k-1}^j(0)$ with $j = 2, 3, \ldots, n$ must be calculated.

ii. Successively determine for $j = 2, 3, \ldots, n$ the new variables:

$$\nu_{k-1}^j(T) = \delta x_{k-1}^j(T) - \sum_{i=1}^{j-1} \left[\delta x_k^j(0) \cdot \delta x_{k-1}^j(T) \right] \cdot \delta x_k^j(0)$$

$$d_k^j = \|\nu_{k-1}^j(T)\|; \quad j = 2, 3, \ldots, n \tag{69}$$

iii. With the previously calculated value determine:

$$\delta x_k^j = \frac{\nu_{k-1}^j(T)}{d_k^j}; \quad j = 2, 3, \ldots, n \tag{70}$$

The calculation process has been implemented by using a Matlab© program throughout the following steps.

1. From Eq.(55) by using the Runge-Kutta or Runge-Kutta-Fehlberg intregration method the state variables $x(t)$ can be calculated.
2. From the values $x(t)$ the jacobian matrix $J[x(t)]$ can be calculated. For example, in a system of order $n = 4$, four vectors of dimension four must be determined. To carry out this, a matrix of 16×16 elements is defined as follows:

$$JJ = \begin{bmatrix} J & Z & Z & Z \\ Z & J & Z & Z \\ Z & Z & J & Z \\ Z & Z & Z & J \end{bmatrix} ; \ J(4 \times 4); \ Z = zeros(4 \times 4) \tag{71}$$

3. The R-K or R-K-F method is applied to the linear system

$$\frac{dz(t)}{dt} = JJ \cdot z(t) \tag{72}$$

and the following values are obtained:

$$\delta x_i(1:4, i) = z(1:4, i)$$
$$\delta x_i(5:8, i) = z(5:8, i)$$
$$\delta x_i(9:12, i) = z(9:12, i) \tag{73}$$
$$\delta x_i(13:16, i) = z(13:16, i)$$

4. From Eq.(73) the Gram-Schmidt orthonormalization is applied to obtain the normalized values $[\delta x_{1n}, \delta x_{2n}, \delta x_{2n}, \delta x_{2n}]$. The following integration step is carried out with the values:

$$z(1:16, i+1) = [\delta x_{1n}; \delta x_{2n}; \delta x_{2n}; \delta x_{2n}] \tag{74}$$

5. From the normalized vector $[\delta x_{1n}; \delta x_{2n}; \delta x_{2n}; \delta x_{2n}]$ the four Lyapunov exponents are calculated as follows

$$S_1 = [d_1; d_2; d_3; d_4]; \quad S_2 = \log_2(S_1); \quad S = S + S2$$

$$NL(1:4, i) = \frac{S}{t(i)} \tag{75}$$

where S is an accumulative auxiliary variable and $t(i)$ is the value of the integration time.

Oscillations in Controlled Processes: Two Experimental Study Cases

R. Femat[1], H.O. Méndez-Acosta[2], and J. Álvarez-Ramírez[3]

[1] Depto. de Matemáticas Aplicadas y Sistemas Computacionales, IPICyT
rfemat@ipicyt.edu.mx
[2] Depto. de Ingeniería Química, CUCEI-UdG
hugo.mendez@cucei.udg.mx
[3] Depto. de Ingeniería de Procesos, UAM-Iztapalapa
jjar@xanum.uam.mx

Summary. The oscillatory behavior in two controlled processes is presented in this chapter. The two study cases can permit further numerical or analytical research that, on the one hand, allows for an evaluation of new criteria on dynamics of controlled systems and, on the other, leads to new pictures featuring nonlinear (possibly chaotic) dynamics. The former is a feedback-controlled digester which is interconnected with heat exchanger via recycle stream. Oscillations in temperature bioreactor is studied in this case. The process is an anaerobic digestion for distillery vinasses and a heat exchanger. The heat exchanger is basically controlled by a Proportional-Integral-Derivative feedback (PID control) while the bioreactor is controlled by a feedback law with uncertainties compensation. In fact, the first case corresponds to the bioreactor discussed in Part II of this book. The latter is open-loop controlled, and concerns to gas-liquid tide column. Such a gas-liquid column can exhibit three classes of dynamical behavior: bubbling (disperse fluid-flow), vortical-spiral type, and turbulent (three-region) regime. Both processes have recycle streams, and their time series are analyzed exploiting Poincaré maps, maximum Lyapunov exponents and power spectrum density.

1 Oscillations in Controlled Systems

The dynamics of controlled systems is an open problem that has recently attracted the attention of scientific community [13]. In fact, oscillatory behavior in chemical systems is an interesting topic (which has been typically studied in autocatalytic reactions, e.g., the Lotka system; see [44] and references therein). Dynamics of controlled systems can be explained in terms of interconnections. Indeed, by analogy with control systems, autocatalytic chemical systems can be described as examples of chemical feedback [44].

In this direction, the recent results have shown that oscillations (including chaotic behavior) can be induced by feedback interconnections in chemical

H.O. Méndez-Acosta et al. (Eds.): Dyn. & Ctrl. of Chem. & Bio. Proc., LNCIS 361, pp. 281–319, 2007.

reaction systems. For instance, oscillatory clusters can be found in homogeneous chemical systems, where the oscillatory behavior can be induced by feedback. Such oscillatory behavior can play an important role in understanding coupled neuron systems [49]. Other results have shown that complex dynamics can be generated in tubular reactors by time delay on energy transport. Such results illustrate the importance of recycling in chemical systems under lag transport [7]. Moreover, it has been reported that pH chemical system can exhibit oscillatory behavior from feedback interconnection [11]. Also homoclinic chaos has been observed in chemical systems [3],[4], even in an monomolecular reaction, A → B, carried out in an exothermic continuous flow stirred tank reactor (CSTR), where the chaotic behavior has been shown to be induced by feedback of temperature measurements [20]. It should be noted from the process control viewpoint, that even though classical feedback interconnections aim the reactor stabilization [40], however, it has been recently reported that the PI-controlled CSTR displays several complex dynamical behavior including periodic oscillation and homoclinic chaos [39]. On the other hand, oscillatory behavior is often deemed inexpedient because of their propensity to cause operational complexities. Nevertheless, it has been demonstrated that the yield of the desired product from an oscillatory autocatalytic reaction, in a CSTR, can be quadrupled with a linear feedback mechanism. Moreover, unsuspected applications can result from the chemical oscillations study. For example, a chaos-based scheme was used in [16] for encoding information into the equation of the Belousov-Zhabotinsky reaction. Here, an arbitrary binary message can be encoded by forcing the chaotic oscillations to follow a specified trajectory. Results in [37] have shown that chemical feedback models can produce a desired symbolic code. That is, from the nonlinear science viewpoint, diverse applications can be found from many models of chemical oscillations. Furthermore, oscillations in chemical reactors are very interesting by themselves.

Now, from its essential notion, we have the feedback interconnection implies that a portion of the information from a given system returns back into the system. In this chapter, two processes are discussed in context of the feedback interconnection. The former is a typical feedback control systems, and consists in a bioreactor for waste water treatment. The bioreactor is controlled by robust asymptotic approach [33], [34]. The first study case in this chapter is focused in the bioreactor temperature. A heat exchanger is interconnected with the bioreactor in order to lead temperature into the digester around a constant value for avoiding stress in bacteria. The latter process is a fluid mechanics one, and has feedforward control structure. The process was constructed to study kinetics and dynamics of the gas-liquid flow in vertical column. In this second system, the interconnection is related to recycling liquid flow. The experiment comprises several superficial gas velocity. Thus, the control acting on the gas-liquid column can be seen as an open-loop system where the control variable is the velocity of the gas entering into the column. There is no measurements of the gas velocity to compute a fluid dynamics regime. In this second process, there is no feedback interconnection for controlling its dynamics. In spite of both bioreactor and gas-liquid column has different interconnection structure and very different nature, both processes depict oscillations with experimental evidence that chaotic behavior can be present into them.

In order to analyze both systems, some techniques from nonlinear science are burrowed. Firstly, a phase portrait is constructed from delay coordinates, a Poincaré map is also computed, FFT is exploited to derive a Power Spectrum Density (PSD); Maximum Lyapunov Exponents (MLE) are also calculated from time series. Although we cannot claim chaos, the evidence in this chapter shows the possible chaotic behavior but, mostly important, it exhibits that the oscillatory behavior is intrinsically linked to the controlled systems. The procedures are briefly described before discuss each study case.

1.1   Space-Phase Reconstruction and Poincaré Map

Delay coordinates are often used in time-series analysis to represent the space-state of a given dynamical system. Thus, time-delay coordinates can be used to reconstruct the attractor, if exists, from the measurements of a single scalar in high-dimensional system. Let us denote $y(t)$ as the reactor temperature measurements and $z(t)$ a m-dimensional vector of time-delay coordinates. The delay coordinates are given by $z(t) = (y(t), y(t-T), y(t-2T), \ldots, y(t-(m-1)T))$ where T is the delay-time and m is the embedding dimension. Thus, if n is the dimension of the dynamical system from where time series were measured, the vector $z(t)$ is generically global one-to-one representation of the system from the output $y(t)$. Thus, such a representation implies that the trajectories defined by vector $z(t)$ are unique in the sense that solutions of the differential equations of the Bioreactor-Exchanger interconnection; i.e., the system (6). For experimental systems, it is not an easy task to obtain an estimated value of the embedding dimension m. However, space-phase reconstruction with low embedding dimension provides information on the geometry features of the reconstructed attractor [23].

1.2   Power Density and Lyapunov Spectra

A spectrum is the distribution of physical characteristics in a system. In this sense, the Power Spectrum Density (PSD) provides information about fundamental frequencies (and their harmonics) in dynamical systems with oscillatory behavior. PSD can be used to study periodic-quasiperiodic-chaotic routes [27]. The filtered temperature measurements $y(t)$ were obtained as discrete-time functions, then PSD's were computed from Fast Fourier Transform (FFT) in order to compute the fundamental frequencies.

The maximum Lyapunov exponent (MLE) was computed to provide major evidence of nonperiodic oscillatory behavior. MLE is one of the most important features in nonlinear science to distinguish chaotic from non chaotic behavior. Essentially MLE measures the distance between attractor orbits along time. When the dynamic of a given system is chaotic the Lyapunov spectrum displays, at least, one positive Lyapunov exponent. A positive MLE quantifies the divergence rate of neighboring states and gives the period of time over which predictions are possible. To compute the MLE is not an easy task, spurious values can be obtained. However, the complete evidence leads us to belief that oscillatory behavior in the bioreactor is realistic. Some improvements to MLE algorithms have been proposed to compute the Lyapunov exponents but a large computing effort is required and they do no guarantee the absence of spurious MLE. In this chapter, the original procedure [50] was programmed to compute MLE. To compute MLE is not an easy task due to spurious values should be avoided. Four parameters must be fixed for computing MLE via Wolf et al's. algorithm [50]: embedding dimension d_e, maximum scale S_M, minimum scale, s_m and evolution time, Θ. Basically, d_e corresponds, in some sense, to the space dimension where orbits were embedded. The embedding dimension was systematically changed and we do not find evidence for changes in attractor structure. Hence, for both cases, the chosen value for i.e., $d_e = m = 3$ satisfies standard criteria [50]. The minimum scale, s_m, is the length scale in which noise is expected to appear and time evolution, Θ, is fixed for computing the divergence measurements and stands the necessary time to renormalize the distance between trajectories. Since minimum scale is related to noise acquirement and the time series have been filtered, it has been fixed at $s_m = 0.001$. Since time evolution, Θ, is fixed for computing the divergence measurements which is necessary to renormalize the distances between trajectories, we have chosen, after exhaustive computing tests, $\Theta = 1.0$. Now, concerning maximum scale, S_M, it stands the estimated value of the length scale on which the local structure of attractor is not longer proved. This parameter resulted critical and mean MLE was computed for several S_M values.

2 The First Case: An Interconnected System of Feedback-Controlled Processes

Biological reactors are sensitive to feedback interconnections toward oscillatory behavior because they comprise (i) recycling streams, (ii) feedback control and (iii) kinetic behavior (which is analogous to autocatalytic reactions since the microorganism involved in the reaction may change its characteristics and generate interactions under certain substrate or product and operational conditions, e.g., reactors of recombinant bacteria, where, under certain conditions, the substrate consumption can lead to aggressive products for bacteria [28]). In this paper we focus on the oscillatory behavior of the temperature in a pilot scale fixed bed bioreactor which is used for wastewater treatment via time series analysis. Oscillatory behavior is induced by interconnection between the bioreactor and a heat exchanger by recycling streams. It is desired that reactor temperature remains constant in order to avoid stress in bacteria within the reactor. If both bioreactor and heat exchanger processes are separately operated, then neither bioreactor nor heat exchanger display oscillatory behavior. We show that, under certain conditions of substrate concentrations, the recycling interaction between the bioreactor and the heat exchanger induces oscillatory behavior in the bioreactor temperature. The dynamical model of the heat exchanger and biological reactor are presented in next subsection. Thereafter, the experimental setup is shown; i.e., some features in regard of experimental equipment and temperature measurements are detailed. Then, results on the oscillatory behavior of the reactor temperature are shown from the time series.

2.1   Dynamical Model of the Interconnected Process

Firstly we shall show that the heat exchanger and bioreactor under separated operation do no display oscillatory behavior. But, if the processes are feedback controlled and interconnected by recycle streams, oscillatory behavior can be induced in the process. The section contains three parts: the first two parts are about the process behavior of separately operated process, whereas the third part is about the interconnected process. As we shall see below, the interconnection of the bioreactor and heat exchanger has two parts: (a) mass and temperature recycle through fluid leaving the bioreactor and (b) the feedback control through the controller for regulation of the substrate concentration and the PID control law for the temperature regulation in the heat exchanger. Such an interconnection stands for a (feedback-controlled) coupled system. Under such situation the dynamic model of the interconnected system comprises the ODE´s of the heat exchanger, the bioreactor and the control loops. Thus, the order of the interconnected system is 11.

From the energy balance, and by considering the total volume of the heat exchanger, the following dynamical model can be obtained for the liquid-liquid heat exchanger

$$\dot{x}_1' = \frac{2}{M_C}\left[F_C(T_C - x_1') + \frac{U_O A_O}{C_{PC}}\Delta T(x'; T_i)\right] \tag{1}$$

$$\dot{x}_2' = \frac{2}{M_P}\left[F_P(T_P - x_2') + \frac{U_I A_I}{C_{PP}}\Delta T(x'; T_i)\right]$$

where the vector $x' = (x_1', x_2') \in \Omega \subset \mathbb{R}_+^2$ denotes the output temperature for coolant and hot fluid (fluid from service), respectively. T_C is the cold temperature entering to the heat exchanger from the bioreactor whereas T_P is the hot temperature entering from service. The heat exchanger is configured in counter-current flow and the service temperature is used to hold x_1' at 308±5 K (see block diagram in Figure 2). $\Delta T(x; T_i)$ is a function and denotes the *mean temperature difference* with $T_i = (T_C, T_P)$ as parameter. U_O and U_I are the external and internal heat transfer coefficient, respectively. In our case, since transfer within the heat exchanger is only between the hot and the cold fluids (i.e., perfect isolation), $U_O = U_I = U(t)$. A_I and A_O are the internal and the external heat transfer surface area, respectively. C_{PC} and C_{PP} are, respectively, the heat capacities of liquid slug in internal and external tubes. F_C and F_P are flow rates of the cold and hot fluids, respectively.

The heat exchanger is operated in countercurrent configuration in order to improve the properties of the heat transfer between fluids. As a consequence, the mean temperature differences can be modeled, within the domain of the physically realizable temperature $x \in \mathbb{R}^2$ [52], by

$$\Delta T(x'; T_i) = \frac{(T_{P,i} - x_1') - (T_{C,i} - x_2')}{\ln\left(\frac{(T_{P,i} - x_1')}{(T_{C,i} - x_2')}\right)}$$

which is called the log mean temperature difference (see [52] and [26] for details in regard to dynamics and heat exchanger operation, respectively). The flow rate F_P was used as the feedback control and computed from proportional-integral-derivative (PID) control law [30]. This feedback control law which is given, in Laplace domain [22], by

$$F_P(s) = \left(K_P + \frac{K_I}{\tau_I s} + \tau_D s\right) \qquad (2)$$

where $s = \omega j, j = (-1)^{1/2}$ and ω stands for frequency, $e = x_1' - x_1'^{ref}$ is the error of temperature control, K_P, K_I, τ_I and τ_D are assumed constants. They, respectively, denote the proportional and integral gains, the reset time and derivative time. A tuning procedure can yield temperature stabilization. In this paper, the control parameters K_P, K_I, τ_I and τ_D can be chosen such that, under no recycle, temperature holds constant at $308 \pm 5K$ [52]. Concerning the dynamical behavior of the heat exchanger without control, it should be remarked that the heat exchanger possesses an unique equilibrium point within (x_1', x_2')-domain for the flow regime physically admissible [27]. Moreover, such an unique equilibrium point is globally stable (i.e., for all the initial conditions contained within the (x_1', x_2')-space, all the trajectories converges to the unique equilibrium point) and its coordinates can be computed from parameter values. Such a features have been formally demonstrated in [52]. Here, the vector field of system (1) is shown for two fixed values of flow rates (the maximum and minimum admissible) in Figure 1a and 1b. As matter of fact, results in [52] imply that heat exchanger, under no recycling, has no oscillatory behavior.

Regarding the biological reactor, one should note that the most important pollution control problems found in wastewater treatment is related to the reduction of the organic matter concentration [8]. This problem is further enhanced in the wine and alcoholic beverages industries whose wastewater (or vinasses) have very complex chemical composition. Many biochemical processes have been proposed to solve this problem being the anaerobic digestion the most promising. In order to describe the dynamical features of the biological reactor we briefly discuss the dynamical behavior of the bioreactor. The

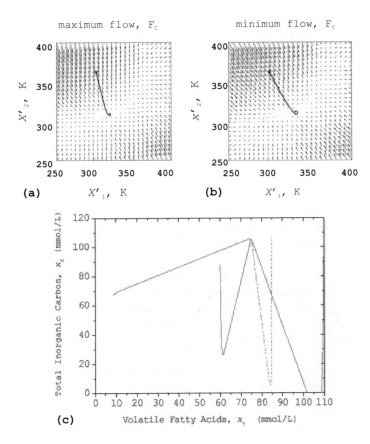

Fig. 1. There is no oscillatory behavior if the system are either separately operated or uncontrolled. Indeed, trajectories converge to an equilibrium point belonging to physically realizable domain. *Above:* Vector field of the heat exchanger under no recycle and for (a) minimum and (b) maximum flow rates. *Below:* (c) 2-dimensional projection of the bioreactor trajectories for several initial conditions.

following dynamical model of the biological reactor describes an anaerobic digestion process. The modeled reactor is a fixed bed continuos bioreactor for vinasses treatment. This model was developed under the following assumptions [8]:

(a1) the hydrodynamical reactor behavior is like a perfectly mixed tank,

(a2) the alkalinity of the wastewater is mainly due to bicarbonate and volatile fatty acids (VFA),

(a3) only two main bacterial population are considered (acidogenic and methanogenic) and

(a4) model validity is for 308 ± 5 K.

In this way, we can assume that the anaerobic digestion comprise basically two steps: acidogenesis and methanization. In the acidogenesis step, organic compounds are fermented into VFA and CO_2 by the acidogenic bacteria while in methanization step the VFA are converted into CH_4 and CO_2 by the methanogenic bacteria. From the assumptions (a1)-(a4), the consideration of the two reaction steps and assuming the gas-liquid equilibrium (see below); the dynamical model can be written as [8]:

$$
\begin{aligned}
\dot{x}_1 &= (\mu_1(x) - \alpha u_{BR})x_1 \\
\dot{x}_2 &= (\mu_2(x) - \alpha u_{BR})x_2 \\
\dot{x}_3 &= (x_{3,e} - x_3)u_{BR} \\
\dot{x}_4 &= (x_{4,e} - x_4)u_{BR} - k_1\mu_1(x)x_1 \\
\dot{x}_5 &= (x_{5,e} - x_5)u_{BR} + k_2\mu_1(x)x_1 - k_3\mu_2(x)x_2 \\
\dot{x}_6 &= (x_{6,e} - x_6)u_{BR} - q_c + k_4\mu_1(x)x_1 + k_5\mu_2(x)x_2
\end{aligned}
\tag{3}
$$

where the state variables respectively stand for: x_1, acidogenic bacteria concentration; x_2, methanogenic bacteria concentration; x_3, total alkalinity; x_4, chemical oxygen demand (COD); x_5 VFA concentration and x_6 total inorganic carbon concentration. COD is the measure of the organic matter in the wastewater. The dilution rate, u_{BR}, is defined as the ratio of the inlet flow rate over the volume of the reactor. $0 \leq \alpha \leq 1$ stands for the mixing degree within the reactor; that is $\alpha = 0$ corresponds to an ideal fixed bed reactor (i.e., no agitation) whereas $\alpha = 1$ corresponds to a CSTR (i.e, perfect agitation). q_c represents the molar flow rate of CO_2 and it can be computed by using Henry's law, namely $q_c = k_{La}([CO_2] - K_H P_{CO_2})$, where k_{La} represents the liquid-gas mass transfer coefficient, K_H the Henry's constant, P_{CO_2} is the partial pressure of CO_2. Thus, if the gas pressure is in equilibrium, then $P_{CO_2} = (\phi - (\phi^2 - 4K_H P_T[CO_2])^{1/2})/2K_H$ where $\phi = CO_2 + K_H P_T + q/k_{La}$ and $[CO_2] = x_6 + x_5 - x_3$ denotes the concentration of the carbon dioxide. Finally, the model (3) considers Monod and Haldane kinetics [36] for the growth of acidogenic and methanogenic bacteria, whose kinetic functions, $\mu_1(x)$ and $\mu_2(x)$, are given by

$$
\mu_1(x) = \frac{\mu_{1,MAX}x_3}{x_3 + K_{1,S}}; \quad \mu_2(x) = \frac{\mu_{2,MAX}x_4}{x_4 + K_{2,S} + (x_4/K_{2,I})^2}
\tag{4}
$$

where the $\mu_{1,MAX}$, $K_{1,S}$, $\mu_{2,MAX}$, $K_{2,S}$ and $K_{2,I}$ are respectively the maximum bacterial growth rate for acidogenesis, the acidogenesis saturation constant, the maximum bacterial growth rate for methanization without inhibition, and the saturation and inhibition constants associated with the substrate x_4. The identified parameters on model (3) are presented in Table 1 (see [8] for details).

The bioreactor has two equilibrium points within the physically realizable domain. Such equilibrium points correspond to washout and operation conditions. For the operation condition (i.e., when degradation of the organic

Table 1. Nominal parameter values of the bioreactor dynamic model

k_1	42.14 (gCOD/gx_1)	K_H	16.0 (mmolCO$_2$/l-atm)
k_2	116.5 (mmolVFA/gx_1)	α	0.5
k_3	268.0 (mmolVFA/gx_2)	$\mu_{1,MAX}$	0.05 (hr^{-1})
k_4	50.6 (mmolCO$_2$/gx_1)	$\mu_{2,MAX}$	0.031 (hr^{-1})
k_5	343.6 (mmolCO$_2$/gx_2)	K_{S1}	7.1 (gCOD/l)
k_6	435.0 (mmolCH$_4$/gx_2)	K_{S2}	9.28 (mmolVFA/l)
k_{La}	0.825 (hr^{-1})	K_{I2}	16.0 (mmolVFA/l)
V_R	1.0 (m^3)	P_T	1.0434 (atm)

matter occurs within the reactor) the coordinates of the equilibrium point are $x_i^* \neq 0 (i = 1, 2, \ldots, 6)$ whereas for the washout condition the equilibrium point has its coordinates at the origin $x^* = (0, 0, \ldots, 0)^T$, and signifies that the bioreactor is unable to reduce the organic matter. In fact, in the washout condition, the substrate concentration $x_{3,e}$ and $x_{4,e}$ at the inlet flow are equal to the outlet flow. Such a condition is reached for very high dilution rate value $\bar{u} < \infty$. The other equilibrium point (operation condition) has coordinates $x^* \neq 0$ belonging to the physically realizable domain. Such a condition can be reached for any value of the dilution rate $0 < u_{BR} < \bar{u} < \infty$ [15]. Formal proofs of stability of open-loop bioreactor is beyond the limits of the present paper and they will be presented elsewhere. However, Figure 1c shows the behavior of the system (3) for several initial conditions for any constant value of the dilution rate $u_{BR} < \bar{u} < \infty$ and the inlet concentrations of $x_{j,e}, j = 3, \ldots, 6$.

Anaerobic digester can be controlled by a feedback control law. The main control objective is to regulate the substrate (COD) concentration x_4 at any desired value $x_4^* \neq 0$ [15]. Indeed, even in face to modeling errors, a control law can be obtained for substrate regulation [35]. Following the ideas reported in [21], since the substrate (COD) concentration x_4 is measured, the dilution rate can be computed by [33], [34]

$$\dot{\hat{x}} = \hat{\eta} + \gamma(x_4)u_{BR} + g_1(x_4 - \hat{x}_4)$$
$$\dot{\hat{\eta}} = g_2(x_4 - \hat{x}_4) \tag{5}$$
$$u_{BR} = (-\hat{\eta} + K_C(x_4 - x_4^*))/\gamma(x_4)$$

where x_4 is the measured concentration of substrate at the outlet bioreactor fluid (see diagrams in Figure 2), $\gamma(x_4) = x_{4,e} - x_4$, $\hat{x}_4$ and $\hat{\eta}$ are the estimated value of the substrate concentration and growth kinetics, respectively. The control parameters g_1, g_2 and K_C were chosen following the criteria in [30] and x_4^* is the reference. As matter of fact, the controller (5) leads the organic matter via COD around the constant reference x_4^* (see Table 3) in the closed-loop experiments. A necessary condition for stability of the controlled

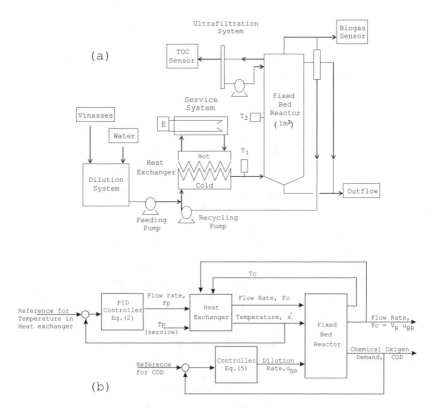

Fig. 2. Diagrams of the processes with instrumentation. (a) Schematic diagram for wastewater digester. Two temperature sensors T_1 and T_2 were used for measurements. Time series were acquired from T_1 and T_2 was used for corroborating no transfer heat to surroundings by comparison with T1 measurements. (b) Schematic for the liquid-liquid heat exchanger. Note that the Bioreactor-Exchanger interconnection involves recycle streams between two feedback controlled processes.

bioreactor (3) under control (5) is that dilution rate u_{BR} is bounded, i.e., $0 < u_{BR} < \bar{u} < \infty$ for the washout value $\bar{u}$.

2.2   The Interconnected Model

Figure 2 show the schematic diagram of both heat exchanger and wastewater digester and the interconnection. As was mentioned in previous subsections, if the bioreactor and heat exchanger are separately operated (i.e., uncoupled system), then they do not exhibit oscillatory behavior for the nominal value of the parameter vector and any value of the dilution rate $0 < u_{BR} < \hat{u} < \infty$. Now, since the bioreactor and the heat exchanger are coupled by the recycle streams and controlled by specific control laws, we need to re-write the models (1) and (3) under recycle and feedback. Figure 2 shows (a) the schematic diagram of the process and (b) the block diagram of the recycle and feedback. Note that the bioreator and heat exchanger are coupled by the recycle of the temperature T_C and the flow rate $Fc = V_R u_{BR}$, where $V_R = 1\text{m}^3$ is the volume of the bioreactor. At same time, the fluid is returned to bioreactor from heat exchanger with flow rate Fc and temperature x_1'. In this sense, system (1) and (3) are coupled by the flow rate Fc, which is affected by the control law (5). In addition, the temperature from the heat exchanger x_1' affects the specific growth rates $\mu_1(x)$ and $\mu_2(x)$ through the parameters $\mu_{1,MAX} = \mu_{1,0} exp(-\Theta_1/x_1')$ and $\mu_{2,MAX} = \mu_{2,0} exp(-\Theta_2/x_1')$, where, in some sense, Θ_1 and Θ_2 stand for activation energy [1]. Thus, the coupled system has the following dynamical model

$$
\begin{aligned}
\Sigma_{HE} & \quad \{\dot{x}' = f_{HE}(x', F_P; Fc(u_{BR})) \\[1mm]
\Sigma_{PID} & \quad \begin{cases} F_P = K_P(x_1' - x_1'^{ref}) + (K_I/\tau_I)x_3' \\ \dot{x}_3' = x_1' - x_1'^{ref} \end{cases} \\[1mm]
\Sigma_{BR} & \quad \{\dot{x} = f_{BR}(x, u_{BR}; \mu_{MAX}(x_1')) \\[1mm]
\Sigma_{CLUE} & \quad \begin{cases} \dot{\hat{x}} = \hat{\eta} + \gamma(x_4)u_{BR} + g_1(x_4 - \hat{x}_4) \\ \dot{\hat{\eta}} = g_2(x_4 - \hat{x}_4) \\ u_{BR} = -\hat{\eta} + Kc(x_4 - x_4^*)/\gamma(x_4) \end{cases}
\end{aligned}
\tag{6}
$$

where the subsystem Σ_{HE} is given by the heat exchanger (1), Σ_{PID} is the control law (2) in the time-domain, Σ_{BR} stands for the bioreactor (3) and Σ_{CLUE} corresponds to controller (5). All parameters are defined above.

The subsystems Σ_{HE} and Σ_{PID} are coupled by feedback and define a space $\Delta_{HE}(x_{HE}) = \Sigma_{HE}(x') \times \Sigma_{PID}(x_3')$ for any $x_{HE} = (x', x_3')$ belonging to the physical domain $D_{HE} \subset \mathbb{R}^2 \times \mathbb{R}$ while the subsystems Σ_{BR} and Σ_{CLUE} are coupled by feedback and define a space $\Delta_{BR}(x_{BR}) = \Sigma_{BR}(x) \times \Sigma_{CLUE}(x_4, \hat{x}_4, \hat{\eta})$ for any $x_{BR} = (x, (\hat{x}_4, \hat{\eta}))$ belonging to the physical domain $D_{BR} \subset \mathbb{R}^6 \times \mathbb{R}^2$. In this manner, the state vector (x_{HE}, x_{BR}) of the wastewater treatment process lies in the physical domain $D_{BR} \times D_{HE}$, which a subset of $\mathbb{R}^2 \times \mathbb{R} \times \mathbb{R}^6 \times \mathbb{R}^2$ (i.e., the order of the full coupled system is 11).

2.3   Details on the Experimental Setup

A 1000 liters pilot scale fixed bed bioreactor was used in the experiments of the anaerobic digestion of vinasses (a waste of the wine industry). Table 2 list the range of compositions and pH of the vinasses used in the experimental runs. The objective of the anaerobic digestion is the reduction of the organic mater concentration in the outflow substrate (COD) by bacteria. Such a process consists of two steps. In first step, volatile fatty acids (VFA) are produced by bacteria whereas CO_2 and CH_4 are produced second step (see Assumption a3). Both acidogenic and methanogenic bacteria are within same reactor. The controller (5) guarantees that substrate regulation is held close to the

Table 2. Composition of the wine distillery wastewater used for experiments

Component	Range
Total COD (g/l)	12-34
Soluble COD (g/l)	7.6-31
Volatile fatty acids (g/l)	1.5-10
Total suspended solids (g/l)	2.4-10
Volatile suspended solids (g/l)	1.2-5.4
Phenol (mg/l)	90-275
pH	4.5-5.2

desired value x_4^* even when substrate concentration is modified at inlet flow (compare Tables 2 and 3) [32]; i.e., the COD in outlet flow holds around reference against load disturbances. The COD was inferred via the Total Organic Carbon (TOC). TOC was measured on-line by a commercial sensor each two minutes and a correlation provides the COD. Values of the mean error are reported in Table 3 for each experiment in this work. Note that, control error is on the order of 10^{-1}g/l.

Table 3. The set-points and mean error for the outlet COD

Experiment	Set-point (g/l)	Mean error (g/l)
E.1a	1.5	0.06409
E.1b	1.0	0.08301
E.1c	1.5	0.04544
E.1d	1.0	0.06079
E.2a	3.0	0.10465
E.2b	3.0	0.01499

Some experiments were performed to observe the effect of substrate concentration in input flow on the bioreactor temperature behavior. All experiments were run, at least, along one week. The main idea behind these runs was to analyze: (a) the effect of reference and inlet concentration and (b) the effect of the parameters of the controller (6). Thus, the following results were chosen for discussing the oscillatory behavior of the bioreactor temperature:

E.1) Effect of COD reference and control parameters: The inlet COD was held constant (diluted substrate with 50% H_2O and 50% vinasees) $x_{4,e}$ = 15.9 g/l. On-line measurements of COD in output flow each 2 min (by means of Zellewer Analytics sensor). Four different COD reference values were chosen to control the COD concentration: (a) x = 1.5 g/l with the following control parameter values: $Kc = $ -0.4 hr^{-1}, $g_1 = 0.7$ hr^{-1} and g_2 = 0.49 hr^{-1}. (b) $x_4^* = 1.0$ g/l with same parameter values than experiment E.1a. (c) $x_4^* = 1.5$ g/l with the following control parameter values: $Kc = $ -0.74 hr^{-1}, $g_1 = 1.08$ hr^{-1} and $g_2 = 1.16$ hr^{-1}. (d) $x_4^* = 1.0$ g/l with same parameter values than experiment E.1c.

E.2) *Effects of inlet COD concentration and control parameters:* The COD reference was held constant. (a) Pure vinasees entered the anaerobic digester and reference value was chosen $x_4^* = 3.0$ g/l for the following control parameter values: $Kc = $ -0.4 hr^{-1}, $g_1 = 0.7$ hr^{-1} and $g_2 = 0.49$ hr^{-1}. (b) Pure vinasees was entered in anaerobic digester and same reference than E.2b for the following control parameter values: $Kc = $ -0.36 hr^{-1}, $g_1 = 0.54$ hr^{-1} and $g_2 = 0.29$ hr^{-1}.

The aforementioned experiments are often used for testing the substrate regulation in anaerobic digesters. Hence, they allow us to study the temperature oscillations under typical operating conditions in this class of biological reactors. In this sense, these results imply, on the one hand, that oscillatory behavior of the bioreactor does not necessarily mean inexpedient process operation. On the other hand, results show that interconnection plays an essential role in chaos induction in process. Such interconnections can be intrinsic (i.e., within the reaction) or not intrinsic (i.e, due to interaction of different process). In our study case, the oscillatory behavior is not intrinsic to the bioreactor or heat exchanger. That is, neither the bioreactor nor the heat exchanger oscillate without interconnection. Thus, the oscillatory behavior can be attributable to the interconnection via recycle of the feedback controlled processes.

2.4   Oscillations Due to the Interconnection of Feedback-Controlled Systems

Now let us focus on the bioreactor temperature. A stream is fed from the reactor to the heat exchanger (see Figure 2b). From the process control viewpoint, the idea is to maintain the reactor temperature constant via the heat exchanger. The temperature reference was chosen at 308 K and the heat exchanger was controlled with a PID control law. A variation of $\pm$ 5K does not represents stress for bacteria inside the anaerobic digester. However, the anaerobic digestion in the bioreactor induces temperature oscillations due to the Bioreactor-Exchanger interconnection. Wastewater regulation from bioreactor is not lost against such time series oscillation (see Table 3). Temperature time series were obtained from on-line measurements by a each two minutes. Such a sampling time is large because the characteristic time in anaerobic digester is on the order of 10^2 min. In order get more information about temperature in the Bioreactor-Exchanger interconnection and to corroborate that the heat transfer to the surroundings is neglectible, two sensors were placed in the experimental set up. One was located at the entrance of the bioreactor and a second was placed at the middle of the bioreactor.

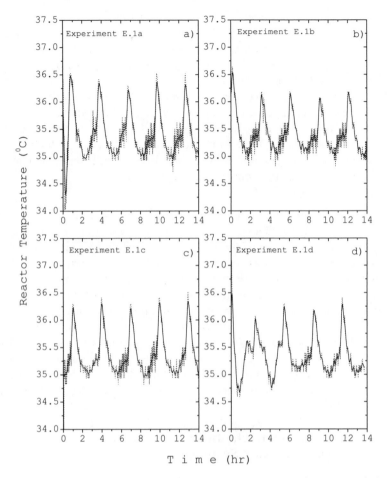

Fig. 3. Filtered signal (solid line) and measured time series (dotted line) for the first experiment E.1 (see text for details). Temperature oscillation were induced by recycle between heat exchanger and biological reactor. The filtered signal remains the oscillatory behavior of the system.

This was also done in order to attribute the temperature oscillations only to the interconnection. Time series were filtered (see solid lines in Figures 3 and 4) by low-pass filter in order to eliminate noise effects in temperature measurements (in Figures 3 and 4, the dotted line and the solid line correspond, respectively, to the temperature measurements and the filtered temperature).

In this study case, although the heat exchanger is a 2-dimensional system (see equation (1), according to results by Luyben [30] the recycle streams increases the order of the model (which is given by the differential equation system). Hence the dimension of the recycle system should be higher than two. We have chosen the embedding $m = 3$ in such a way that the space-phase

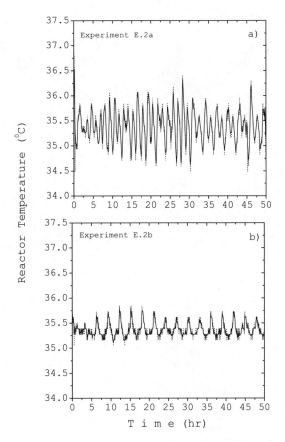

Fig. 4. Filtered signal (solid line) and measured time series (dotted line) for the first experiment E.2 (see text for details). Temperature oscillation were induced under recycle between heat exchanger and biological reactor.

reconstruction preserves, according to [21], the geometry features of the original attractor and, according to [1], the dimension of the system is increased due to the recycle streams. The delay time has been chosen equal to the sampling rate, i.e., $T = 2$ min. Such a choice in delay time was made to avoid the noise effects which possibly remained after the filtering step. The chosen values for embedding dimension and delay time satisfy standard conditions [19]. In this manner, the time-delay coordinates of the reconstructed space-phase becomes $z(t) = (z_1(t), z_2(t), z_3(t)) = (y(t), y(t - T), y(t - 2T))$.

Figure 5 shows the 3-dimensional reconstructed attractors and their projections on canonical planes. The reconstructed phase portraits do not exhibit a defined structure, i.e., it is not toroidal or periodic. As matter of fact, the oscillatory structure is only observed in the Poincaré map. The Poincaré map is often used to observe the oscillatory structure in dynamical systems. The

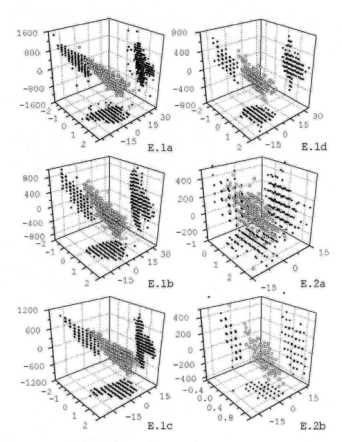

Fig. 5. Reconstructed space phase (z_1, z_2, z_3) from time series in Figure 3 and 4. The attractors seem ordered in layers (see projections in canonical planes). The attractor for concentrate vinasses (experiment E2.b, see text for details) is smallest. z_1, z_2, z_3 are dimensionless.

idea behind the Poincaré map is to get stroboscopic measures of the trajectory with respect to lower-order manifold $\Sigma(z)$ [23]. The stroboscopic measures are obtained as the trajectories go through the Poincaré section. Thus, an inherent description of the oscillations is provided by crosses of trajectories on the manifold. We have chosen the plane $\Sigma(z) = z_3 = 0$ as the Poincaré section of the reconstructed space phase.

Figure 6 displays the Poincaré maps for all experiments. Note that even the projections in canonical planes (see Figure 5) seem ordered in layers. That is, a toroidal structure can be seen form the Poincaré surface. That is, small amplitude oscillations were detected in time series (see Figures 3 and 4) for all experiments. The typical behavior of aperiodic (possibly chaotic) oscillations can be confirmed is one takes a look at the corresponding Poincaré section

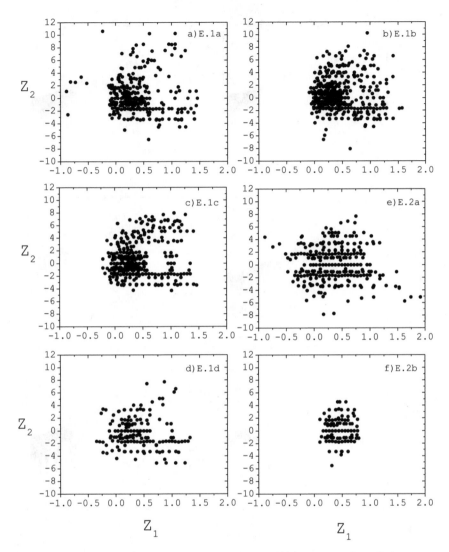

Fig. 6. Poincare maps. The section was chosen $\Sigma(z) = z_3 = 0$ and the crosses indicate no periodic oscillation. Once again, the smallest attractor corresponds to experiment E2.b. z_1, z_2, z_3 are also dimensionless.

$\Sigma(z) = 0$ (see the "hole" around coordinates $(0.6,0)$ at the plane (z_1, z_2) for experiments E.1). Nevertheless, it is not clear if oscillatory behavior is deterministic or stochastic.

Figure 7 shows the PSD of filtered time series in Figures 4 and 5, respectively. Note that the PSD comprises low fundamental frequencies, i.e., less than 10 Hz. Low fundamental frequencies were expected because dynamics of

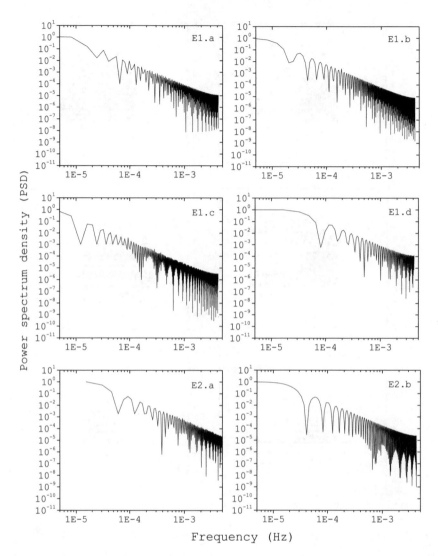

Fig. 7. Power spectrum density. The measured time series comprise several fundamental frequencies. Since frequencies have low-order (< 0.03 Hz) noise effect can be neglected. Note that if the values reference outlet substrate and the control gains decrease (experiment E.1d), then the number of fundamental frequencies in PSD decreases. This leads us to belief that there is a suitable values such that system displays limit cycle. However, this behavior was not experimentally found.

the process is slow (of the order on hours). Although PSD is not a sufficient to conclude that temperature oscillations are chaotic. However, since FFT displays a continua spectrum, it is a very good evidence that many fundamental frequencies are contained in the oscillations of the measured temperature,

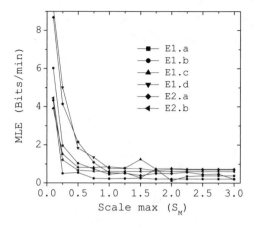

Fig. 8. Mean value of Maximum Lyapunov exponent versus Maximum scale (which is the most important for computing MLE). A positive value indicates that, at least in one direction, trajectories in space phase are diverging.

which is a characteristic of chaos. However, PSD is not sufficient to conclude chaos. Results on the Lyapunov spectra are shown in Figure 8 for $\Theta = 1$, $s_m = 0.001$ and $d_e = 3$. Note that for all cases, the mean of the MLE´s remain positive for all S_M values. Finally, Figure 9 shows the time evolution of the MLE for $\Theta = 1$, $s_m = 0.001$, $d_e = 3$ and $S_M = 1.5$. Notice that MLE's hold positive for the most experiments.

2.5   Remarks on the First Case

In this paper the temperature oscillations in a biological reactor have been studied. Such a bioreactor is used to deal with wastewater treatment from wine industry; in this sense the bioreactor is part of a coupled process. The parts of the process do no oscillate when they are separately operated. Here, we have shown experimental evidence of nonperiodic oscillatory behavior which can be attributable to interconnection of the bioreactor with a heat exchanger via recycle streams. Both bioreactor and heat exchanger are feedback controlled. The heat exchanger is interconnected with bioreactor in order to hold the bioreactor temperature at $308 \pm 5K$. Such a temperature allows the vinasses conversion by bacteria. The results show, on the one hand, that, since chemical oxygen demand is held controlled around desired values, temperature oscillation does not affect the reduction of the organic matter from the wastewater treatment by bioreactor. On the other hand, according to evidence form Poincaré map, PSD spectrum and computation of MLE, fluctuations of reactor temperature are chaotic. Former implies that oscillatory behavior can be induced if two dynamical system, chaotic or not, are interconnected, i.e., if two dynamical system share, at least a portion of information by means of links, then chaotic behavior can be displayed by total system. Latter result

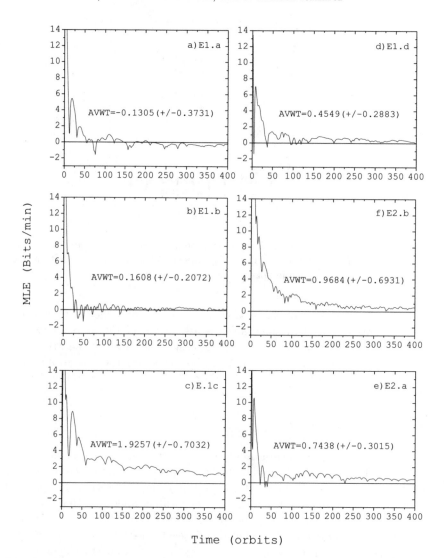

Fig. 9. Maximum Lyapunov exponents along the orbits in the space phase for scale maximum $S_M = 1.5$ (compare with Figure 8). Even for experiment E2.b (which has the smallest attractor) the maximum Lyapunov exponent holds positive. The average values of the Lyapunov exponents where computed by neglecting the transitory effects (AVWT means Average Value without Transitory effects).

mean that presence of chaotic behavior does not imply inexpedient operation of chemical or biochemical process. The experimental evidence which involves nonperiodic behavior and possibly chaotic.

3   Oscillations in an Open-Loop Controlled System

A set of experiments on gas-liquid motion in a vertical column has been carried out to study its dynamical behavior. Fluctuations volume fraction of the fluid were indirectly measured as time series. Similar techniques that previous section were used to study the system. Time-delay coordinates were used to reconstruct the underlying attractor. The characterization of such attractor was carried out via Lyapunov exponents, Poincaré map and spectral analysis. The dynamical behavior of gas-liquid bubbling flow was interpreted in terms of the interactions between bubbles. An important difference between this study case and former is that gas-liquid column is controlled in open-loop by manipulating the superficial velocity. The gas-liquid has been traditionally studied in the chaos (turbulence) context [24].

The chaotic behavior is an interesting nonlinear phenomenon which has been intensively studied during last two decades. The deterministic techniques have been used to understand the dynamical structure in several nonlinear systems [5], [6], [42]. Particularly, the two-phase flow systems present nonlinear dynamical behavior which can be studied by means of chaos criteria behavior [2], [17], [25], [31], [45], [51]. Two-phase flows they provide a rich variety of cases whose dynamics lead to oscillatory patterns. The following published results are example:

(a) The dripping of a faucet, studied by Martien et al. [31], where substantial evidence for a broad range of dynamical behavior in the drip (i.e., period doubling, transaction to chaos and hysteresis) was found.

(b) The chaotic flows in rotating lid cavities, reported by Hadid [25], where two-dimensional two-phase flow inside a closer rotating cavity reveals a sequence of complex flow structures that occur when the rotational Reynolds number varies.

(c) The motion of a solid through fluid, studied numerically by Aref and Jones [2]. In this work, the chaotic motion of solid through an ideal fluid was found by compelling and consistent evidence. the authors suggest that chaotic motion can be detected experimentally by monitoring the trajectory in space of a fixed point in the body.

(d) Tritton and Edgell [45] reported a experiment sets of bubbling from a submerged orifice. Their results show observation of ("two-level", large-scale and limited) chaos. Besides, they found periodic and period doubling behavior.

(e) Zardi and Seminara [51] studied pulsating bubbles subjected to acoustic waves. They analyzed the competition mode from erratic drift effect of bubbles. Draho et al. [17] studied the fractal behavior of pressure fluctuations in a bubble column. Agreement between experimental data and simulated series of Brownian motion was found.

(f) Chaotic motion in dynamics of the bubble formation has been studied by Tufaile and Sartorelli [12], [46], [47]. They observed bifurcation in a bubble gun as function of the pressure. In their experiments, when pressure is fixed, the used the amplitude of sound wave as the second (open-loop) control [12].

On the other hand, the flow structure of bubbling has been the subject of a deal of academic research [10], [29], [48]. Some experiments lead to the following qualitative results. (i) A gross circulating flow pattern is observed under dispersed bubble and coalesced bubble regimes [10] (in bubble columns, these regimes are so-called homogeneous, churn-turbulent and slugging). In homogeneous bubbling regime the coalescence can be neglected whereas coalesced bubble regime can exhibit 3 or 4 regions of flow [10], [29], [48]. (ii) The gross circulation comprises an upward flow the column core and a downward flow along the wall with the inversion point (zero axial liquid velocity). (iii)The coalesced bubble regime may include 4-region flow. A central plume arise a homogeneous bubbling while spiral-vortical region yields a transition to chaos. (iv) However, in columns of width lass than 0.15 m, the central plume region becomes indistinguishable from the fast bubble flow region yielding a 3-region flow pattern [29] (turbulent bubbling).

3.1   Previous Results on Gas-Liquid Flow Structure

Recently, visualization techniques (video-digitalization of light reflection) have been used to develop correlations between flow structure and column dimensions and operation conditions in 2-D and 3-D bubble columns and three-phase fluidized beds [29]. Additionally, analogous description of the hydrodynamics of fluidized beds and bubble columns has been reported [9]. It must be pointed out that bubble columns find applications as reaction and separation units in chemical, mining, pharmaceutical and biochemical industries. Some examples of industrial operation are: chemical systems [9], biotechnology [14] or detecting and preventing of cavitation effects in physiologic microchemistry [17]. Hence, the dynamical analysis of bubbling systems is a very important problem. In what follows, before the description of the experimental setup, we discuss a brief review of literature. Thus, three different regimes have been commonly identified for bubble columns. Namely, disperse, churn-turbulent and slugging [18]. However, in order to establish the conceptual context of gas-liquid vertical flow structure, is pertinent to make a brief review of recent numerical simulations and experiments.

It has been pointed out by numerical experiments that pulsating bubbles subject to acoustic waves can exhibit chaotic behavior [51]. A second-order model for the pulsating bubbles which is governed by slow variations in amplitude was analyzed in [51]. The effect of parameters such as amplitude and frequency of the external wave was found to induce chaotic behavior.

On the other hand, the complex phenomenon of gas-liquid bubbling has been recently analyzed via experimental velocity profiles measured from hot-film anemometry [45] or light reflection [10], [29], [48]. Several oscillatory patterns (periodic and nonperiodic oscillations, including chaos and period doubling behavior) were reported by Tritton and Edgell in [45]. The authors concluded that: In bubbling from a submerged orifice, different oscillatory characteristics can exist. These are due to the fact that as flow gas rate increases, the spacing between bubbles decreases. Hence the motion (and formation) of each bubble may be influenced by the presence of the previous one. This result can be seen as axial interaction of subsequent bubbles. This is, the modes of a bubble raising from the submerged orifice induce an external wave (perturbation) onto the previous one.

The image-digitalization technique has been used to video-record the light reflection in vertical bubble columns and obtain correlations between flow structure and column dimensions as well as operation conditions. The first quantitative analysis of 2-D bubble columns and its transition to vortical-spiral regimes was reported [29]. The authors reported that, any given operation conditions, the vortical-spiral flow can be formed for 3-flow regions. The main results reported in [29] are the following: (i) To any column width the variation of the vortex size as a function of gas velocity is given by $v_s = \alpha[1 - \exp(-g_v/\beta)]$ (where v_s is the vortex size, α is a parameter which depends on the column width, only, and g_v is the gas velocity). (ii) If the gas velocity $g_v < 1.0$ cm/s and the column width is greater than 0.15 cm, then dispersed regime is present. (iii) If the gas velocity increase around 3 cm/s, the two fast bubble flow regions merge to form one central bubble stream that moves in a wave-like manner and the central plume region disappears. In others words, the 4-flow region of coalesced regime becomes 3-flow region. The above results provide a qualitative picture of the macroscopic flow structure and its dependence on the column dimension and gas velocity.

More recently, it has been reported a set of experiments in a 2-D bubble column where only the gas velocity was varied [48]. The results obtained from video-digitalization of light reflection show the following: As the gross circulation of the liquid phase occurs, the migration of the bubbles away from the walls causes bubble coalescence and the creation of the fast bubble flow region which stretches the vortical flow[1] and descending flow regions. From this result, we can observe that, in 2-D gas-liquid flow, the effect induced by the oscillations of both coalesced and "faster" bubbles modifies the dynamic behavior on the bubbling flow.

Additionally, macroscopic flow structure of 3-D bubble columns were studied [10]. The results reported can be resumed as follows: (a) In disperse regime, the bubbles rise linearly and the liquid flow falls downward between the bubble stream. (b) If gas velocity increases, the gas-liquid flow presents a vortical-spiral flow regime. Then, cluster of bubbles (coalesced bubbles) forms the central bubble stream moving in a spiral manner and 4-flow region can be identified (descending, vortical-spiral, fast bubble and central flow region). Figure 10 shows an illustrative schemes of the results found in [10].

---

[1] The vortical flow regime consists of multiple vortices located along the side walls at the axial direction.

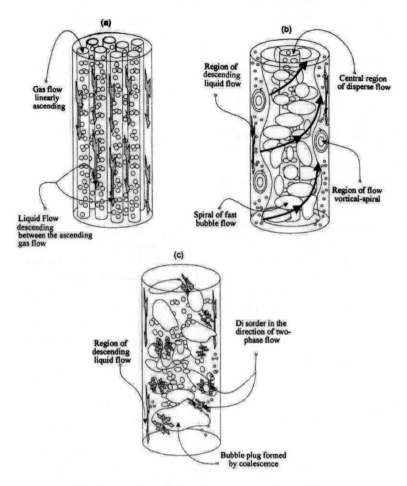

Fig. 10. Schematic representation of the regimes from Chen et al. [10]. (a) Homogeneous regime, (b) four-flow region regime and (c) three-flow region regime.

Although the results obtained in [10], [29], [48] can not provide quantitative information of the interaction of the vortical-spiral flow, the image-digitalization results lead us to the following questions. Can the interactions between bubbles and flow regions induce chaotic behavior even if the regime is not slugging? Which is the transition to chaos in vertical bubble columns? Is the embedded attractor chaotic? In order to investigate these interactions and its influence at the route to the chaotic flows structure on bubble columns, we design the experimental apparatus which is described in the next section.

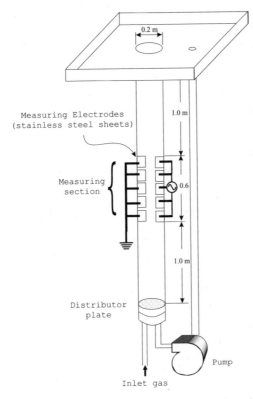

0.2 m

Measuring Electrodes
(stainless steel sheets)

1.0 m

Measuring
section

0.6

1.0 m

Distributor
plate

Pump

Inlet gas

Fig. 11. Experimental apparatus. The flow is fully developed in the measuring section.

3.2   Experimental Setup for Gas-Liquid Column

It is well known that experimental studies on gas-liquid bubbling are not an easy task. Several design parameters as verticality, type of distributor and superficial fluid velocity may affect the flow regimes [29], [41], [48]. Besides, phenomenon such as circulation, coalescence or break-up are very difficult to study experimentally [18]. We have chosen the following column as experimental apparatus (see Figure 11). Air-in-water dispersion were used in the experimental runs. Table 4 shows the superficial gas velocity and the average volume fraction of the liquid. The core of the experiment was a Plexiglas column of 0.2 m of diameter and 2.6 m of axial length above the distributor plate. Verticality condition was assured (tilt is less than 0.2 degrees). The column was divided in three sections. The both below and upper sections are of 1.0 m of longitude which guarantees that in middle part the flow is fully developed. Thus, the middle section is used to scan the flow, this part is so-called "measuring section". In this way, the entrance and disengagement effects can not affect the measurements. The measuring section was adapted with a set of five pairs of guard electrodes with a separation of 0.003 m between adjacent electrodes. These electrodes were stainless steel rectangular sheets, 0.157×0.075 m, provide with a bolt passing through the column wall. The bolts were wired to the electronic unit for signal processing. An electrical current of 31250 Hz was induced by an oscillator to prevent the polarization of the electrodes that would otherwise be induced on the water molecules. It most be pointed out that: the induced electrical current is used to avoid the polarization of the water molecules. A modulation of the recorded signal is performed.

Table 4. Global operating conditions for runs. Superficial liquid velocity for all cases was 0.0197 m/s. Runs (a), (b) and (c) are presented in this work.

Superficial gas velocity (m/s)	Average liquid volume fraction ε_{av}
0.0045	0.9373
0.0061	0.9144
0.0076	0.8955
0.0094	0.8730 (a)
0.0109	0.8598
0.0154	0.8398 (b)
0.0258	0.8236
0.0340	0.7934 (c)
0.0437	0.7678

The signals were recorded as electrical potential in millivolts. The well known Maxwell's formula and an adjustable empirical coefficients were used to obtain the equivalent volume fraction of liquid [43]. Since it is known that kinematic waves exist only in the frequency of some few Hertz, hardware low pass filter with 20 Hz cutting frequency was included for each channel in the electronic unit. The filter was tuned at 82.66 Db. Data were acquired by a computer at 100 Hz. The following comments regarding the above apparatus description are in order:

(a) The dimension of the experimental column preserves the scaling laws respect to the equipment used in [29] and [48]. Thus, the results are comparative.
(b) The column diameter is equal to 0.2 m, so that 4-region coalesced flow regime may be obtained. In this way, the bubble interactions can lead to the vortical-spiral and the possible route to chaos can be studied.
(c) Only the superficial gas velocity was varied. The another design variables (such as verticality, type of distributor, etc.) are constant.
(d) Neither gas density nor bubbles velocity were measured.

In seek of simplicity, three experiments are reported (see Table 4). We choose the most significative cases to illustrate the features of the attractor raised from the bubbling. These cases are labeled as follows: (a) $\varepsilon_{av} = 0.87301$, (b) $\varepsilon_{av} = 0.83884$ and, (c) $\varepsilon_{av} = 0.79344$ (ε_{av} represents the average liquid volume fraction). In the first one, disperse flow regime is performed. The estimated size of the bubbles is on the order of 5-8 mm. In the second case, 4-region coalesced bubble regime is present. In the last one, 3-region coalesced bubble regime was obtained. The size of bubbles was not estimated in neither cases $\varepsilon_{av} = 0.83884$, $\varepsilon_{av} = 0.79344$. The corresponding vortex size are presented in the Table 5. The vortex size in the first case is very small whereas the vortex size at (b) and (c) cases is similar.

Table 5. Vortex size for the runs reported, calculated from the results in [29]

Run	Vortex size	Operating regime
(a)	<0.08	Disperse flow
(b)	0.10	Four-region coalesced flow
(c)	0.15	Three-region coalesced flow

The Figure 12 shows the fluctuations of the liquid volume fraction as time series. Note that if ε_{av} decreases (or, correspondingly, the void fraction increases) the time series display a more erratic behavior. Such effect may be imputed to the following facts: (i) The number of bubbles per unit of volume increases as ε_{av} decreases. Then, the bubbles modes can be influenced by the presence of the others. That is, the interaction of closer bubbles is coupled. (ii) The two-phase flow contains local flow structures of different size. In others words, the flow regions in the bubbling can be seen as a system of coupled oscillators. (iii) The oscillators are coupled due to the interactions of the flow regions. Each flow region can be seen as an oscillator which is perturbed by external forces (i.e., the others flow regions). In this sense, it is possible to find chaotic behavior in the vertical bubble column.

3.3 Application of the Nonlinear Techniques

In this second study case, since the time series has been measured from a bubble column and it can be seen as a 3-D velocity pattern, we have chosen a 3-dimensional reconstruction ($m = 3$) with $T = 10.0$ ms (which is equal to the sampling rate). This choice of the time-delay was made to retain the main features of the attractor whereas avoiding noise effects not eliminated in the filtering step. We have been renamed the coordinates as follows: $z_1(t) = x(t)$, $z_2(t) = x(t - T)$ and $z_3(t) = x(t - 2T)$. Figure 13 shows the 3-dimensional reconstructed attractors and their projections on the canonical planes of the time series displayed in Figure 12. It is easy to see the following: (a) For the case where $\varepsilon_{av} = 0.87301$ (homogeneous bubbling), the attractor is not

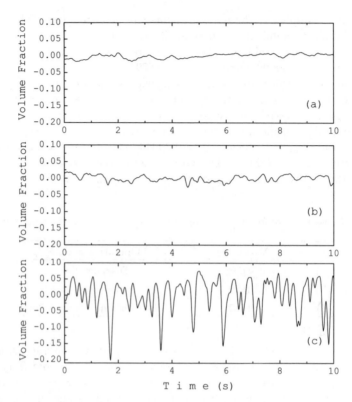

Fig. 12. Time series fluctuations of volume fraction of liquid. (a) ε_{av}=0.87301 (disperse flow regime), (b) ε_{av}=0.83884 (4-region flow regime) and (c) ε_{av}=0.79344 (3-region flow regime).

periodic. However, it exhibits a layer distribution of the states (see Figure 13a and the projections on the canonical planes). In this sense, the bubbling attractor presents an almost-periodic behavior which can be attributed to the order in the rising bubbles (see scheme in the Figure 10a). (b) For ε_{av} = 0.83884 (4-flow regions regime) the layer distribution is loosen up (see Figure 13b and the projection on (z_1, z_2)-plane). However, the almost regular behavior is not lost for vortical-spiral bubble flow (a comparison of the projections on the planes (z_1, z_2) and (z_2, z_3) of the Figures 13a and 13b). (c) For ε_{av}= 0.79344 (3-flow region regime), the bubbling attractor displays two regions: (i) great concentration of winding orbits of small amplitude and (ii) disperse orbits of larger amplitude. The former is attributed to the slug due to the formation of plugs-bubbles and the latter is attributed to the vortical oscillations. Notice that the layer distribution is fully lost (Figure 13c).

    In a similar manner than previous case, we have chosen the plane $\Sigma(Z) = z_3 = 0$ to construct a Poincaré section of the attractors in Figure 13. We tracked the orbit $z(t)$ and recorded the pairs (z_1, z_2) at which $z_3 = 0$. Due to

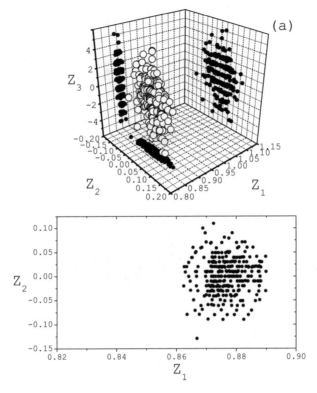

Fig. 13. (a) Phase portrait of the time series in Figure 12a, where z_1, z_2, z_3 are dimensionless

the fact that the orbit $z(t)$ is given only at discrete-time points, we used spline interpolation to approximate the Poincaré map for the time series displayed in Figure 12. In some sense, the distance between the smaller and the bigger regions is a measure of the rate between small and larger amplitude of the orbit. Figure 14 shows the Poincaré map of the cases (a) disperse flow, (b) 4-flow region regime and (c) 3-flow region regime. Note that the attractor of the disperse flow regime is the smallest one whereas the attractor of the 3-flow region regime is the largest one. Then, the relative attractor size indicates if the gas velocity increases then the amplitude of the orbit increases. This fact is attributed to the following: As the gas velocity increases the slugging behavior is present; in others words, the central plume in the 4-flow region regime disappears due the formation of plug-bubbles and the structure of the bubbling changes. The mapping on the Figures 14a and 15b are similar (the orbits are very close). However, the map in Figure 14a shows that the orbits is practically immersed in a "short line". The short line is generated by the layer distribution. On the other hand, in the Figure 14c (3-flow region regime), one can see the great concentration region of winding orbits at the right side

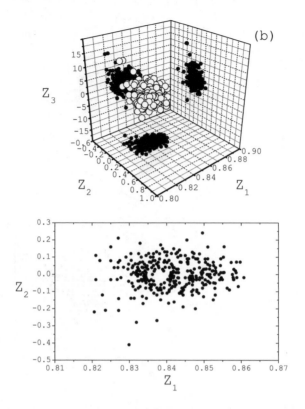

Fig. 13. (b) Phase portrait of the time series in Figure 12b

of the graphic. It must be mentioned that the this numerical technique was implemented in two-phase flow [45], where the authors photographed bubbling from an orifice with stroboscopic light under several operation conditions. A comparison between the Poincaré maps in Figure 14 and the pictures taken by Tritton and Edgell in [45] shows similar structure.

Essentially, MLE is a measure on time-evolution of the distance between orbits in an attractor. When the dynamics are chaotic, a positive MLE occurs which quantifies the rate of separation of neighboring (initial) states and give the period of time where predictions are possible. Due to the uncertain nature of experimental data, positive MLE is not sufficient to conclude the existence of chaotic behavior in experimental systems. However, it can be seen as a good evidence. In [50] an algorithm to compute the MLE form time series was proposed. Many authors have made improvements to the Wolf et al.'s algorithm (see for instance [38]). However, in this work we use the original algorithm to compute the MLE values.

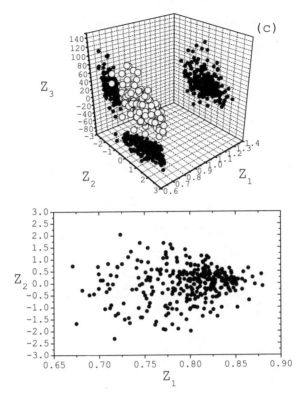

Fig. 13. (c) Phase portrait of the time series in Figure 12c

There are four parameters which must be fixed to use the MLE algorithm: Embedding dimension, d_e, maximum scale, S_M, minimum scale, S_m and evolution time, Θ. Basically, d_e is the attractor dimension where the orbits were embedded, S_M is the estimate value of the length scale on which the local structure of the attractor is not longer being proved. S_m is the length scale in which noise is expected to appear. Θ is fixed for compute of divergence measurements which is the necessary time to renormalize the distances between trajectories (for more details see [50]).

Since minimum scale is related with noisy measurement, it has been fixed at $S_m = 0.001$. Besides, since the measured scalar can be seen as a gas velocity pattern at the three directions, we have chosen the embedding dimensions as $m = d_e = 3$. In order to investigate if the structure signaling an attractor emerges, the embedding dimension was systematically increased. However, we do not find evidence of changes in the attractor structure. Then, we have two freedom degrees represented by maximum scale and evolution time. Figure 15 shows the MLE versus S_M for an evolution time $\Theta = 5$. Note that for $S_M = 1.15$ the MLE of all cases are positive. Thus, we have chosen this value for

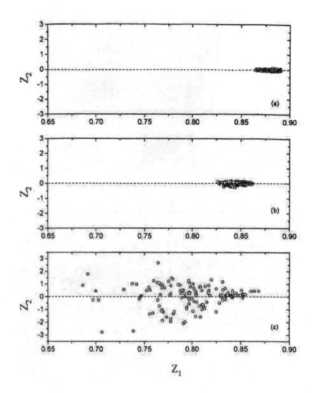

Fig. 14. Poincaré map at $z_3(t) = 0$. (a) For the disperse regime, the attractor crosses the Poincaré section in a short line, (b) for the four-flow region regime, the width of the short line increases, and (c) for the three-flow region regime, the attractor contains two regions (i) winding orbits and (ii) disperse orbits.

S_M which implies that the length of the attractor is two times similar than maximum scale. Consequently, the structure of the attractor is not longer than the maximum scale value (see Figure 15c). It must be pointed out that for $S_M > 5$ the estimated values of the MLE are less than zero for all cases.

We have the following parameters values: $d_e = 3$, $S_M = 1.15$, $S_m = 0.001$ and $\Theta = 5$. Figure 16 shows the evolution of the estimated value of the MLE. Orbits mean the inverse of time needed for computation of MLE at fiducial point [50]. Notice that MLE achieves a stationary value which is an evidence of the characteristic behavior of the class of time series in Figure 12. For $\varepsilon_{av} = 0.87301$ (disperse flow regime), $\lambda_M < 0$ for all time whereas for $\varepsilon_{av} = 0.82884$ there exists a transition, nevertheless the stationary value of $\lambda_M < 0$. For $\varepsilon_{av} = 0.79344$ (3-region flow regime) $\lambda_M > 0$ every time. This is an evidence that the bubbling attractor in 3-flow region regime may be chaotic.

The power spectrum density (PSD) is a widely used tool to find fundamental frequencies (and their harmonics) in dynamical systems with oscillatory

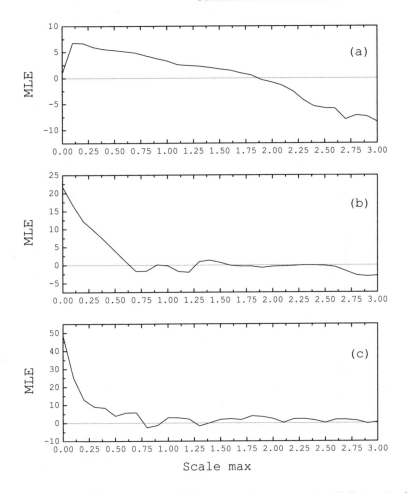

Fig. 15. Maximal Lyapunov exponent versus maximum scale. If $S_M > 5$, then $\lambda_M < 0$ for all cases.

behavior. It is used to study periodic-quasiperiodic-chaotic routes [27]. In our study case, Power spectrum was implemented by means of fast Fourier transform method. A typical size of the acquired data is about 2000, which suffices to find the fundamental frequencies of any given time series. Figure 17 shows the normalized power spectrum of the time series displayed in Figure 12. Note that, if ε_{av} decreases (i.e., superficial gas velocity increases) the number of fundamental frequencies in the power spectrum increases. This is an evidence that modes of interacting bubbles induce a more complex oscillatory pattern. Of course, if ε_{av} decreases the interactions between bubbles increases because they are closer to each other.

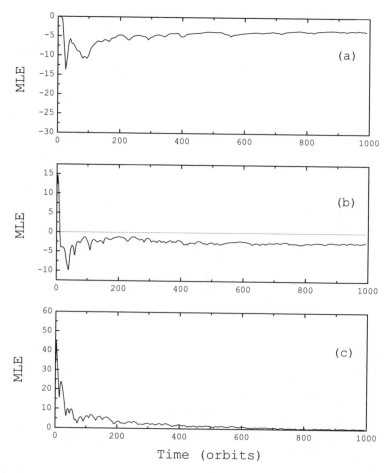

Fig. 16. Maximal Lyapunov exponent versus time evolution. (a) ε_{av}=0.87301, almost-periodic behavior, ε_{av}=0.83884, almost-periodic behavior and (c) ε_{av}=0.79344, chaotic behavior.

3.4 Remarks on the Second Case

The flow structure in a tall vertical bubble column was analyzed using deterministic techniques. The characterization of the two-phase flow structure was realized in the middle section where the flow is fully developed (measuring section). Three cases under different volume fraction of liquid were shown. The results can be resumed as follows:

(a) ε_{av}=0.87301 (*homogeneous flow regime*). The fluctuations of volume fraction of liquid has small amplitude and low frequency. The reconstructed attractor is quasiperiodic and presents certain order (the layer distribution). Note that the power spectrum exhibits two fundamental frequencies.

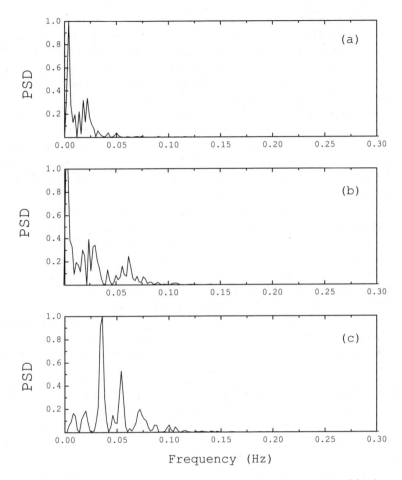

Fig. 17. Power spectrum. As the gas velocity increases the number of fundamental frequencies also increases.

Since the estimated MLE is negative, $\lambda_M < 0$, we can say that this case displays as essential regular dynamical behavior. The Poincaré map displays an orbit set contained in a short line. The above results can be imputed to the following fact: The bubbles rise in an almost-linearly pathway and the liquid phase falls downward between bubbles streams (see scheme in the Figure 10a). This means that the bubbles interactions are feeble. In this way, the modes induced by one bubble stream can not affect another one.

(b) $\varepsilon_{av}= 0.82884$ (*4-flow region regime*). The orbits of the attractor are not contained in a short line. Although the size of the reconstructed attractor is similar than case (a), the bubble streams interactions are stronger due that space between bubble streams is smaller than the first one. This induces a transition from disperse flow to 4-flow flow regime (see scheme in Figure 10b). The 4-flow region regime presents more fundamental frequencies than case (a). Additionally, erratic drift displayed in the fluctuations of liquid volume fraction can be attributed to continuous generation of multiple vortex cells are continually generated. The generated vortex are contained in the wave motion of the fast bubble region. The behavior of these vortices is dynamic and its formation appears independent of each other. A central plume lies in the core of the column. Into the central plume the bubble rise almost-linearly. This is, the velocity vector of liquid and gas are in vertical order. The MLE λ_M converges to -5.0 and the layer distribution of the bubbling attractor have not been fully lost. As in the case (a) regular quasiperiodic motion was found.

(c) $\varepsilon_{av} = 0.79344$ (*3-flow region regime*). The bubble streams interactions are evidently strongest because they are very close each other. The 3-flow region regime appears due that the central plume disappears (see scheme in Figure 10c). The instantaneous velocity vectors of liquid and gas are more erratic (because vortical-spiral flow regime disappears). The orbits of the reconstructed attractor are more disperse. However, there is a region of high concentration of winding orbits into the attractor due the formation of vortices of similar size than case (b) (see Table 5). The following facts lead us to conclude that the reconstructed attractor is chaotic: (i) $\lambda_M > 0$ every time, (ii) there are more than three fundamental frequencies and (ii) the layer state distribution has been fully lost.

As a summary for the second study case, the bubble streams interactions induce the presence of complex oscillatory phenomenon [51]. For instance, since the bubble rise almost-linearly into the plume, the presence of the central plume induce an almost-periodic behavior. We shown that the number of fundamental frequencies (which are in some sense induced by the modes of the bubble streams) increases when the superficial gas velocity increases yielding a route to chaos (periodic- quasiperiodic-chaotic behavior).

References

[1] B. Andrés-Toro, J.M. Girón-Sierra, J.A. López-Orozco, F. Fernández-Conde, J.M. Peinado, and F. García-Ochoa. *Math. and Comp. in Simulations*, 48:65–74, 1998.

[2] H. Aref and S.W. Jones. Chaotic motion of a solid throught ideal fluid. *Phys. Fluid. A*, 5(12):3026–3028, 1993.

[3] F. Argoul, J. Huth, P. Merzeau, A. Arnéodo, and H.L. Swinney. Experimental evidence for homoclinic chaos in an electrochemical growth process. *Physica D*, 62(1-4):170–185, 1993.

[4] A. Arnéodo, F. Argoul, J. Elezgaray, and P. Richetti. Homoclinic chaos in chemical systems. *Physica D*, 62(1-4):134–169, 1993.

[5] H. Bai-Lin. *Chaos II*. World Scientific, Singapore, 1990.

[6] G.L. Baker and J.P. Gollub. *Chaotic Dynamics: An introduction*. Cambridge Univ. Press, Cambridge, 1990.

[7] M. Berezowski. Effect of delay time on the generation of chaos in continuous systems. One-dimensional model. Two-dimensional model - tubular chemical reactor with recycle. *Chaos, Solitons & Fractals*, 12(1):83–89, 2001.

[8] O. Bernard, Z. Hadj-Sadok, D. Dochain, A. Genovesi, and J.P. Steyer. Dynamical model development and parameter identification for anaerobic wastewater treatment process. *Biotechnol. Bioeng.*, 75(4):424–438, 2001.

[9] A. Biesheuvel and L. van Wijngaarden. Two-phase flow equations for a dilute dispersion of gas bubbles in liquid. *J. Fluid Mech.*, 148:301–318, 1984.

[10] R.C. Chen, J. Reese, and L.S. Fan. Flow structure in a three-dimensional bubble column and three-phase fluidized bed. *A.I.Ch.E. Journal*, 40(7):1093–1104, 1994.

[11] T. Chevalier, A. Freund, and J. Ross. The effects of a nonlinear delayed feedback on a chemical reaction. *J. Chem. Phys.*, 95:308–316, 1991.

[12] E. Colli, V.S.M. Piassi, A. Tufaile, and J.C. Sartorelli. Bistability in bubble formation. *Phys. Rev. E*, 70(6):066215, 2004.

[13] F. Colonius and W. Kliemann. *The Dynamics of Control*. Birkhäusser, Boston, 2000.

[14] N. Devanathan, M.P. Dudukovic, A. Lapin, and A. Lübbert. Chaotic flow in bubble column reactors. *Chem. Eng. Sci.*, 50(16):2661–2667, 1995.

[15] D. Dochain and G. Bastin. *Modelling and Control of Biotechnological Processes*. Pergamont Press, Oxford UK., 1986.

[16] M. Dolnik and E.M. Bollt. Communications with chemical chaos in the presence of noise. *Chaos*, 8(3):702–710, 1998.

[17] J. Drahoš, F. Bradka, and M. Punčochář. Fractal behaviour of pressure fluctuations in a bubble column. *Chem. Eng. Sci.*, 47(15-16):4069–4075, 1993.

[18] L.S. Fan. *Gas-Liquid Fluidization Engineering*. Butterworth, Stoneham, MA, 1989.

[19] J.D. Farmer, E. Ott, and J.A. Yorke. The dimension of chaotic attractors. *Physica D*, 7(1-3):153–180, 1983.

[20] R. Femat. Chaos in a class of reacting systems induced by robust asymptotic feedback. *Physica D*, 136:193–204, 2000.

[21] R. Femat, J. Alvarez-Ramírez, and M. Rosales-Torres. Robust asymptotic linearization via uncertainty estimation: Regulation of temperature in a fluidized bed reactor. *Comput. Chem. Eng.*, 23:697–708, 1999.

[22] R. Femat, J. Capistran-Tobias, and G. Solis-Perales. Laplace domain controllers for chaos control. *Phys. Lett. A*, 252(1-2):27–36, 1999.

[23] J. Guckenheimer and P. Holmes. *Nonlinear Oscillations, Dynamical Systems, and Bifurcations of Vector Fields*. Springer-Verlag, Berlin, 1990.

[24] E. Guyon, J.P. Nadal, and Y. Pomeau. *Disorder and Mixing*. Kluwer Academic Publishers, Netherlands, 1987.

[25] A.H. Hadid. Chaotic flow in rotating lid cavities. *Phys. Fluid. A*, 5(8):1939–1946, 1993.

[26] F.P. Incropera and D.P. DeWitt. *Fundamentals of Heat and Mass Transfer*. Wiley and Sons, USA, 1990.

[27] H. Lamba and C.J. Budd. Scaling of lyapunov exponents at nonsmooth bifurcations. *Phys. Rev. E*, 50(1):84–90, 1994.

[28] J. Lee and W.F. Ramirez. Mathematical modeling of induced foreing protein production by recombinant bacteria. *Biotechnol. Bioeng.*, 39:635–, 1992.

[29] T.J. Lin, J. Reese, T. Hong, and L.S. Fan. Quantitative analysis and computation of two-dimensional bubble columns. *A.I.Ch.E. Journal*, 42(2):301–318, 1996.

[30] W.L. Luyben. *Process Modeling, Simulation and Control for Chemical Engineering.* McGraw-Hill, New York, 2nd. edition, 1990.

[31] P. Martien, S.C. Pope, and R.S. Shaw. The chaotic behavior of the leaky faucet. *Phys. Lett. A*, 110(7-8):399–404, 1985.

[32] H.O. Méndez-Acosta. *Control Robusto de la Digestión Anaerobia en el Tratamiento de Aguas Residuales de la Industria Vitivinícola (in Spanish).* PhD thesis, Universidad de Guadalajara, Mexico, 2004.

[33] H.O. Méndez-Acosta, D.U. Campos-Delgado, R. Femat, and V. González-Álvarez. A robust feedforward/feedback control for an anaerobic digester. *Comput. Chem. Eng.*, 31:1–11, 2005.

[34] H.O. Méndez-Acosta, R. Femat, and D.U. Campos-Delgado. Improving the performance on the COD regulation in anaerobic digestion. *Ind. Eng. Chem. Res.*, 43(1):95–104, 2004.

[35] O. Monroy, J. Álvarez Ramírez, F. Cuervo, and R. Femat. An adaptive strategy to control anaerobic digesters for wastewater treatment. *Ind. Eng. Chem. Res.*, 35(10):3442–3446, 1996.

[36] J. Nielsen and J. Villadsen. *Bioreaction Engineering Principles.* Plenum Press, New York, 1994.

[37] K. Otawara and L.T. Fan. Increasing the yield from a chemical reactor with spontaneously oscillatory chemical reactions by a nonlinear feedback mechanism. *Comput. Chem. Eng.*, 25:333–335, 2001.

[38] U. Parlitz. Identification of true and spurious Lyapunov exponents from time series. *Int. J. Bifur. and Chaos*, 2(1):155–165, 1992.

[39] M. Pérez and P. Albertos. Self-oscillating and chaotic behaviour of a PI-controlled CSTR with control valve saturation. *J. Process Control*, 14:51–57, 2004.

[40] M. Ratto. A theoretical approach to the analysis of PI-controlled CSTRs with noise. *Comput. Chem. Eng.*, 22(11):1581–1593, 1998.

[41] R.G. Rice, D.T. Barbe, and N.W. Geary. Correlation of nonverticality and entrance effects in bubble columns. *A.I.Ch.E. Journal*, 36(9):1421–1424, 1990.

[42] H.G. Schuster. *Deterministic Chaos.* VCH Publishers, 1993.

[43] A. Soria. *Kinematics Waves and Governing Equations in Bubble Columns and Three Phase Fluidized Beds.* PhD thesis, University Western Ontario, Canada, 1991.

[44] J.I. Steinfeld, J.S. Francisco, and W.L. Hase. *Chemical Kinetics and Dynamics.* Prentice-Hall, USA, 1989.

[45] D.J. Tritton and C. Edgell. Chaotic bubbling. *Phys. Fluid. A*, 5(2):500–502, 1993.

[46] A. Tufaile and J.C. Sartorelli. Chaotic behavior in bubble formation dynamics. *Physica A*, 275(3-4):336–346, 2000.

[47] A. Tufaile and J.C. Sartorelli. Bubble and spherical air shell formation dynamics. *Phys. Rev. E*, 66(5):056204, 2002.

[48] J.W. Tzeng, R.C. Chen, and L.S. Fan. Visualization of flow characteristics in a 2-D bubble column and three-phase fluidized bed. *A.I.Ch.E. Journal*, 39(5):733–744, 1993.

[49] V.K. Vanag, L.F. Yang, M. Dolnik, A.M. Zhabotinsky, and I.R. Epstein. Oscillatory cluster patterns in a homogeneous chemical system with global feedback. *Nature*, 406(6794):389–391, 2000.

[50] A. Wolf, J.B. Swift, H.L. Swinney, and J.A. Vastano. Determining lyapunov exponents from a time series. *Physica D*, 16(3):285–317, 1985.

[51] D. Zardi and G. Seminara. Chaotic mode competition in the shape oscillations of pulsating bubbles. *J. Fluid Mech.*, 286:257–276, 1995.

[52] A. Zavala-Río, R. Femat, and G. Solis-Perales. Countercurrent double-pipe heat exchangers are a special type of positive systems. In *First Multidisciplinary International Symposium on Positive Systems: Theory and Applications (POSTA 2003)*, volume IEEE LNCIS 294, pages 385–392, Roma, Italy, August 2003.

Printing: Mercedes-Druck, Berlin
Binding: Stein+Lehmann, Berlin

Lecture Notes in Control and Information Sciences

Edited by M. Thoma, M. Morari

Further volumes of this series can be found on our homepage:
springer.com